AF213367

Studies in Systems, Decision and Control

Volume 635

Series Editor

Janusz Kacprzyk, Systems Research Institute, Polish Academy of Sciences, Warsaw, Poland

Editorial Board

Dmitry A. Novikov, Institute of Control Sciences (Director), Russian Academy of Sciences, Moscow, Russia

Peng Shi, School of Electrical and Mechanical Engineering, The University of Adelaide, Adelaide, SA, Australia

Jinde Cao, School of Mathematics, Southeast University, Nanjing, China

Marios Polycarpou, KIOS Research and Innovation Center of Excellence, University of Cyprus, Nicosia, Cyprus

Witold Pedrycz, Department of Electrical and Computer Engineering, University of Alberta, Edmonton, AB, Canada

Fatima Anwar, Process Engineer, Applied Materials (United States), Santa Clara, CA, USA

The series "Studies in Systems, Decision and Control" (SSDC) covers both new developments and advances, as well as the state of the art, in the various areas of broadly perceived systems, decision making and control–quickly, up to date and with a high quality. The intent is to cover the theory, applications, and perspectives on the state of the art and future developments relevant to systems, decision making, control, complex processes and related areas, as embedded in the fields of engineering, computer science, physics, economics, social and life sciences, as well as the paradigms and methodologies behind them. The series contains monographs, textbooks, lecture notes and edited volumes in systems, decision making and control spanning the areas of Cyber-Physical Systems, Autonomous Systems, Sensor Networks, Control Systems, Energy Systems, Automotive Systems, Biological Systems, Vehicular Networking and Connected Vehicles, Aerospace Systems, Automation, Manufacturing, Smart Grids, Nonlinear Systems, Power Systems, Robotics, Social Systems, Economic Systems and other. Of particular value to both the contributors and the readership are the short publication timeframe and the world-wide distribution and exposure which enable both a wide and rapid dissemination of research output.

Indexed by SCOPUS, DBLP, WTI Frankfurt eG, zbMATH, SCImago.

All books published in the series are submitted for consideration in Web of Science.

Andrii Marchenko · Zbigniew Koruba
Editors

Decarbonization of Transport Energy Installations in the Context of Sustainable Development Strategies

Springer

Editors
Andrii Marchenko
National Technical University "Kharkiv Polytechnic Institute"
Kharkiv, Ukraine

Zbigniew Koruba
Department of Computer-Based Technology and Arm Engineering
Kielce University of Technology
Kielce, Poland

ISSN 2198-4182 ISSN 2198-4190 (electronic)
Studies in Systems, Decision and Control
ISBN 978-3-032-05883-6 ISBN 978-3-032-05884-3 (eBook)
https://doi.org/10.1007/978-3-032-05884-3

© The Editor(s) (if applicable) and The Author(s), under exclusive license to Springer Nature Switzerland AG 2025

In September 2015, the UN Summit on Sustainable Development was held as part of the 70th session of the UN General Assembly. The 16 components of sustainable development and 169 main tasks were approved.

This work is subject to copyright. All rights are solely and exclusively licensed by the Publisher, whether the whole or part of the material is concerned, specifically the rights of translation, reprinting, reuse of illustrations, recitation, broadcasting, reproduction on microfilms or in any other physical way, and transmission or information storage and retrieval, electronic adaptation, computer software, or by similar or dissimilar methodology now known or hereafter developed.
The use of general descriptive names, registered names, trademarks, service marks, etc. in this publication does not imply, even in the absence of a specific statement, that such names are exempt from the relevant protective laws and regulations and therefore free for general use.
The publisher, the authors and the editors are safe to assume that the advice and information in this book are believed to be true and accurate at the date of publication. Neither the publisher nor the authors or the editors give a warranty, expressed or implied, with respect to the material contained herein or for any errors or omissions that may have been made. The publisher remains neutral with regard to jurisdictional claims in published maps and institutional affiliations.

This Springer imprint is published by the registered company Springer Nature Switzerland AG
The registered company address is: Gewerbestrasse 11, 6330 Cham, Switzerland

If disposing of this product, please recycle the paper.

Preface

Oh man! In vain to wait for the knowledge.
Of the entire depth of the laws of the Universe.
Incomprehensible for an electron is Everest.
The more we know, the wider the circle of ignorance.

Dear reader!

Thank you very much for your interest in our collective monograph: 'Decarbonisation of Transport Energy Installations in the Context of Sustainable Development Strategy'. Our ideas and hopes, which are reflected on every page of the monograph, are the result of the painstaking work of a team of authors from the USA, Germany, Austria, the Czech Republic, Lithuania, Poland, and Ukraine, as well as the diligent support of a large number of people who care about today's problems. We sincerely hope that this monograph will be useful for students and postgraduates, young scientists and scientists who already have significant achievements in educational and scientific activities, and for all innovators of sustainable development.

We are far from thinking that the publication of this monograph will solve all the accumulated educational, scientific and scientific-technical problems of sustainable development in transport. This is only a small part of the great efforts required from the scientific and educational community of society on the path to scientific support for sustainable development.

Sustainable development, decarbonisation, green technologies—all these global challenges of today concern each of us and are relevant to the survival of our civilisation. The high relevance of solving these important problems is evidenced by the memoranda and strategies adopted at the state and international levels over the past 30–40 years. Just recently, the United States Department of Energy released an industry decarbonisation strategy that aims to cut CO_2 emissions by 58% by 2040 and 87% by 2050. At the same time, climatologists in many countries emphasise that humanity is on the brink of global warming, and some of them say that humanity has already crossed that line.

Through their scientific work, the international team of authors of the monograph, using the example of transport energy installations, not only analyses the state of the problem, proposes strategies and practices for solving it, but also collectively demonstrates the need for joint action to solve this global problem. At the same time, our editorial epigraph in the style of Omar Khayyam, a wise man who lived 1000 years ago, allows us to assert, on the one hand, that we do not yet know the full extent of the problem of global warming, and on the other hand, to hope that humanity has a chance for the future. It was with this generally positive attitude that we began our work on this monograph.

We are aware that solving these problems requires the use of a large number of known and new strategies and technologies, accompanied by scientific support. Of course, scientific support is not a 100% guarantee of optimal management decisions to overcome these pressing challenges to sustainable development. However, it is much more productive than the populist and erroneous management decisions on this issue that are currently being made and, as a rule, only exacerbate the climate problems of civilisation. In our opinion, this requires separate research and joint publications by unbiased scientists and government officials.

In conclusion, we, the editors of this monograph, hope, dear reader, that the innovative strategies and technologies proposed by our international team of authors will not only receive your constructive assessment, for which we are grateful to you regardless of whether it is positive or negative, but will also encourage you to express your personal position on this topical issue in lectures, scientific works of students and postgraduates, scientific research of scientists, scientific conferences and publications. Your position and your constructive actions are extremely important to all of us!

We wish you success in your work!

Below is a brief description of the monograph and its contents

A preliminary analysis of 16 Sustainable Development Goals has showed a high dependence on communications and transport for the successful achievement of each of these areas. We are convinced that none of these areas can be fully implemented without transport and communications technologies. At the same time, issues such as decarbonisation, widespread use of hydrogen technology, Industry 4.0 and Industry 5.0 strategies have become relevant. The editors of the monograph proposed that scientists from different countries prepare publications on these scientific issues. The publications prepared by the authors present their personal views on solving sustainable development issues related to energy installations in transport. This is a great work by a large team of professionals. We, the editors, are infinitely and sincerely grateful to each author and each team of authors for their hard work in the interests of education and science. A special feature of the monograph in this sense is the example of effective collaboration between scientists from different countries on topical issues of sustainable development, which, we are convinced, should be continued in other collaborations.

The main principle proposed by us, the editors of the monograph, in its formation was the principle of free expression of the authors' opinions in the text of each publication, and the task of the editors and the team of experts and support staff was only to adhere to the general concepts of this monograph. In the opinion of the editors of the monograph, it is the authors' position that is important and should be presented to the scientific community, and then, in accordance with scientific traditions, each reader will be able to find their own rational solution. At the same time, the reader becomes directly acquainted with the position of the author of the publication. For this purpose, the principle of free expression of the authors' opinions in the publications was proposed. Moreover, leading experts from seven countries in Europe and America were invited to join the team of authors.

The publications present the results of the authors' research on solving traditional problems in engine building, as well as problems related to the life cycle of hybrid and electric transport power plants. It is emphasised that innovations in transport power plants should reduce CO_2 emissions from fossil fuels and other harmful substances, as well as reduce energy consumption per unit of work. In the context of the implementation of decarbonisation technologies, new criteria for assessing the perfection of internal combustion engines have been proposed. The research focused on power plants for land, sea (river) and air transport. The authors' research tools include Industry 4.0 and Industry 5.0 strategies, modelling technologies, including modelling for evaluating management decisions, strategies for creating an effective innovation environment, and training highly qualified specialists, innovators and leaders to solve sustainable development problems.

In fact, all publications by the authors of the monograph can be divided into two parts. The first part deals with general technologies and strategies, while the second part deals directly with specific technologies and strategies that directly relate to transport energy installations. The editors included publications on staffing strategies for sustainable development through the training of highly qualified specialists, innovators and leaders in the first part. The second publication concerns modelling technologies, including modelling for the evaluation of management decisions, as an important tool for scientific support of sustainable development and countering populism in management strategies. The third publication concerns strategies for creating an environmentally innovative environment in universities—the basic scientific and educational sources for all 17 areas of sustainable development.

The second part deals directly with strategies and technologies for the sustainable development of land, sea (river) and air transport energy installations, including the use of effective Industry 4.0 and Industry 5.0 tools.

The publications present the results of the authors' research on solving traditional problems in engine building, as well as problems related to the life cycle of hybrid and electric transport power plants. It is emphasised that innovations in transport power plants should reduce CO_2 emissions from fossil fuels and other harmful substances, as well as reduce energy consumption per unit of work.

We, the editors, hope that the publications provided in the monograph by 17 international teams of authors on sustainable development in transport power plants will be supplemented by many other scientific publications and presentations at scientific conferences.

In conclusion, the editors of the monograph would like to express their sincere gratitude to everyone who participated in the preparation, writing and publication of the monograph. Our joint efforts have been wonderfully realised in the form of this monograph. We are convinced of its usefulness for education and science on the path to solving the problems of sustainable development.

Many thanks to our colleagues at the National Technical University 'Kharkiv Polytechnic Institute' (Ukraine, Kharkiv) and Kielce University of Technology (Poland, Kielce). Your support inspired us to work effectively throughout the creative and practical journey of this monograph. The professionalism of the authors of the monograph and the scientific achievements published in this scientific edition are undoubtedly based not only on the support of our colleagues—scientists, but also on the traditions of our scientific schools, the knowledge and experience passed on to us by our teachers and mentors.

Special thanks to our families. We worked hard at all stages of the monograph's life cycle, thereby devoting less attention and effort to our families. But you supported and inspired us. In our work, this only gave us strength and faith in success. We are aware that without your positive emotional infusion, tolerance and contribution to the realization of our world, this monograph would have been impossible.

Our work was carried out in extremely difficult and unfavourable conditions of war. Air raid sirens, which were frequent and often lasted more than a day, rocket and bomb strikes, human suffering and destruction, and periodic lack of telephone and internet connections hindered and negatively affected our work. It was only thanks to the resilience of the team of authors and the comprehensive support of all authors that this monograph was created. We are deeply convinced that this monograph works for sustainable development, while war works against sustainable development and the well-being of society.

The monograph is our victory on the path to sustainable development!

With the greatest respect!

Prof. Andrii Marchenko
Vice—Rector for Science and Research
Kharkiv Polytechnic Institute
National Technical University
Kharkiv, Ukraine

Prof. Zbigniew Koruba
Rector of the Kielce University of Technology
Kielce, Poland

Contents

The Critical Role of Leadership in the Era of Industry 5.0

Mukuni Joseph and Lapuzina Olena

Abstract This paper examines the emergence of Industry 5.0 and its implications for leadership development. It highlights the need to update curricula to include sustainable development skills alongside traditional leadership training. Industry 5.0 emphasizes collaboration between humans and technology, demanding leaders who can navigate this paradigm and integrate sustainability into their strategies. The paper explores the drivers of Industry 5.0, including its focus on humancentric approaches, and proposes frameworks for fostering competencies aligned with these objectives. By addressing the evolving requirements of Industry 5.0, the study offers insights into developing leaders capable of managing sustainability-focused transformations in modern organizations.

1 Introduction

Over the past decades, we have seen unprecedented technological advances made by humanity. Today, the world around us is replete with inventions that were hardly imagined a decade or two ago. Who would have thought barely several decades ago that we would have a mobile telephone that could tell you pretty much anything that you would like to know, such as time, the weather, news, directions to a destination, and answers to questions on any subject matter? The cellphone is also a video camera and storage device for digital information, among several other functions. The invention of this single technological device with multiple functions was a result

M. Joseph
Department of Teaching and Learning, Career and Technical Education, Virginia Tech, Blacksburg, VA, USA
e-mail: mjoseph7@vt.edu

L. Olena (✉)
Department of Social System Control Pedagogy and Psychology named after I. A. Ziaziun, National Technical University, "Kharkiv Polytechnic Institute", Kharkiv, Ukraine
e-mail: Olena.Lapuzina@khpi.edu.ua

© The Author(s), under exclusive license to Springer Nature Switzerland AG 2025
A. Marchenko and Z. Koruba (eds.), *Decarbonization of Transport Energy Installations in the Context of Sustainable Development Strategies*, Studies in Systems, Decision and Control 635, https://doi.org/10.1007/978-3-032-05884-3_1

of man's desire for convenience. It is more convenient to carry a single device than to carry several devices, each serving only one or two functions.

The cellphone is just one example of recent technological advancements. In the world of work, productivity continues to be enhanced by technology-driven inventions such as robots and smart machines that are, in a sense, increasingly becoming part of the 'workforce'. Klaus Schwab, a famous German economist, Doctor of Technical Sciences, professor, has long been talking about the coming of a new Industrial Era, arguing that in terms of scale and complexity it will be something amazing and will fundamentally change many things in our reality [1].

Industry 4.0 came as predicted. It transformed human life, the nature of work, and interpersonal relationships through automated work processes, artificial intelligence, material science, genetic engineering, the Internet of Things (IoT), biotechnology, 3-D printers, and many other technological tools, processes, and products [2–4]. Industry 4.0 can be credited for increased efficiency and mass production, leading to maximum profit. However, its emphasis on profit at the expense of people and the environment has generated criticism and triggered another revolution which is referred to as Industry 5.0 [5, 6].

This paper discusses Industry 5.0 from the perspective of leadership and administration of enterprises. The paper describes the characteristics of 5.0, identifies its implications on leadership, and then suggests a pedagogical response to the leadership challenges brought about by 5.0.

2 What is Industry 5.0?

Industry 5.0 is a concept for future industry that involves the integration of people and robots into production processes. This concept is a continuation of previous industrial concepts such as Industry 4.0, which included automation and digital transformation of production.

In Industry 5.0, people and jobs will work together, complementing each other in the production process. Jobs will involve routine and dangerous tasks, and people will engage in creative and complex tasks that require human intuition and experience.

The main idea of Industry 5.0 is that the person remains the center of the production process, and robots become his assistants. This allows us to ensure high quality products and save jobs for people. Unlike Industry 4.0, which was more concerned with the transformational aspects of the impact of modern digital technologies and changing business models, Industry 5.0 is about the following transformations, among many others: (1) full connection with the Green Economics, leading to better sustainability and circular economy (double digital and green transition); (2) creating systems that are resilient to pandemics and other various modern challenges.

Industry 5.0 is a promising concept to improve product quality and ensure sustainable industrial development in the future. In recent years, industry has undergone significant changes, made possible using Industry 4.0 technologies. However, they are now talking about the next stage—Industry 5.0, which involves even greater automation and the use of artificial intelligence; provides industry frameworks (development frameworks) that combine competitiveness objectives; focuses on the impact of alternative modes of governance models leading to permanence and sustainability; makes workers more meaningful and empowered through the use of digital devices and thus embodies a human-centric approach to the use of technology; builds a path towards a sustainable economy; and introduces indicators that show, for each industry ecosystem, progress towards sustainability, prosperity and continuity.

Innovative Industry 5.0 is described by the European Union as an industry that:

(1) does not operate at high productivity but rather aims to meet the needs and interests of society;
(2) puts the well-being of the worker first and uses new technologies as a priority to ensure the prosperity of society and the planet as a whole;
(3) to implement a policy of transition to the implementation of innovations in the economy that would be focused on people;
(4) to provide actions at the new stage of development that would complement those already adopted in Industry 4.0 [7–9].

Because Industry 4.0 is driven by the profit motive in the application of technology to industrial processes and practices, Industry 5.0 augments Industry 4.0 by bringing into focus values of environmental sustainability, human-centricity, and industrial resilience.

The European Union's conceptualization of 5.0 is understandable because when enterprises focus solely on efficiency, mass production, and maximum profits, some advancements in workplace smart technologies can be detrimental to the wellbeing of workers, clients, the community, and the environment [9]. Already, there is a concern that the interaction between workers and machines is creating psycho-social problems among workers. For instance, Majerník has suggested that: "It will be necessary to address the potential stresses of human cooperation with machines and robots and to develop methodologies for measuring and evaluating mental workload and the emergence of mental illness" [10]. Doyle-Kent has raised concerns about environmental degradation resulting from recent industrial revolutions [11]. These concerns call for good leadership in industry.

Figure 1 shows the centrality of leadership in industry's response to a public need through an interaction between the workforce and technological tools/processes to produce goods and services meeting the needs of the public while ensuring a healthy and sustainable interface with the physical and social environment. The interaction of the elements in Fig. 1 has the following features:

1. Public Need: This is the raison d'être of industry's existence. Production processes of goods and services take place to meet the needs of society.
2. Industry and Commerce: In satisfying a public need, industry and commerce must consider the impact of production processes on the physical and social environment to ensure that there is no harm to the environment.

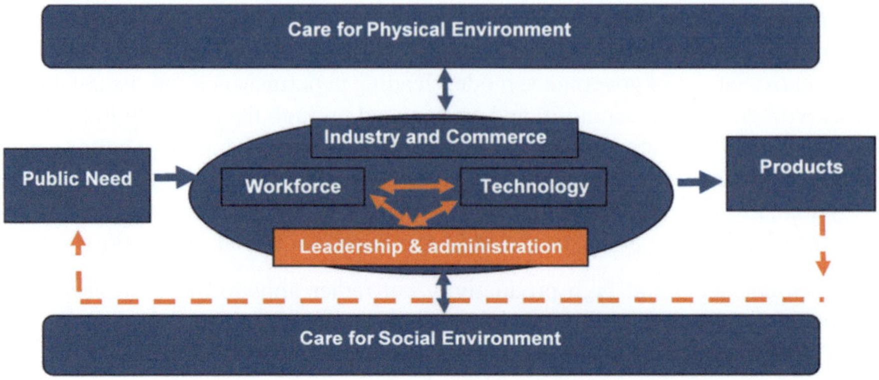

Fig. 1 Conceptualization of leadership in 5.0 industry

2.1. Workforce: Industry 5.0 requires a workforce with not only the right hard skills but also soft skills that include a caring attitude to fellow workers, clients, and the environment.
2.2. Technology: The tools, processes, and outcomes of production must be driven by technology that causes no physical or mental harm to workers and has no negative impact on the environment while contributing to efficiency, mass production, and maximum profit.
2.3. Leadership/Administration: Industry 5.0 requires ethical leaders that establish and maintain teams which are skilled, knowledgeable, and motivated; and foster healthy collaboration between human and machine workers, with a caring attitude to the environment. A good leader also develops strategies for making the enterprise resilient in the wake of shocks in the social and physical environment.

(a) Products: The outcomes of technological processes must meet clients' needs without causing damage to society and to the physical environment.
(b) Physical Environment: Industry 5.0 seeks to protect air, wildlife, trees, natural vegetation, rivers, lakes, and the ocean from possible adverse effects of technology to sustain life on planet earth.
(c) Social Environment: The use of technology must not have any negative impact on relations with social partners, culture, organizational structures, internal customs, gender equity, ethnicity, and other social factors that contribute to the wellbeing of the society which the enterprise serves.

The balance between increased use of smart technological tools and processes for maximum profits on one hand, and safeguarding the wellbeing of workers, clients, and the environment on the other hand, requires changes in the repertoire of knowledge and skills of leaders.

The next section explores the implications of 5.0 on leadership, administration, and governance in industry.

3 Leadership, Administration, and Governance in Industry 5.0

The history of the world community shows that the system of economic relations, the level of economic development is not something separate, independent of society [12]. Highly developed economic relations both in the country and between countries are motivated by the requirements of society: socially, informationally, culturally, and morally. Professional leaders play a very important role in meeting the needs of society. Industry 5.0 requires leaders that are creative and practical in meeting society's needs while fostering the ideals of sustainable development in industry and commerce [13].

Today, key problems of social progress are associated with leadership, competitive strategies, management, creative approach to collective work, and the latest technologies. Only one who works on the problems of tomorrow and, thus, prepares himself and his organization for new tasks, will be a successful leader in the era of Industry 5.0 [14].

In his professional activity, a modern leader in technical and managerial issues is faced with many problems. The selection and justification of strategies that determine the successful development of an organization over a long period of time is complex. In this regard, the formation of a charismatic leader of the twenty-first century, who possesses not only knowledge and a systemic, analytical, predictable style of thinking, but also has a high moral ideology, is one of the most important tasks of modern higher education [15].

Sustainable development is a complex, important, and constantly changing problem of the present. Only through the combined efforts of professionals-leaders in their fields is it possible to effectively solve the pressing problems of sustainable development. That is why scientists around the world are using Environmental, Social, and Corporate Management (ESG) in social management, that means the following: "environment", "social sphere", "corporate management". This is an approach about making management decisions considering the impact of corporate activities on the ecology and society. ESG is a business strategy that implies a careful attitude to nature, employees, and society. It was developed by the former Secretary General of the UN Kofi Annan. The goal of this approach is to rebuild the company's work in such a way as not to harm the surrounding (social and physical) environment [16].

ESG is a set of standards on which the company's work is evaluated. They are used by investors to check potential investments. The environmental standard defines how the company protects nature and the environment. The social standard determines how the companies relate to personnel, customers, and society. The corporate criterion relates to the company's management, how effectively it runs the company, what wages it pays employees, and how the rights of shareholders are respected [17].

ESG criteria appeared relatively recently. This is a kind of answer to what is happening with the environment, including global warming, and social inequality between the poor and rich. All this was a help to the emergence and distribution of ESG new criteria around the world.

In addition to financial statements, a business enterprise is evaluated on its contribution to the stable development of the economy. Today, for the company to be considered successful, it needs to demonstrate not only stable development, but also a good reputation. The correspondence of ESG significantly increases the reputation of the company compared to competitors and makes the brand more recognizable in society [18].

Let us consider in detail each ESG criterion.

(a) *Ecological criterion (E):* The company's contribution to the preservation of the environment is evaluated. This includes whether nature pollutes its activity, whether greenhouse gases pollute the atmosphere, or the company has special filters in its activities, or whether natural resources are used, and in principle environmental norms are observed.
(b) *The social criterion (S)* reflects the company's attitude to personnel, suppliers, customers, and partners. This also includes the health of workers and safety, the use of children and slave labor, professional development of workers, harmful working conditions, compliance with human rights, and responsibility to customers for the quality of goods [14]. A well-thought-out ESG strategy helps the organization in attracting and retaining competent employees, increasing staff motivation, and increasing labor productivity.
(c) *The management criterion or corporate governance criterion (G)* is associated with the effectiveness of the leadership, the validity of the wages of managers, the rights of shareholders, the quality of the audit, fraud, and corruption. The modern world has long been thinking that we will leave the inheritance for future generations. It was these considerations that led to the emergence of new initiatives that formed the basis of the idea of sustainable development. Sustainable development can be defined as development that meets the needs of modern society without harming the environment and future generations.

Sustainable development was officially announced in September 2015 after approval at the 17th Summit, where 17 Sustainable Development Goals (SDG) were determined, which entered into force in January 2016. These SDGs, known as global goals, are a world-round call for action aimed at solving various global problems, the desire to eradicate poverty and inequality, protect the environment and promote the world and prosperity of all. These SDGs are complex and interconnected, and together they are aimed at ensuring the balance of three aspects of economic development, namely economic, social, and ecological.

Given these interconnected problems, companies are increasingly integrating the practice of sustainable development into their activities. In connection with the development of responsible investments, environmentally responsible companies operating in accordance with ESG principles attract more investors. The real results of the company can be evaluated using quantitative and qualitative principles and indicators of ESG.

Strategic approaches to the implementation of the concept of sustainable development, which were first developed by Kofi Annan, require the active participation of international companies and leading organizations. Now leading global companies include the fundamental principles of sustainable development in the basic documents of their activities. This phenomenon has become popular in recent years, especially in Western countries. In the early 2000s, there were only 20 ESG companies in the United States. By 2020, their number increased to almost 800.

A business that claims to have a good assessment of ESG must comply with development standards in three categories: social, management and environment. The principles of the environment determine how much the company takes care of the environment and how it tries to reduce negative ecological impact. Social principles show the company's attitude to personnel, suppliers, clients, partners, and consumers. To comply with the standards, the business must work on working quality, monitoring gender balance, and investing in social projects.

Principles of management relate to the quality of company management: transparency of reporting, management salary, healthy situation in offices, relations with shareholders, and anti-corruption measures. From the pressure of investors and banks, to modern companies, it is unprofitable to have a low ESG rating. In 2020, EY conducted a survey among institutional investors. As a result, 98% of respondents said that they strictly monitor the rating companies of ESG, which are considered for investment [17].

Here are some examples of ESG strategies among technological and financial companies in the world.

(d) Mastercard

Last November, Mastercard pledged to reach zero carbon emissions by 2040 (previously set by 2050). The company has also made progress in reducing the gender gap in wages: employees now earn 93 cents for each dollar earned by employees on average, compared to 92.4 centers in 2020. And recently, the global payment system has declared plans to tie bonuses of employees to initiatives in environmental, social, and corporate governance (ESG).

(e) PayPal

The Fintech Company has developed an ESG strategy based on four pillars: support for corporate culture, construction of an inclusive global economy, reduction of environmental impact and responsible and ethical business.

(f) Apple

The technological giant said that in 2021, 20% of all materials used in Apple products were redesigned. Also, since 2015, Apple has reduced the amount of plastic in the package by 75% and plans to completely remove the plastic from its packaging by 2025.

(g) Bank of America

One of the largest American banks last year declared the allocation of $ 1 trillion in support of green funding by 2030. The Bank also plans to achieve zero carbon emission by 2050.

ESG in Ukraine

To determine the transparency index (ESG) of Ukrainian companies, authoritative organizations such as the Professional Association of Corporate Governance (PACG) and the Center for CSR Development assess the indicators of their transparency in the corporate governance system and corporate social responsibility of these companies. Based on the study, the ESG index of transparency of companies of Ukraine 2020 was presented.

Sustainable development of companies involves tolerant activities in the areas of corporate governance, labor relations, the environment, consumer relations, fair business practices and much more. The relevance of ESG principles is constantly increasing, and this must be considered when creating a strategy for any company.

The importance of sustainable development in the business environment is increasing every year. More and more market participants are aware of the importance of sustainable development for their activities. Ukrainian companies are also influenced by this global trend of change and are moving in the direction of achieving sustainable development, establishing several necessary measures, and adhering to them.

ESG goals

(a) Open the path to Ukrainian entrepreneurs to implement ESG.
(b) Provide tools to evaluate companies' readiness for ESG.

Adherence to environmental, social and governance approaches, which includes sustainable leadership, is essential for ethical business in Ukraine. Especially in our difficult times, when the Russian state is waging war on the Ukrainian nation, the Ukrainian economy, and Ukrainian business. To implement the above, it is necessary to train and educate Ukrainian specialist leaders—professionals who are capable of making bold decisions based on the principles of sustainable development.

This section has explored issues that impact the competencies of effective leaders in the era of 5.0. The next section discusses what schools, colleges, and universities need to do to prepare leaders for the responsibilities arising from what has been said above concerning Industry 5.0.

4 Pedagogical Implications of Industry 5

The leadership training curriculum in schools, colleges, and universities needs to include characteristics and qualities of sustainable development leadership. The characteristics and qualities of sustainable development leaders today include several components, namely:

1. Strategy of sustainability leaders who:
   - have a clear vision based on sustainability principles;
   - support progress and inspire others to achieve success.
2. Demonstrate examples of behavior based on sustainability values.
3. Form a culture of sustainability in the company or organization.
4. Teamwork.
5. Invest in education and development of leadership skills of sustainability professionals.
6. Create systems for measuring progress in sustainability.

These aspects should be recognized through a collective common understanding in academic circles and appropriate contextual principles should be created for the formation of leadership skills of modern intellectuals—scientists, researchers, professors, and teachers—representatives of the scientific community of higher education institutions. The intellectual leadership of teachers and researchers in higher education institutions is a topic of great interest for pedagogical research and practice.

Intellectual leadership is a field of complex processes of designing, providing, criticizing, teaching, coaching, researching, mentoring, generating ideas, envisioning, encouraging, rethinking, managing, representing, consulting, evaluating, and achieving. The strategy of intellectual leadership is a strategy of personal development not only of the student, but also of the entire scientific and pedagogical staff.

For teachers, intellectual leadership is defined as the need for constant intense intellectual work, self-improvement and development, and a valuable attitude to the profession. In the scientific activity of the teacher, intellectual leadership is manifested in conducting research on current scientific problems, the number of publications, the citation index, the popularity of the scientist, the contribution to the development of science, etc.

In the pedagogical activity of the teacher, intellectual leadership is manifested in the constant updating of the content of education in accordance with the latest achievements of science and technology, in the involvement of students in scientific work, in the use of intellectually developing, creative, and problem tasks in classes. For students, intellectual leadership is manifested not only in the desire for scientific work, but also in professional orientation.

Professionalism, skill, and achieving success in any field require certain intellectual qualities. So, in this case we are talking about deep knowledge, broad erudition, critical skills, and desire for personal and professional growth, intellectual work, and scientific knowledge of the student's world. It is important to have a developed sense of personal responsibility for the results of education and awareness of the need to work throughout the life. Thus, for a modern university in the context of globalization, it is important to develop a development strategy aimed at intellectual leadership of both students and academic staff [18].

In modern market conditions, the successful development of society and the achievement of competitive advantages cannot be ensured without innovative approaches to education and management, the basis of which is the development and implementation of intellectual leadership and innovations of various types. All this is based on modern concepts of sustainable development Industry 4.0 and Industry 5.0.

As mentioned in the previous section, Industry 4.0 is a concept of digital transformation of production, which implies the use of modern innovative Internet technologies in factories, plants and other enterprises, where robotic machines, equipment and other devices can interact with each other and people. All this will allow for more efficient and high-quality production.

The main Industry 4.0 idea is that through digital technologies it is possible to achieve automation, optimization, and increased productivity of production, which allows increasing product quality and reducing production costs. Industry 4.0 is a promising concept to improve production efficiency and ensure sustainable industrial development in the future.

To actively implement Industry 5.0, it is necessary to introduce several changes at the level of business and education. Generations of businessmen raised on the theories of the free market and competition, where profit was the basis, must change to others, where people-centered approaches dominate, and CSR (corporate social responsibility) is not a screen for making profit, but one of the postulates of any business. The special focus of Industry 5.0 should be on job creation and retention, and the formation of workers in new market sectors.

In this regard, a question arises—*How would learning institutions prepare engineering leaders of various specialties to work in the conditions of Industry 5.0 and sustainable development?* First of all, it is necessary to provide high quality education in the field of engineering and technology. Modern education should be focused on developing creativity and innovation, as well as providing knowledge in the field of digital technologies.

The second aspect is practical experience, which is necessary to acquire skills in working with modern innovative technologies used in various modern industries. Dual training, practical-oriented technologies in education—all this is extremely necessary for specialists in the era of Industry 5.0.

The third aspect is the development of communication and leadership skills. Engineering leaders must be able to communicate with different people and manage teams. Therefore, it is necessary to include courses on communications and leadership in the training program. It is also important to remember the importance of sustainable development. Engineering leaders must understand the consequences of their actions and know how to reduce their negative impact on the environment. Therefore, it is necessary to include courses on ecology and sustainable development in the training program.

As a result, training engineering leaders of various specialties during the period of sustainable development of Industry 5.0 requires an integrated approach (It is necessary to provide high quality education, practical experience, development of communication and leadership skills, and remember the importance of sustainable development. This approach will allow successful work in the conditions of Industry 5.0 and ensure sustainable development of industry.

Let's consider the US experience in training engineering leaders of various specialties during the period of sustainable development of Industry 5.0. The experience of the United States, as one of the leading countries in industry and technology, is very valuable in preparing engineering leaders of various specialties during the period of sustainable development of Industry 5.0.

One of the key aspects of preparing engineering leaders is the high quality of education. There are many high-quality technological universities in the US, such as the Massachusetts Institute of Technology (MIT), the California Institute of Technology (Caltech), Stanford University, Virginia Tech, and many others. These universities provide high quality education in engineering and technology, focused on developing creativity and innovation.

In addition, there are many programs in the United States that help undergraduate and graduate students gain hands-on experience with modern hardware and software used in industry. For example, the Co-op program at MIT allows students to gain hands-on experience working for technology companies. The program at Virginia Tech allows students to gain hands-on work experience in technology companies and gain thought leadership skills. Its title is "Intellectual Leadership and Technology Practice" (ILTP). Through this program, students can work in companies involved in software development, security and data technologies, innovative technologies, etc. Undergraduate and graduate students gain hands-on experience working in these companies and learn thought leadership skills that will help them succeed in their careers. In addition, the ILTP program provides students with the opportunity to gain knowledge about technology and innovation in technology, which will help them become competitive in the job market.

In addition, the United States places great emphasis on developing communication and leadership skills. Many US universities offer courses in communications and leadership. For example, Virginia Tech offers courses in communications and leadership that will help students develop effective communication and people management skills. Some of them are as follows:

1. **Communication Strategy:** this course helps students understand how to communicate effectively with others and develops communication skills in a variety of situations.
2. **Leadership in education:** this course helps students to develop leadership and educational skills.
3. **Conflict management:** this course helps students to understand how to effectively resolve conflicts and find compromise solutions.
4. **Ethics of Leadership: t**his course helps students to understand how ethical principles can affect leadership and management.
5. **Personality development:** this course helps students to develop the skills of independent development and introspection and prepares them to become effective leaders.

These courses help students develop communication and leadership skills that will be useful in any area of life.

Finally, the United States places a strong emphasis on sustainability, particularly in the area of renewable energy. Many universities and companies are actively working on developing technologies that reduce environmental impacts and have energy efficiency programs. Virginia Tech pays great attention to the creation of sustainable development and to the reduction of negative impact on the environment.

Some of the programs and technologies used at the university are:

1. **Green campus:** Virginia Tech is actively working on the creation of a green campus, which means reducing emissions into the atmosphere, energy and water consumption and waste reduction.
2. **The use of solar panels:** Virginia Tech installed solar panels on the roof of the building to provide energy for some rooms.
3. **The use of LED lighting:** Virginia Tech has established LED lighting in rooms, which helps reduce electricity consumption.
4. **The use of electric vehicles:** Virginia Tech uses electric cars to transport students and employees, which helps reduce emissions into the atmosphere.
5. **Waste sorting:** Tech provides the opportunity to sort waste in all rooms, which helps to reduce the number of dead wastes.

These programs and technologies help Virginia Tech to reduce the negative technological impact on the environment and continue the program of sustainable development of society.

Virginia Tech University, located in Virginia, USA, occupies a leading position in the field of technical leadership during sustainable development and Industry 4.0. One of the key components of their strategy is the Center for Sustainable Development.

Virginia Tech has a Sustainability Center that aims to promote the development of technologies and projects that reduce negative impacts on the environment and improve resource efficiency, as well as:

- studying and developing new technologies with sustainability in mind;
- developing energy-efficient technologies;
- using renewable energy sources;
- developing technologies to reduce emissions of harmful substances and much more.

The Center for Sustainable Development also actively cooperates with industry partners and State Institutions to develop and implement new technologies. This allows students and teachers of the university to work on real projects and gain valuable experience in interaction with partners.

In addition to research and development, the Center for Sustainable Development also offers various programs and events for students and the public. For example, the Center organizes seminars, conferences and other measures that discuss Sustainable Development and innovation.

In addition, Virginia Tech University offers various training programs for technical leaders for sustainable development and Industry 4.0. Some of them are described below:

1. The training program at the master's level of energy engineering and Sustainable Development. This program allows students to gain knowledge and skills necessary for the development and implementation of energy technologies and Sustainable Development projects. It includes energy efficiency courses, renewable energy, Sustainable Development and other topics.
2. Teaching at the master's level in the field of engineering and technological systems. This program allows students to gain knowledge and skills necessary for the development and implementation of technological systems, taking into account sustainable development. It includes courses for managing technological systems, sustainable development, economics and other topics.
3. The training program at the master's level in information technology and sustainable development. This program allows students to gain knowledge and skills necessary for the development and implementation of information technology, taking into account sustainable development. It includes programming, databases, sustainable development and other topics.
4. The training program at the master's level in mechanical engineering and sustainable development. This program allows students to gain knowledge and skills necessary for the development and implementation of projects for the construction of machines, taking into account sustainable development. It includes courses on mechanics, material science, sustainable development and other topics.

Virginia Tech in the United States is one of the leading educational institutions that prepare leaders of various specialties during the period of sustainable development of Industry 5.0, taking into account the needs of the modern labor market. This approach is based on the principles of sustainable development and innovation, which are key success factors in the modern world. In this context, the education school offers a wide range of programs covering various specialties, such as engineering, electronics, information technologies and others.

The main goal of the Virginia Tech is to teach qualified specialists who can work in the conditions of Industry 5.0 in areas that include the widespread use of artificial intelligence, automation and other innovative solutions. Virginia Tech is actively working with industry partners and state institutions to develop and implement new technologies.

In general, Virginia Tech is an important component of the university strategy for learning leaders of various specialties during the period of sustainable development of Industry 5.0. Their programs and research help to ensure high-quality training of specialists, which is an important task in the modern world.

Thus, the US experience shows that in order to train leaders-engineers of various specialties during the period of sustainable development, Industry 5.0, it is necessary to ensure high quality education, practical experience, the development of communication and leadership skills, as well as remember the importance of sustainable development. The experience of Virginia Tech in leaders' training during the period of sustainable development, Industry 5.0, is extremely important not only for the United States, but also for Ukraine, especially during Russian military invasion. This invasion lasts to this day and has a significant negative impact on the environment and the general psychological state of the population of Ukraine.

After the start of Russian aggression, Ukraine was forced to transform and rethink its long-term plans, in which leadership strategies in all spheres of activity occupy a critical place. Now, despite the war, new styles and models of leadership are being developed in various fields, based on the values of sustainable development, such as humanity, responsibility and justice.

Passion, courage, consistency, and responsibility are those virtues that have gained wide recognition both in the war and in the conditions of business competing. Let's look at some of them.

Consistency can be defined as predictability, when your words match your actions. People trust leaders whose actions are consistent and reasonable.

Responsibility is another important feature of a modern leader.

It is important to note that the responsibility of a leader-manager should be not only to the team, subordinates, but also to society.

In order to effectively use new technologies, you need to **change your mind-thinking**. In the modern era, one of the critical skills is to have a new digital mindset, which allows you to be at the forefront of the most important changes in all areas of our lives today and have a competitive advantage.

Another important task of a leader is to train leaders in **decision-making skills**, as well as other innovative skills that will be in demand in the near future.

Leaders must choose the right strategy of action. It is necessary to inform all team members about the leader's strategic intentions, this leads to the formation of a strong team and the achievement of goals.

Sun Tzu, a Chinese philosopher, said: "If you know the enemy and know yourself, you need not fear the result of a hundred battles. If you know yourself but do not know the enemy, for every victory you win, you will also suffer a defeat. If you know neither the enemy nor yourself, you will be defeated in every battle" [19].

In the context of rapidly changing economic and social events, the uncertainty of the choice of actions creates a variety of opportunities for the leader: it is important to be able to quickly change tactics when circumstances change, use external factors and create winning scenarios.

Today, international cooperation between universities is important for the formation of leadership during the war in Ukraine and helps to ensure an effective exchange of knowledge and experience in the problem of leadership.

Participants in international cooperation exchange ideas and best practices in the field of leadership, which helps students and teachers gain new knowledge and skills. They also work together on projects aimed at improving the situation in Ukraine during the war.

International cooperation also helps access to new technologies and resources, which is very useful for solving complex problems that arise during the war. At the same time, universities can cooperate with international organizations, scientific centers and other institutions to provide access to the necessary resources.

In our opinion, the international cooperation between the Virginia Tech University and NTU "KhPI" contributes to the effective training of future specialists and improving their activities to ensure effective guidance and solutions to problems related to leadership especially during hostilities in Ukraine. In order to solve these important problems in NTU "KhPI" in cooperation with Virginia Tech University, international conferences, practical classes on various topics, which is associated with the leadership of modern specialists of the period of Sustainable Development.

In the classroom, students conduct discussions about real problems arising in the practice of professional leaders of different countries; independently develop situational tasks related to the activities of the head of the enterprise, an engineer-research, an economist, and also conduct role-playing games etc., using the most modern interactive pedagogical technologies.

Conducting international conferences, round tables, seminars with Virginia Tech is of great interest for Ukrainian students, scientists, graduate students, teachers. Their activity in classes is growing, and independent training is significantly increasing, students and graduate students are not only getting the necessary skills, but also put these skills in practice by participating in the practical activities of enterprises and participating in international conferences and projects.

5 Conclusion

The realities of Industry 4.0 are still unfolding. For example, the full impact of Artificial intelligence (AI) on the workplace is yet to be fathomed, but from what is known so far, advancements in technology appear to have the capacity to transform not only the nature of work but also the relationships among workers and between workers and machines. Industry 4.0 has the potential to alienate workers and cause damage to the environment because of the emphases on the profit motive to the exclusion of the wellbeing of man and the environment. To address these concerns, and in the spirit of sustainable development, the European Union has proposed Industry 5.0, an approach that promotes all that is good with Industry 4.0 and minimizes the unintended consequences of technological advancements, by putting the welfare human beings and environmental protection at the center of industrial processes and products.

This article argues that realizing the intentions of Industry 5.0 requires leaders whose repertoire of knowledge, skills, and dispositions must include both the ability to balance the pursuit of efficiency, mass production, and maximum profit with ethical considerations for the well-being of others and the environment. The paper has suggested what could be included in the curriculum of leadership development programs, and why. The paper has also given examples of learning institutions where competencies needed for sustainable development are offered.

References

1. Schwab, K.: The Fourth Industrial Revolution. Crown Publishing Group, New York (published 2017) (2016). ISBN 9781524758875. Retrieved 29 June 2017. Digital technologies... are not new, but in a break with the third industrial revolution, they are becoming more sophisticated and integrated and are, as a result, transforming societies and the global economy. https://www.thriftbooks.com/w/the-fourth-industrial-revolution_klaus-schwab/9911596/?utm_source=google&utm_medium=cpc&utm_campaign=pmax_high_vol_scarce_under_$10&utm_adgroup=&utm_term=&utm_content=&gad_source=1&gclid=Cj0KCQiAst67BhCEARIsAKKdWOmv-wMJ3FWI_DyG55Y4e7bIoTM--XVc7Qden-DNgP8Ko2r2CgtRCO8aAoJ-EALw_wcB#edition=9365475&idiq=15988187
2. Darling-Hammond, L., LaPointe, M., Meyerson, D., Orr, M.T., Cohen, C.: Preparing School Leaders for a Changing World: Lessons from Exemplary Leadership Development Programs. Stanford University, Stanford Educational Leadership Institute, Stanford, CA (2007)
3. Kuckertz, A.: Bioeconomy transformation strategies worldwide require stronger focus on entrepreneurship. Sustainability **12**, 2911 (2020). https://doi.org/10.3390/su12072911
4. Mukuni, J.: Behold the Fourth Industrial Revolution and how to keep pace with workplace competencies in an ever-changing world of work!. *The* Changing Landscape of Workplace and Workforce [Working Title] (2023). IntechOpen. https://www.intechopen.com/chapters/1163981
5. Breque, M., De Nul, L., Petridis, A.: Industry 5.0: Towards a sustainable, human-centric and resilient European industry. European Commission, Directorate-General for Research, and Innovation (2021). https://doi.org/10.2777/308407
6. Ghobakhloo, M., Iranmanesh, M., Mubarak, M.F., Mubarik, M., Rejeb, A., Nilashi, M.: Identifying industry 5.0 contributions to sustainable development: a strategy roadmap for delivering sustainability values. Sustain. Prod. Consum. **33**, 716–737 (2022). https://doi.org/10.1016/j.spc.2022.08.003
7. European Commission (n.d.). Research and innovation. https://research-and-innovation.ec.europa.eu/research-area/industrial-research-and-innovation/industry-50_en
8. (2020) European Commission: Directorate-General for Communication, Circular economy action plan—For a cleaner and more competitive Europe, Publications Office of the European Union, https://doi.org/10.2779/05068
9. (2020) European Commission: Directorate-General for Environment, Leading the way to a global circular economy—State of play and outlook, Publications Office, https://doi.org/10.2779/013167
10. Majerník, M., Daneshjo, N., Malega, P., Drábik, P., & Barilová, B.: Sustainable development of the intelligent industry from Industry 4.0 to Industry 5.0. Adv. Sci. Technolo. Res. J. **16**(2) (2022). https://doi.org/10.12913/22998624/146420
11. Doyle-Kent, M., Kopacek, P.: Industry 5.0: Is the manufacturing industry on the cusp of a new revolution?. In: Proceedings of the International Symposium for Production Research, pp. 432–441. Springer International Publishing (2020). https://doi.org/10.1007/978-3-030-31343-2_38
12. Hightower, A.M., Delgado, R.C., Lloyd, S.C., Wittenstein, R., Sellers, K., Swanson, C.B.: Improving student learning by supporting quality teaching (2011). Retrieved from: http://www.edweek.org/media/eperc_qualityteaching_12.11.pdf
13. Sosik, J.J., Gentry, W.A., Chun, J.U.: The value of virtue in the upper echelons: a multisource examination of executive character strengths and performance. Leadersh. Q. **23**(3), 367–382 (2012). https://doi.org/10.1016/j.leaqua.2011.08.010
14. Matos, Pedro, ESG and Responsible Institutional Investing Around the World: A Critical Review (August 7, 2020). CFA Institute Research Foundation Literature Reviews, May 2020, ISBN: 978-1-944960-97-1, Available at SSRN: https://ssrn.com/abstract=3668998 or https://doi.org/10.2139/ssrn.3668998
15. Mohan, G., Seijts, G., Miller, R.: Does leader character have a gender?, J. Bus. Ethics Springer **188**(1), 169–186 (2023). https://doi.org/10.1007/s10551-022-05313-9

16. OECD: The circular economy in cities and regions: Synthesis report, OECD Urban Studies, OECD Publishing, Paris (2020), https://doi.org/10.1787/10ac6ae4-en
17. EYGM: How will ESG performance shape your future? (2020). https://assets.ey.com/content/dam/ey-sites/ey-com/en_gl/topics/assurance/assurance-pdfs/ey-global-institutional-investor-survey-2020.pdf
18. Roth, M.A., Price, J.K.: The critical role of leadership for education transformation with successful technology implementation. In: Huang, R., Kinshuk, Price, J. (eds.), ICT in Education in Global Context. Lecture Notes in Educational Technology. Springer, Berlin, Heidelberg (2016). https://doi.org/10.1007/978-3-662-47956-8_10
19. Szczepanski, K.: "Sun Tzu and the Art of War". Asian History. Archived from the original on 22 January 2016. Retrieved 13 March 2016. 4 February 2015. https://www.military-quotes.com/Sun-Tzu.htm

System Theory Computing Elements Comparison: Two Modelling Examples Towards Sustainability

Bernhard Heiden, Bianca Tonino-Heiden, and Volodymyr Alieksieiev

Abstract This chapter defines key system theory elements for computational modelling, highlighting the challenges of system dynamics and proposing a framework using dimensions like holisticity, realisticity, and orgitonality. Comparative analysis of modelling paradigms and system simulation programs, alongside the Turing machine concept, underscores their applicability to sustainability focused projects. Two modelling examples, Witness and value stream analysis, illustrate how system translations reveal unique strengths and limitations of each approach. The chapter concludes with broader applications in computational sustainability and future research directions to enhance model integration for addressing complex system challenges.

1 Introduction

In Industrial Engineering and Management, the future prediction of systems is of key importance. To use computer simulation is a means of increasing the predictability of future system performance, as well as the understanding of the "mechanics" of such systems (see, e.g. [1, 2]). Computers are a means of mechanisation of thinking (mt) and by applying practical thinking concepts. One early concept of mt is that of the Aristotelean logic, which Leibniz has refined and later Boole to adapt it to computational implementation. Beginning from the first computational programming

B. Heiden
Industrial Engineering and Management Studiengang, Carinthia University of Applied Science, Villach, Austria
e-mail: B.Heiden@fh-kaernten.at

B. Tonino-Heiden
Institute of Philosophy, University of Graz, Graz, Austria

V. Alieksieiev (✉)
Institute of Production Management and Technology, Hamburg University of Technology, Hamburg, Germany
e-mail: volodymyr.alieksieiev@tuhh.de

© The Author(s), under exclusive license to Springer Nature Switzerland AG 2025
A. Marchenko and Z. Koruba (eds.), *Decarbonization of Transport Energy Installations in the Context of Sustainable Development Strategies*, Studies in Systems, Decision and Control 635, https://doi.org/10.1007/978-3-032-05884-3_2

languages, logic plays a key role in these "machine" languages, meaning that logical operators, <and>, <or>, <not> and <if then> are integrated structurally in the syntax of every computational language, which is now called predicate-logic of the first order (see, e.g. [3]).

With regard to system dynamics, that is, computer-modelling time-dependent systems in our understanding here, there exist now, over time, a series of approaches and programs, like Witness, Anylogic, Stella, StochSD and many others. On the other side, there can be some meta-system approaches that try to close the gap between "small" and "big" systems or those that are at least different in size, complexity and abstraction scale. Such systems can be regarded as, e.g., hierarchical whole and part systems by the holon theory [4] or as different cybernetic units or orgitons by the orgiton theory [5].

A. *Goal of This Work.*

The goal of this work is to compare the elements of several different program systems and modelling heuristics. The investigated programs and heuristics will comprise several types (see, Fig. 2), (a) one, used in production and general systems, is Witness (see, e.g. [6]), (b) another one that is interdisciplinarily used in general dynamics, with two representatives, Stella by ISEE Systems [7] and StochSD by Leif Gustaffson [8], (c) a method for valuing production processes, the value stream analysis [9], (d) a heuristic of describing universal machines done by the concept of the Turing machine [10, 11] and (e) the funnel modelling method in production logistics [12–16].

The goal is then further to find out what and whether there is a common language and what is the difference and how those different program system and heuristic types are translatable in terms of those basic elements, and to give at least a basic explanation of a principal translation matrix and a meta-system view.

B. *Research Questions.*

The research questions are:

(1) What could be a qualitative or heuristic method to increase system simulation software's predictability and counteract the idealism dilemma of system simulation?
(2) What could be a common typology and framework for different programs and heuristics for
    (a) generic production systems software,
    (b) system dynamics software,
    (c) production value stream modelling,
    (d) the Turing machine and
    (e) the funnel model systematic in production logistics with regard to their basic elements?

The basic assumptions and restrictions of these questions are that the given programs can represent those program types and heuristics (a-e) and are sufficiently different from each other type.

C. *Method of this Work and Results Sketch.*

We make a qualitative analysis of the system elements that we investigate, and then construct two frameworks by logic natural language argumentation. The first framework is intended to give a qualitative orientation, the second gives a translation matrix of system theoretic basic elements.

D. *Content of the Work.*

Section II presents a first qualitative framework for ordering systems with regard to three dimensions in the context of complexity. Types and elements of the investigated software, and the heuristics, value stream analysis, the Turing machine and the funnel modelisation method are given structurally. In Section III, we complete the system-elements and -types matrix and embed it in a possible translation systematic. In Section IV we give examples and how to translate them to simulation models schematically and as a general strategy. In Section V, we then discuss the principle problem of the system simulation dilemma, give an approach to gain greater system predictability, and hint how and where to be more cautious in system modelling. In Section VI, we summarise the results, give conclusions and an outlook.

2 Elements and Operating Modes of System Theory

First, one crucial precondition shall be stated. According to Weizenbaum [11], a program of a universal Turing machine, as computers are (ibid. p. 93), is a means of transforming a machine into another one by a program (e.g. ibid. p. 139). A program then has the function to transform one computer into another. Concerning modelisation, a system is transformed from an observed system into a model of the observed system. This can be regarded as a mirroring of a system in an abstract dimension and, above all, an abstract dimension of it. With regard to orgiton theory we say it is of a higher orgitonal state. When we regard Fig. 1, we see an idea of how this leads to a language and theory mirroring of cybernetic systems of computers as well as that of thought. Some systems differ from other systems that they are "faster" in increasing their orgitonal higher state, or in other words that have more backcouplings and connections or relations to other systems than others. These Turing machines can be of different kinds, meaning that they are implemented differently, like computers or humans.

System theory can be understood as a concept of modelisation or mirroring of a base system that is mirrored (c.f. also Fig. 1 and the mirrorplane). Two mirrors, then, like the two types, aim at simulating these realities. So, the hypothesis here is that one mirror can be translated into another. If this would be possible in the general case for "all" models, this would be the general case, which might be impossible to reach ever. So, at least, we then could say at the end of our investigation that it is possible to give a sufficient generalisation of system dynamics programs so that, e.g. type one can be transformed into type two with regard to (a) a to be specified modelisation

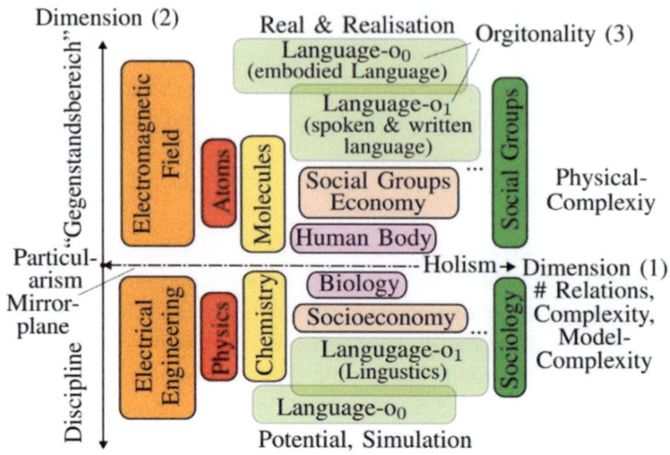

Fig. 1 Three Dimensions of System Complexity; (1) Holisticity, (2) Realisticity, and (3) Orgitonality or Cyberneticality

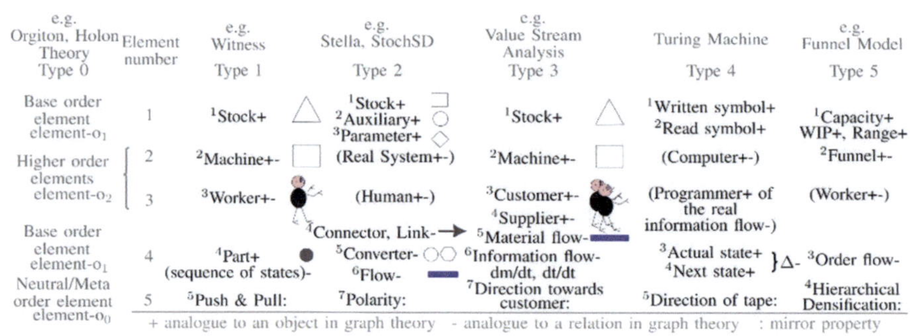

Fig. 2 System Dynamics (SD) Program Elements for different types of Programs (Software), the Concept of the Turing Machine (Hardware and Software) and a possible framing of meta systems ("Unifying" System). The uppercase indexed numbers for each type represent the "elements" in each specific denoted system used there

and (b) the investigated systems. In the ideal case of the perfect transformation, this can be tested by testing the consequences of the modelisations and the difference in their outcome concerning the calculation results. With the diction of the Turing machine, this would be a "program" to transform one Turing machine into another, meaning a type 1 program into a type 2 program and vice versa. If there might be a difference in details, this could then be a specific advantage or disadvantage of the chosen program or heuristics, which can also be expressed as a unique selling property (USP), according to the economic diction.

A. *Witness Program Elements and Modes*

We first define the basic elements of the Witness program, which means that all models can be composed of the elements in principle, which are given as follows and are depicted in Fig. 2:

(1) Stock. The stock is a cumulating term where items are stored virtually.
(2) Machine. The machine is the core element in the simulation. For information processes mainly the time used is the key operating parameter. For production machines, also other parameters are of importance.
(3) Worker. The worker can do a specific task in a machine.
(4) Part. The part is the product that is operated in the machine. It can be regarded numerically as a number associated with a machine for a certain time. If this number is associated with a material, it is a material process. This process (in Witness here denoted as part) can also be or represent "anything else" or an abstracted process, like e.g. information. In this case, the process denotes then an information process.
(5) Push & Pull. To get a complete working model, we need additionally some basic modes, which are listed below:

    (a) There must be a complete flow operationally closed from input to output.
    (b) There is either a pull or push application of the system.

In both the above cases, a logical connection of two elements is given, in the form of a logical "sentence", by this implementing, in principle, an agent's like behaviour that classical (Boolean) logical operations can also supplement alternatively principally.

B. *Stella and StochSD Elements*

Analogously, we can define the basis elements of Stella and StochSD Element below and in Fig. 2:

(1) Stock (analogously to A.). According to [17, p. 10], a stock can be regarded synonymously as, e.g. state variable, compartment or level.
(2) (Real System). In the following, the round brackets indicate that it is an implicit elements property. The overall approach models the real system, or the reality. This means that the approach is toward chosen aspects or attributes of a holistic system, where only those properties are looked at. For this to be "true", the properties have to be relevant to the chosen problem, which means that it is also connected with "all" the problems of a holistic system. One problem is here, e.g. how to choose the relevant attributes of the system, theoretically as well as empirically. Practical solutions to this problem are that the "experts" (in the observed field) shall give here relevant information due to their experience in the field. Another is to gain experience through a cybernetic learning process by applying the systematic toolbox.
(3) (Human). The human is also implicitly in the system. The regarded system deals with whole systems the human is predominantly in two parts there. As part of the whole system, as one ingredient. On the other side, there is the human that

manages, e.g. a company or a state politically, and so "models" the whole as a system according to his to be defined "goals". The goal can not only be seen as a priori but as prescriptive, as it "influences" the system, potentially with a self-fulfilling prophecy property, hence must be regarded with caution. In our opinion, the main goal should be to get a more rational decision, independent of setting the goals, which is itself a somehow "societal" decision of the company or state of the given thought example.

(4) Connector, Converter, Flow. There are three elements related to flow or relations in this software type. All relate, in principle, to a relation.

   (a) Connector, Link. A connector or Link in StochSD (see, e.g. [17]) "transfers information from one primitive to another", where primitive denotes in StochSD an element, similarly denoted to the elements in Fig. 2 here. It is synonymously to information and signal.
   (b) Converter. A converter is a tabularised relation between two variables inside the system. So, it means a mapping relation of two variables.
   (c) Flow. This denotes a material flow vs. information flow.

(5) Polarity. The polarity is the concept of the direction of information or causation flows. It determines in a cybernetic loop the direction of increase (growth) or decrease (decay). So, it is an important feature to formulate stability-related properties of cybernetic loops.

C. *Value Stream Analysis.*

(1) Stock. A stock is analogously to the previously defined stocks in A. and B. Here the stock has the meaning of storage of a product in production. This can be a building, a warehouse, or also another machine like a conveyor belt.
(2) Machine. A machine here denotes predominantly a production machine, producing goods.
(3) Customer, Supplier. The customer is seen as the driving force of production. His wish shall be fulfilled. He is at the beginning of the action chain. The supplier is on the other side of it or also of the supply chain.
(4) Material Flow, Information Flow. There are two main types of flow, that of information and that of material flow.
(5) Direction Towards Customer. The fifth element can be regarded as the (fixated) direction towards the customer or the orientation of all the actions towards the "wish" of the customer.

E. *Turing Machine.*

The Turing machine can be thought of as a quintuple of elements according to [11, p. 92] (cf. also [10]): written and read symbol, actual and next state, and the direction of the tape, for the storage of and reading from the symbols. The basic idea of it is that it can imitate another machine with regard to any "effective procedure" (in German "effektives Verfahren").

(1) Written and Read Symbol. This can be interpreted as a storage content. Here this are the written and read symbols.
(2) (Computer). The computer is the implicitly supposed machine that the Turing machine represents.
(3) (Programmer). The programmer is the human that feeds the computer. It could also be the computer in a higher orgitonal state, in the case that it is programming itself.
(4) Actual and Next State. The declaration of an actual and a next state can be interpreted as the difference of these states, and by this as a "flow" especially if regarded as sequential in time events.
(5) Direction of Tape. The direction of the tape can be regarded as the meta-information of the Turing machine.

F. *Funnel Model.*

The funnel model enables the general logistical modelling of a production system as a funnel. Herewith it is possible to apply different modelling scales [14, 15]. In the following, the main elements of the funnel model are presented (cf. also [12, 13, 16]) to compare it with the previous presented models and programs.

(1) Maximum capacity: maximal possible funnel vent size and Work-In-Progress (WIP): orders in the funnel and range.
(2) Funnel: as previously noted, funnel symbolizes production systems of different scope, e.g. a single working station or a whole production.
(3) Input or output order rate: number of orders which are entering or leaving the funnel in a certain time period.
(4) (Worker). The worker is implicitly part of the production process. He or she fulfils the order with the production machine.
(5) The structure of the funnel is implicitly densifying or ordering orders sequentially.

F. *Meaning of the Investigated Different "Types" in the Original Context.*

In Fig. 2, we have depicted the different types 1–5. We have also given an index number, which denotes the typical order in the specific context. E.g. in the value stream analysis, there are six typical elements, which are represented by the six uppercase indices in Type 3: Stock, Machine, Customer, Supplier, Material flow, and Information flow (see, e.g. [9]).

3 Heuristic Meta-System Theory Computing Concept

In the following, we want to sketch a concept of how system-theoretic modelling can be translated from one type into another (see Fig. 2). When we regard the conception of the Turing machine, it can be considered as a machine that translates one computer into another by imitating the computer by the other computer. One software for a system dynamics simulation can analogously be regarded as a Turing machine, although in a somewhat different meaning. The symbolic representation are then the system's base elements that represent the analogon to the Turing machine. We can measure this in grades of increasing complexity. We have sketched this with regard to orgiton theory [5], which denotes cybernetic loops of informational, material, and energy units named as orgiton (by appending -o) or orgitons (by appending -os) so that we can define units, as element-os that have increasing complexity, when shifting from lower to higher complexity, in our case by indices 0,1,2. An element-o_0, or all the elements in row 5 in Fig. 2 denote here a basic element, also a pre-orgiton, or a precondition for the overall system. In the Turing Machine, e.g. the direction of the tape, indicates some outer indicating order parameter. Similarly, push & pull, polarity or customer-oriented denote "directions", as accompanying information of the overall process. The element-o_2s are represented by rows 1 and 4 in Fig. 2. This relates basically to two basic categories, which can also be found in graph theory (see, e.g. [18]), nodes (row 1) and edges or relations (row 4). In knowledge graphs, these can also be shifted. This feature is in the here given systematics of programs (evidently) not supported. This would then mean that rows 1 and 4 could be regarded as interchangeable with regard to elements, depending on the context. Rows 2 and 3 in Fig. 2 represent more complex elements. We have indicated with round brackets () that those elements are explicitly left out in the specific type, which means that the programs or the heuristics representing the types do not go into detail concerning their base element structure in this sort of modelling. In the Turing machine, or system dynamics programs like Stella or StochSD, to give some examples, the humans, and the real system are principally left out. This observation might be connected to the thoughts and critics of Max Horxheimer of the "instrumentelle Vernunft" (in English rationality operated or done using machines or instruments) [19], and congruent with what one colleague, Joachim Werner, said recently: "They have forgotten the humans in operations- and system dynamics-research in the beginnings". On the other side, e.g. in Witness and Value Stream Analysis, humans and machines are explicitly considered.

There also exist further developments towards Artificial Intelligence (AI) approaches to include humans by agents, e.g. with the program Anylogic, that regard "intelligent interaction" of humans more explicitly. In Witness, on the other side, the classical approach implements human intelligibility by logical boolean structures or predicate logic of first order, as indicated above.

4 Systemic Computing Elements Comparison—Examples

In the following we give two examples. One more basic for a general production workflwo in Example I. The other in for the application of a modern transport system in Example II.

A. *Example I*

To illustrate the system theory comparison of computing elements done in Section II and III, a fictive production line has been modelled using (a) a Witness simulation and (b) a value stream mapping. The modelled production line consists of three processes (see Fig. 3): two of them are fully automated (a([2]machine), b([2]machine)), and a worker (a([3]worker), b([2]machine)) carries out the third process. In addition, a buffer (a([1]stock), b([1]stock)) is located between the first and the second working station. The material flow (a([4]part), b([5]material flow)) between machines is organized according to the push principle (a([5]push), b([7]direction towards customer, pull from customer)).

The models of the presented example are shown in Fig. 3, where Fig. 3a shows the Witness model and Fig. 3b depicts the value stream mapping of the sample production line.

For example, a worker is assigned to a particular machine or flows (cf. Fig. 2 and 3). Machines, the core active elements in the Witness simulation, are similar to processes in the value stream mapping. Although the processes in value stream mapping are not obligatorily carried out with machines, they are considered primarily as value-added actions that correlate with the function of machines in Witness. The element "buffer" can have different symbols in a Witness simulation, it fulfils in congruence with the value stream mapping a passive function, i.e., it can not push or pull. Some differences in both modelling approaches can be addressed in the example

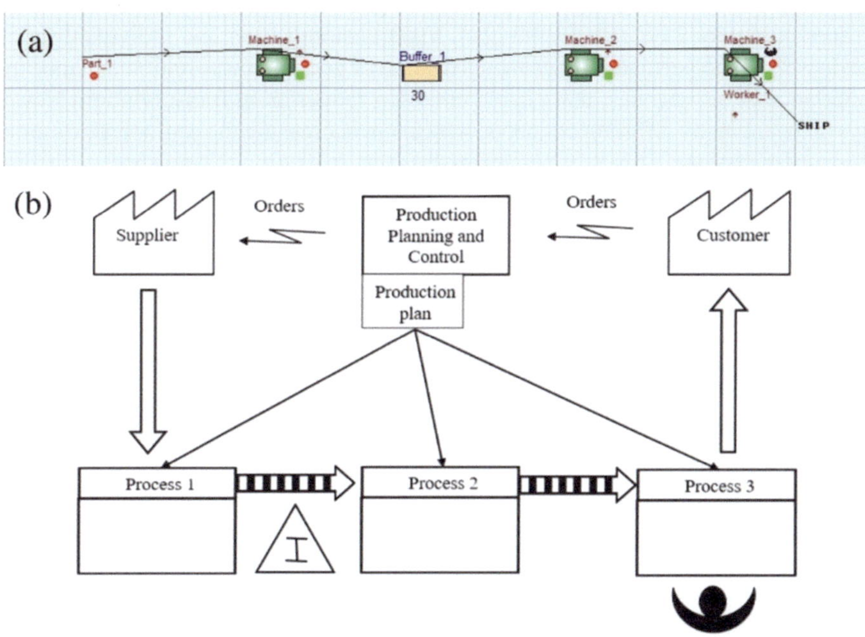

Fig. 3 **a** Model of a sample production line in Witness, **b** Model of a sample production line in value stream mapping (for Excel template source see: https://www.orgaimprove.com/)

of interaction between active and passive elements. So, in the Witness simulation, once the buffer is added to the process chain, the pure push interaction between elements can not be established since the items have to be at least pulled from the buffer. In addition, the items in Witness have to be pulled "out of world", i.e. from outside of the system, whereas in value stream mapping, they are commonly depicted as delivered by a supplier. The customer side, typically the last step in value stream mapping, can be shown in Witness, e.g. as an additional element (machine or buffer) or is supposed to be outside the considered system. In the last case, the command "push to ship" has to be executed (see also Fig. 3a). External and internal information flows (e.g. orders, production plans) that are depicted in value stream mapping can not be directly depicted in Witness; some parts of it (e.g. production quantity) can, however, be input in the corresponding elements (e.g. machines). In general, both modelling approaches, the Witness simulation and the value stream mapping, show good similarity so that the systems depicted in one of them can be translated to the other one, as long as the application conditions of specific modelling approaches are met. It can enable, among others, to use the advantages of both approaches without significant effort.

B. *Example II*

Transportation plays a crucial role in globalisation and product customisation processes, enabling the supply and delivery of materials and finished goods (cf. [20]). The transport sector has a significant CO_2 footprint. For road transportation it grew by approximately 21% since 1990 [21]. Mass customisation and individualisation trends well as the permanently growing volume of goods transportation enhance the importance of approaches, that enable emission reduction in transportation. Nowadays there are many different concepts, which aim decreasing the ecological footprint of transport systems. This Section targets the systematisation of some basic concepts using the theory of systems. For this purpose, the Section is structured as follows: Firstly, some fundamentals of theory of systems in the interpretation of Spencer-Brown are presented (Section IV-B1). Afterwards this approach is transferred for analysing and comparing different sustainable transport concepts (Sections IV-B2 and IV-B3).

(1) *Interpretation of a system according to Spencer-Brown*: Spencer-Brown defines a system as a so-called marked state surrounded by its environment (unmarked state) and with the border of the system as an important systems appearance parameter (see Fig. 4). The system environment itself is located, according to the author, in a so-called unmarked space (real world).[22] Following this concept, basically any system can be described (cf. [23]) for social systems or [24] for logistic systems). The following Sections apply this concept for classifying the sustainability approaches using a vehicle as a transport system example. For this purpose, Section IV-B2 describes firstly the sustainability concepts, which change the system itself for reducing the CO_2 footprint. Afterwards, Section IV-B3 addresses approaches, which aim to redesign the system's environment.

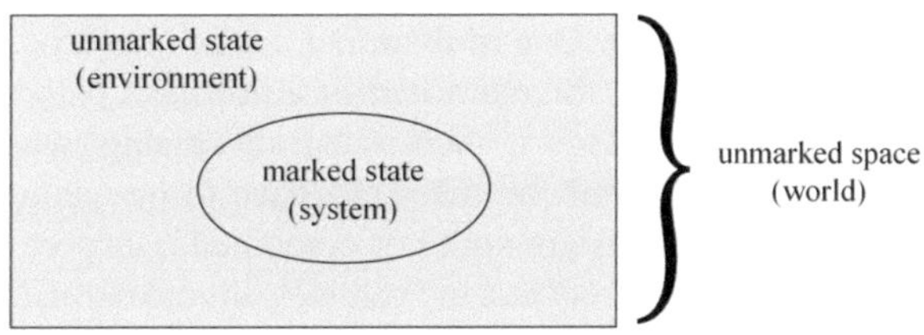

Fig. 4 System, environment and world according to [22]

(2) *Sustainable transport concepts—system's change*: When exploring a transport system as a system, it has firstly to be defined, what are its borders and environment (cf. [22] and [24]). As the borders of a transport system its physical borders can be understood. Environment of a transport system could comprise e.g., other systems, which surround the considered transport system (e.g., other transport, logistic, social systems, etc., see Fig. 5). This Subsection considers firstly the concepts, which suggest changing the system for increasing the sustainability performance of different scales. Depending on the scalability of a concept, it can have a positive impact either on the system's environment (small scalability), on the system and the environment or on the overall sustainable performance of the "world" (see Section IV-B1). The following example illustrates this interdependency: a car manufacturing company, which has developed a unique concept for significantly reducing the carbon footprint of their car can either make this solution open for other companies or keep it confidential it e.g., as a Unique Selling Point (USP). From the point of system's theory, the first case can lead to scaling this solution and its bigger impact, so that not only the company as a system itself, but also "real world" can profit from it. Using a vehicle as an example of a system, its change towards increasing the ecological sustainability could be e.g., recyclable materials or sustainable drive concept (e.g., electric drives). While this change could be technologically difficult, it is usually easier to scale, if it is accessible. Accessibility in this case illustrates also a certain kind of border, e.g. a price border for customers or implementation border for other producers. The change of "environment" or "world" results from changing the system itself and spreading the solution for changing the surrounded systems (environment).

(3) *Sustainable transport concepts—change of environment*: Another opportunity for sustainable development is changing the environment of a system organisationally. Nowadays there are many organisational concepts for sustainable

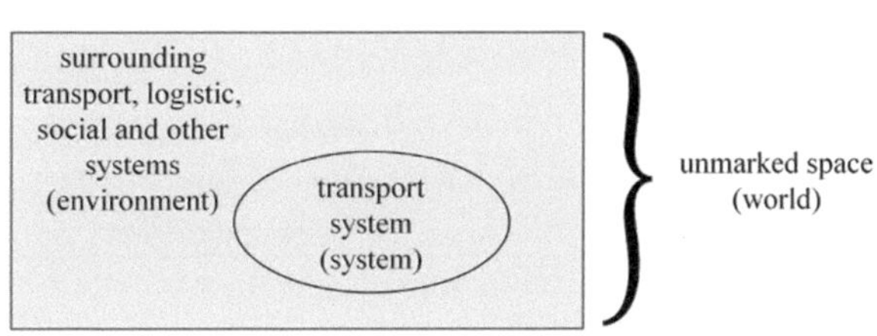

Fig. 5 Transport system abstraction according to [22]

transportation. One of them is e.g., a concept of combined transport, that aims at conducting the main part of a transport path using an ecologically friendly means of transport, such as a train or ship, whereas only the short distances before and after the main run have to be conducted using trucks. Transshipment terminals are nodes of combined transport chain, where loading units are transshipped between the vehicles. Figure 6 illustrates the principle of combined goods transportation between points A and B. The goal of this concept is to unite sustainability advantages of trains and ships with the flexibility of trucks.[25, 26] The main difference of this concept in comparison to the system's change concept is that this concept designs the sustainable conditions for available systems without changing the systems itself. With other words, the concept of combined transport enables reducing CO2 not by redesigning the vehicles technically, but by using the advantages of different vehicle types, such as their size (trains and ships can e.g., transport many loading units at the same time, reducing the corresponding carbon footprint per loading unit) and flexibility of trucks (cf. [26]). Whereas by a technical change of the system's accessibility and when technological difficulties arise that impact the success of the system, then changing the environment conditions for better suited ones as well as the information exchange and coordination between different systems becomes increasingly important. Using the example of a combined transport, a good information exchange between all participants is crucial (cf. [27]). Combining both approaches enables the use of both advantages. The implementation of combined approaches requires however longer time and bigger efforts, alone due to the bigger scope required to implement the change of the "environment" conditions. Nevertheless, these combined approaches would be the most effective and are hence strategically crucial.

(4) System Dynamics Model: The system described in this section can be described by one of the systems in Fig. 2, as a system dynamics model. A modelisation of Example II might then start to model the system and the environmental part. These systems can then be further differentiated by subsystems describing the transport, like e.g. in the Witness or value stream case. Depending on the environmental situation, the system might then be more and more complex, as systems may be nested together to yield a more complex and realistic system, regarding the as relevant regarded causal relationships and feedback loops. In the practical context it is of high relevance to use empirical data and real physical laws as background knowledge. An example for a system dynamics model performed in this fashion can be found according to [28], which calculates, to the best empirical evidence, the main effects that lead to global warming,

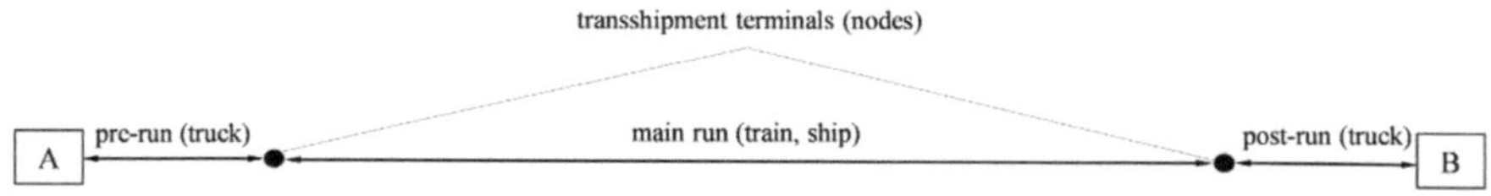

Fig. 6 Principle of combined transport (cf. [26])

depending on the currently available empirical and theoretical knowledge. The known disadvantages of such huge models is that the models are to be said "always wrong" (because of the intrinsic modelisation problem, to find the model that represents reality as it is, which is virtually impossible), but on the other side, they allow for understanding the complex dynamics and relevant mechanisms. Especially, when empirical and theoretical knowledge supports the model, dependencies become clearer. This is even more true, as is here the case, when the usability of the model allows that non-experts can easily use the cockpit of environmental influence levers in these simulations. With this user driven "playing around" dimension of active participation, not only the motivation, the memory, but also the competence gained increases, as the simulation is data driven and hence evidence based.

5 Discussion of Holism Under Consideration of Social Dynamics and Socioeconomics

Socioeconomics is a scientific field that is interdisciplinary and emergent. As such, it attempts to combine and unite areas of different origins. One approach to achieving this is the socio-dynamic approach. This is based on second-order cybernetics, which generally describes systems. Social dynamics, understood as social mechanics, can thus be described with flow and memory variables. The different "wholes", in the sense of relational structures or systems, can be modelled scientifically, as Popper also expresses, in such a way that wholes in the sense of all properties cannot be described scientifically, whereas wholes in the sense of relations can [29, p. 68 f.]. These relational structures or systems, therefore, describe systems of different sizes with different degrees of aggregation from subsystems, just as John von Neumann did with the foundation of socio-economic game theory [30] concerning value projections regarding the information unit money. He thus projected economic activity onto the quantitative and mathematical relationship structure as a theory. This approach, which was very fruitful due to its mathematisation, ultimately led to many other scientific methods, e.g. the mathematical optimisation with the min–max approach and the Nash equilibrium of game theory. In [30], John von Neumann also formulated the term dominance so that if $A > B$, A is dominant over B, which takes on precise meaning in the quantitative expression of a theory. In just such a sense, social dynamics can be understood in the sense of and going back to Jay Forrester [31] in the form of dynamic systems, i.e. time-dependent systems, each in the sense of an internal mechanics of an underlying cybernetic or computational model. Dominant modes reduce the order so that they can be simplified in specific situations while implicitly retaining the full complexity of the system. On the other hand, the discredited model approaches of Georg W. F. Hegel, which are characterised by a dialectical methodology, describe also part-whole relationships and their evolutionary development (see also [32]).

This means, (a) that this might be a late rehabilitation of whole-part systematologies by computer modelisations, (b) that the problematic features of such structures, can be better overcome by computational systems, by correction of "wrong" views associated with such a modelling, and (c) that we still have to be very cautious not to tap in in the trap of to much believing in those sorts of "idealisations", which can be translated into modelisations of reality. The last point (c) is of ever increasing importance, as we need to adapt to new paradigms given by such systems. Modern AI systems would be of this type. Although we did not include AI systems explicitly here, this is clearly included as by the following steps of cybernatisations or by increasingly complex systems and realities.

The problem of holistic systems remains if they are modelled mathematically as mathematical system dynamics and qualitative and philosophical system dynamics. A starting point for criticism of Hegel, particularly by Popper and also Topitsch, is the excessive idealisation of his philosophical system. As a particular point may be here, the absolute interpretation. Another approach of Topitsch, here concerning Habermas [33, p. 104 f.], which might be attributed to his holism, the argumentation is "theological", which means in the specific situation, that it lacks empirical verifiability (in German "Überprüfbarkeit", vs. verification), which renders the system arbitrary. This is a serious problem and attack on the system dynamics approach, which means that empirical evidence has to be considered carefully in general with these systems, as here is the crucial lever to counterargument. So, this is another point (d) to make those cybernetic holistic systems reliably is to include empirical data, and also to (e) decrease the distance with regard to back-couplings of the knowledge (see, e.g. [34]), where, e.g. one form is to increase "real-time" system behaviour.

In view of the social upheaval that we are currently facing from AI systems, which in reality represents a different form of increased system dynamics, and the points of criticism outlined above, it could be essential to further investigate and, above all, critically examine part-whole systems at a higher level, i.e. possibly also with more robust scientific-methodological tools.

Holistic systems in the context of system dynamics are often related to the logistic equation (see, e.g. [35]), which is a typical growth function showing towards limitations, where each is related to self-reference and exponential functions. The exponential function results from a specific mechanism that has often been observed and was used to describe such systems, as it is a limiting function. The Lyapunov coefficient λ (see, e.g. [36]) is closely related to this property and is defined for chaotic systems by the exponential function in the form of $e^{\lambda \cdot t}$. The basic idea here can be summarised that (e) the complexity dimensions decreases under some circumstances, which can be a good exercise for modelisations. So to investigate by computational models, whether, when and how a problem dimension reduces is a method, that can be used always as a safe harbour to improve the understanding of the observed system in its qualitative context.

6 Summary and Outlook

This work has compared different system simulation tools concerning their elements and the Turing machine concept.

For the qualitative ordering, or prerequisite of system dynamics modelling, we have proposed in Fig. 1 a model in three dimensions: (1) holisticity denoting the scale of complexity with regard to the axis particularism to holism or the relation whole to part, (2) realisticity, or the real to model interaction field, and (3) orgitonality for the ordering inside a domain, concerning cybernetic or orgitonal relationships.

We have formulated one translational framework for the given tools that allows, in principle, for the translation of the system conceptually. For this, we have developed five types as columns and five elements as rows in Fig. 2, that can be transformed principally into each other.

Application examples have illustrated such a possible translation of the models with regard to Witness and the Value Stream Analysis, in a first model more in detail, in a complex one more as a general approach towards modelisation principles.

Finally, we have given methodological reasoning on using system dynamic simulation and improving systematic accuracy and trust in the models to be investigated and discussed limitations. Five characteristics indicate the usefulness of whole-part system modelisations: (a) computational modelling, (b) increasing error correction, (c) increased awareness of being cautious with such models, (d) including increasingly empirical data, (e) learning to understand by modelling qualitative system properties.

As an outlook, we want to further investigate system dynamics software and other heuristics and develop the systems depicted here with examples. Another point will be to increase the complexity and investigate the consequence concerning knowledge, especially in modelling system dynamics. We also invite researchers in the field to further develop the given translation matrix and fill in with their own examples to realise a more resilient and unified world according to the osmotic paradigm of working together effectively.

References

1. Towell R.D.: System dynamics—background, methodology and applications—Part 1: Background and methodology. Computi. Control Eng. J., 201–208 (1993)
2. Towell R.D.: System dynamics—background, methodology and applications—Part 2: applications. Comput. Control Eng. J., 261–268 (1993)
3. Mates, B.: Elementare Logik—Prädikatenlogik der ersten Stufe. 2nd ed. Vandenhoeck und Ruprecht in Göttingen (1978)
4. Koestler, A.: The Ghost in the Machine, Danube Hutchinson, London (1976)
5. Heiden, B., Tonino-Heiden, B.: Diamonds of the orgiton theory. IJET **15**(4), 185–190 (2023), https://doi.org/10.7763/IJET.2023.V15.1244
6. Teichert, C.: "WITNESS—Logistik", Mit Innovationsmanagement zu Industrie 4.0. Ed. by P. Granig, E. Hartlieb, and B. Heiden, Springer Gabler Verlag, Wiesbaden, pp. 34–42 (2018)

7. ISEE Systems, available at: https://www.iseesystems.com/.
8. Gustafsson, L.: StochSD Home Page (2022). Jan. 30, 2022. available at: https://stochsd.sourceforge.io/.
9. Erlach, K.: Wertstromdesign—Der Weg zur schlanken Fabrik, 2nd edn. Springer Verlag, Berlin, Heidelberg (2010)
10. Turing, A.M.: On computable numbers, with an application to the Entscheidungsproblem. Proc. London Math. Soc. 2–42, 230–265 (1936)
11. Weizenbaum, J.: Die Macht der Computer und die Ohnmacht der Vernunft. 15th ed. suhrkamp taschenbuch wissenschaft, 369 pp. (2020)
12. Lödding, H.: Handbook of Manufacturing Control. Springer, Berlin, Heidelberg (2013). https://doi.org/10.1007/978-3-642-24458-2
13. Lödding, H.: Verfahren der Fertigungssteuerung. Grundlagen, Beschreibung, Konfiguration, 3rd ed. VDI-Buch. Berlin, Heidelberg: Springer, Berlin, Heidelberg (2016). ISBN: 9783662484593
14. Bechte, W.: Steuerung der Durchlaufzeit durch Belastungsorientierte Auftragsfreigabe bei Werkstattfertigung. VDI-Z (70,2). phdthesis, 1980. Düsseldorf: University of Hannover (1984)
15. Nyhuis, P., Wiendahl, H.-P.: Fundamentals of Production Logistics. Springer, Berlin, Heidelberg (2009)
16. Wiendahl, H.-P.: Fertigungsregelung, Logistische Beherrschung von Fertigungsabläufen auf Basis des Trichtermodells : 42 Tabellen. Hanser, München (1997)
17. Gustafsson, L.: StochSD User's, Manual and Tutorial. File: StochSD User's Manual 2022-01-01.docx. Uppsala, Sweden, Jan. 1, 78 pp. (2022), available at: https://stochsd.sourceforge.io
18. Dasgupta, S., Papadimitriou, C., Vazirani, U.: Algorithms. The McGraw-Hill Companies, Boston (2008)
19. Horkheimer, M.: Zur Kritik der instrumentellen Vernunft. Aus den Vorträgen und Aufzeichnungen seit Kriegsende, 5th ed. Fischer Taschenbuch (1997)
20. Grzelakowski, A.S.: Transport conditions of the global economy. In: Transport Economics and Logistics 80, pp. 75–84 (2018), ISSN: 2545-1642, https://doi.org/10.26881/etil.2018.80.08
21. (2023), Statistisches Bundesamt, Carbon dioxide emissions by road transport, available at: https://www.destatis.de/Europa/EN/Topic/Environment-energy/CarbonDioxideRoadTransport.html (visited on 05/15/2024)
22. Spencer-Brown, G.: Laws of Form, 6th edn. Bohmeier, Joh, Leipzig (2014)
23. Luhmann, N.: Soziale Systeme. Grundriß einer allgemeinen Theorie, 1st ed. Suhrkamp Verlag AG (2002)
24. Heiden, B., Alieksieiev, V., Tonino-Heiden, B.: Scalable logistic cell RFID witness model. Proceedings of the 5th International Conference on Internet of Things, Big Data and Security—Volume 1: IoTBDS. IOTBDS 2020 Conference. SCITEPRESS—Science and Technology Publications, pp. 420–427 (2020), https://doi.org/10.5220/0009490204200427
25. (2001), United Nations. Terminology of Combined Transport, New York and Geneva
26. Kaffka, J.: Kombinierter Verkehr. In: Verkehrs- und Transportlogistik, Ed. by U. Clausen and C. Geiger, Berlin, Heidelberg: Springer (2013)
27. W. Böse, J., Voß, S.: Informationsmanagement im Kombinierten Verkehr. Informationsmanagement im Verkehr, Heidelberg: Physica Verlag, pp. 269–322 (2000)
28. (2024), System Dynamics Society. How Did En-ROADS Get 755,000 users? Lessons on Modeling, Interface Design, and Facilitation, May 15, 2024. available at: https://systemdynamics.org/event/how-did-en-roads-get-755000-users-lessons-on-modeling-interface-design-and-facilitation/
29. Popper, K.R.: Das Elend des Historizismus, 4th edn., p. 157. Mohr, J.C.B (1974)
30. von Neumann, J., Morgenstern, O.: Spieltheorie und wirtschaftliches Verhalten, Ed. by F. Sommer, Trans. by M. Leppig, 2nd ed. Würzburg: Physika-Verlag (1967)
31. Forrester, J.W.: Industrial Dynamics, 4th edn. M.I.T. Press, Cambridge, Massachusetts, USA (1965)

32. Topitsch, E.: Die Sozialphilosophie Hegels als Heilslehre und Herrschaftsideologie. Ulm: Luchterhand,. 103 pp. (1967)
33. Topitsch, E.: Im Irrgarten der Zeitgeschichte, p. 178. Dunker and Humblot, Berlin (2003)
34. Heiden, B., Volk, M., Alieksieiev, V., Tonino-Heiden, B.: Framing artificial intelligence (AI) additive manufacturing (AM). In: Procedia Computer Science. 14th International Symposium "Intelligent systems" (INTELS'20), 14.-16. Dec. 2020 (Lomonosov Moscow State University, Online), Vol. 186, Moscow, Russia: Elsevier B.V., pp. 387–394 (2021), https://doi.org/10.1016/j.procs.2021.04.161
35. Soldov, A., Ochkov, V.: Differential Models, Springer (2005)
36. Hilborn, R.C.: Chaos and Nonlinear Dynamics—An Introduction for Scientists and Engineers. Oxford University Press, New York (1994)

The Synergy of University Innovation Ecosystem, Education and Science as a Key Factor in Ensuring the Implementation of Sustainable Development Strategies

Kryvobok Ruslan and Shmatko Nataliia

Abstract The study addresses pressing global challenges outlined in the UN's Sustainable Development Goals (SDGs) of 2015, emphasizing the need for innovative, context-specific solutions across the 17 key areas. The authors highlight the critical importance of training specialists capable of implementing these strategies and fostering innovation. Such training should integrate the university innovation ecosystem with education and science, prioritizing sustainable development goals. For Ukraine, facing the challenges of the Russian war, the strategy involves two stages: innovative recovery and subsequent innovative development, both requiring skilled innovators. The study examines the innovation ecosystem of the National Technical University 'Kharkiv Polytechnic Institute' (NTU 'KPI'), focusing on initiatives like the Spark Startup Centre, the Research Department, and the University Science Park. Using a systematic approach, the authors evaluate the role of this ecosystem in achieving SDGs through methods of generalization, systematization, and comparative analysis. Empirical data from surveys conducted among participants in NTU 'KPI' projects, including a startup school and the All-Ukrainian Festival of Innovations, confirm the curriculum's effectiveness in fostering skills from basic memorization to creative application, per Bloom's Taxonomy. Successful practices, such as engaging schoolchildren and students in innovation activities with local and international partners, are highlighted. The study offers practical recommendations for enhancing university ecosystems to support innovation among youth and researchers, essential for training future professionals.

K. Ruslan (✉)
Research Departmen, National Technical University "Kharkiv Polytechnic Institute", Kharkiv, Ukraine
e-mail: ruslan.krivobok@khpi.edu.ua

S. Nataliia
Department of Management, National Technical University "Kharkiv Polytechnic Institute", Kharkiv, Ukraine
e-mail: nataliia.shmatko@khpi.edu.ua

© The Author(s), under exclusive license to Springer Nature Switzerland AG 2025
A. Marchenko and Z. Koruba (eds.), *Decarbonization of Transport Energy Installations in the Context of Sustainable Development Strategies*, Studies in Systems, Decision and Control 635, https://doi.org/10.1007/978-3-032-05884-3_3

Introduction. In the current context of global challenges defined by the UN Sustainable Development Goals (SDGs) of 2015, the search for innovative solutions that take into account the local context and contribute to the achievement of 17 key areas of sustainable development is of particular relevance. Education and science play a key role in training specialists capable of effectively implementing these solutions. An important element of this process is the integration of the innovation ecosystem of universities with educational and scientific processes, which allows for the formation of competencies necessary for achieving the SDGs. The adaptation of these approaches to the conditions of Ukraine, which faces numerous challenges, mainly caused by russian aggression, deserves special attention.

In the conditions of war, Ukraine is faced with a significant outflow of personnel, a worsening demographic situation and a reduction in the total number of scientific personnel, which complicates the development of science and technology. At the same time, low state funding of science limits the opportunities for conducting fundamental research, which is the basis for the formation of an innovative economy. Today, STEM and STEAM educational and scientific ideology, which covers the fundamental areas of Education and Science in Physics, Chemistry, Biology, Mathematics and Engineering, serves as a basis for the creation of innovative technologies and provides the prerequisites for the general innovative development of any country. In view of this, the innovative recovery and further development of Ukraine require the creation of an effective system for training highly qualified specialists who are able not only to generate new ideas, but also to successfully implement them.

Problem statement. The problem under consideration is the insufficient integration of educational and scientific activities with real practices of today and innovative development, which hinders the training of specialists capable of effective activity in the conditions of global challenges. The research is focused on developing recommendations for improving university ecosystems to promote the development of innovative competencies.

Research objective. The purpose of this study is to determine the role of the university's innovation ecosystem in the context of ensuring sustainable development by training specialists capable of creating and implementing innovations. Research objectives:

- analyze the current state of financing and staffing of university education and science;
- determine the role of STEM and STEAM education and science in the innovative development of the state;
- substantiate the role of the university's innovation ecosystem in the context of ensuring the country's sustainable development;
- conduct an empirical study of the impact of the university's innovation ecosystem on the formation of skills and creative thinking of participants in educational programs.

The subject of the study is educational, scientific and innovative processes integrated into the university ecosystem. The object of the study is the innovation ecosystem of the National Technical University "Kharkiv Polytechnic Institute" (NTU "KhPI"), which includes its individual elements, such as Startup Centre Spark, the research part and LLC "Science Park NTU "KhPI".

1 Analysis of the Current State of Funding and Staffing of Science

The science intensity of GDP in Ukraine has been significantly decreasing over the past decade, which indicates a systemic lack of investment in research and innovation activities. Over the period from 2012 to 2022, this indicator has more than halved: from 0.67% to 0.32% of GDP. The science intensity of GDP in Ukraine is 0.4%, which is one of the lowest indicators and corresponds to the level of Chile, inferior even to Romania (0.5%) and Latvia (0.6%). Ukraine is 5 times behind the average indicator of OECD countries (2.0%) and 10 times behind leaders such as Israel (4.3%), Sweden and Japan (3.3% each) (Fig. 1). This indicates a critically low level of investment in science, which limits the country's innovative development opportunities and deepens its lag in the global scientific and technological space.

Insufficient funding for science and low intensity of research activities in Ukraine create serious obstacles to the personnel support of innovative sustainable development of Ukraine. For example, in industry, the number of scientific and technical personnel in 2022 decreased to 2100 people, while in 2021 this figure was 2377 people [3, 4]. After 2022, this figure deteriorated significantly.

Given the announced data, the problem of financing science in Ukraine is becoming systemic, which significantly affects the country's ability to develop innovations and remain competitive in the global environment. Although an increase in science funding by 20% in 2024 to 2.2 billion hryvnia is a positive signal, it does not

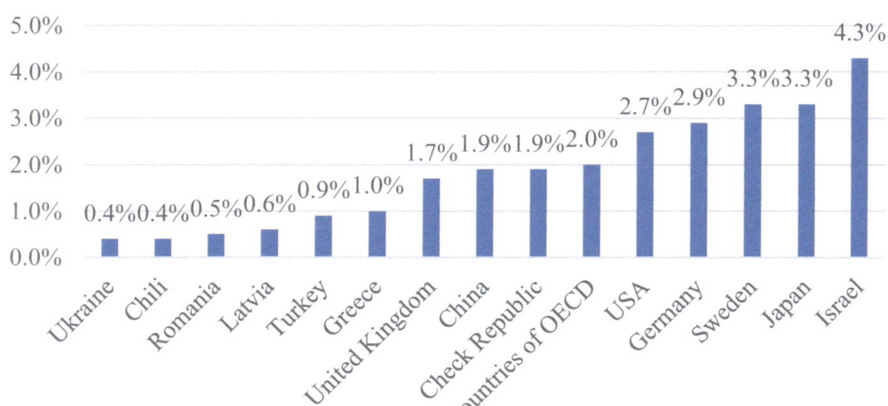

Fig. 1 Comparison of the knowledge intensity of GDP in countries around the world (2022) [1, 2]

change the general trend of decreasing relative indicators of the science intensity of GDP, which are projected at 0.17% for 2024 [5, 6].

This indicator is critically low, especially compared to EU countries, whose average level in 2021 was 2.26%. For comparison, in Sweden this indicator reached 3.35%, in Germany—3.13%, and even in countries with lower indicators, such as Latvia (0.6%) or Romania (0.5%), the level significantly exceeds the Ukrainian one.

Failure to meet the legally defined norms of science financing raises a reasonable question about the prospects for obtaining significant scientific results in Ukraine. According to Article 48 of the Law of Ukraine "On Scientific and Scientific and Technical Activities", the level of state funding of the scientific sphere should be at least 1.7% of GDP. However, since the adoption of this law in 2015, this norm has never been met, which largely determines the lag of domestic science from global development trends [6].

Moreover, in the context of global competition and rapid technological progress caused by the fourth and fifth industrial revolutions, insufficient funding for science leads not only to technological lag, but also to the loss of opportunities for increasing economic efficiency, improving the quality of life of the population and developing innovative solutions for strategically important industries. The lack of proper support for scientific research limits the potential of the national economy, reduces its competitiveness in the international market and complicates integration into the global scientific and technological space.

There is a decline in human resources. Along with insufficient funding, there is a critical decline in human resources in industry and research. From 2010 to 2022, the number of workers employed in industry almost halved: from 3,291.9 thousand to 1,808.6 thousand people [3, 5]. The war exacerbated these negative trends, leading to a significant outflow of labor resources through mobilization, migration, and internal displacement of the population. In 2022, the number of people employed in industry decreased by 16.5% compared to 2021.

The quality of labor potential is deteriorating, in particular due to the shortage of young specialists in important engineering, design and technological specialties. Universities are faced with problems of training and attracting young personnel, which threatens the prospects of innovative development of the country. In conditions of war, the shortage of personnel has become one of the main problems for 43% of enterprises, according to surveys by the Civil Organization "Institute of Economic Research and Policy Consultations". Insufficient funding for science and the low level of knowledge-intensiveness of GDP directly affect the development of key industries that shape the future of the state by realizing the possibilities of its existing innovative potential. Therefore, one of the key systemic factors for overcoming economic problems is the proposed STEM and STEAM strategies of education and science, which are the basis for training highly qualified personnel capable of developing and implementing the latest technologies. In the context of global competition, investments in STEM education are not only educational, but also a strategic economic priority that requires special attention from Ukraine.

2 Conceptual Provisions of Sustainable Development in the Synergy of Education and Science

Sustainable development requires a comprehensive approach that involves the integration of economic, social and environmental aspects, as well as the mobilization of the innovative potential of society. In this context, universities play a key role as centers of knowledge, scientific research and training of future personnel. However, the effective implementation of the 17 UN Sustainable Development Goals requires new approaches to innovation management, the development of the education system and human resources for science. In this regard, there is a need to outline conceptual provisions that determine the strategic guidelines and framework of this study.

Provision 1. Transport is the basis for achieving the Sustainable Development Goals.

The implementation of the UN Sustainable Development Goals is impossible without an effectively functioning transport system that ensures the mobile efficiency of resources, people and information in an innovative model of private–public partnership of education, science, business and government institutions. Transport infrastructure contributes to economic integration, social cohesion and access to basic services, such as education and health. Transport plays a particularly important role in achieving goals related to sustainable production and consumption, climate change and urban mobility. Therefore, the development of an effective sustainable development strategy requires taking into account transport aspects as a fundamental component.

Provision 2. The STEM and STEAM paradigm of education and science is the driving force of innovative development.

Achieving the goals of sustainable development is possible only if scientific and technological progress is ensured, which is based on the training of personnel of the new paradigm of innovators. STEM and STEAM education and science (science, technology, engineering and mathematics) forms the competencies necessary for the development of technological and social innovations capable of solving global challenges. It contributes to the formation of critical thinking, creativity and the ability to interdisciplinary interaction. Therefore, investments in STEM and STEAM education and science are a necessary condition for the training of innovators who can ensure the implementation of sustainable development strategies.

Provision 3. The features of sustainable development in Ukraine are more profound in connection with the consequences of the war, economic instability and social transformations.

These factors complicate the processes of socialization, economic recovery and human capital development. The problem of finding innovators and training personnel capable of adapting the world experience of sustainable development to national realities is particularly acute. This requires the formation of a special strategy that takes into account the specifics of the socio-economic situation in the country.

Provision 4. Universities are a source of innovation for the implementation of the Sustainable Development Goals.

Universities play a key role as a source of new innovators for achieving the 17 Sustainable Development Goals by developing relevant competencies in young people. They act as centers of scientific research, generators of new knowledge and platforms for interdisciplinary interaction. Innovative ecosystems of universities ensure the transfer of knowledge to the real sector of the economy, contributing to the development of startups and technology clusters. Thus, universities become catalysts for sustainable development through the synergy of education, science and entrepreneurship.

Provision 5. Effective implementation of the sustainable development strategy is possible only under the conditions of ensuring the synergy of science, education and innovation.

The relationship between fundamental research, applied developments and educational processes creates the prerequisites for the formation of an innovative economy. This requires updating the content of education, focused on the development of creative thinking, entrepreneurial skills and social responsibility. The synergy of these components allows to ensure a continuous cycle of innovation from fundamental knowledge to its practical application.

Provision 6. Sustainable development requires global cooperation and strategic partnerships between universities, research centers, business and government institutions, systems of legislative and current state management and regulation.

Integration of local initiatives into global networks allows for the exchange of experience, resources and best practices. This contributes to the acceleration of innovation processes and the effective resolution of global challenges. International scientific projects play a special role, contributing to the mobility of knowledge and the formation of a global innovation environment.

Thus, the conceptual foundations of the sustainable development strategy, based on the synergy of university innovations, education and science, define strategic guidelines for achieving the 17 UN Sustainable Development Goals. In particular, they emphasize the importance of transport, STEM education, innovation ecosystems, global cooperation and digital transformation. Special attention is paid to the specifics of the Ukrainian context, which necessitates the development of adaptive sustainable development strategies. The proposed provisions create a methodological basis for further empirical research and the formulation of hypotheses aimed at solving current problems of sustainable development.

3 Determining the Role of STEM and STEAM Education and Science in the Innovative Development of the State

The role of STEM and STEAM education and science (science, technology, engineering, mathematics) in the innovative development of the state is fundamental, as it contributes to the formation of highly qualified human resources capable of generating new ideas and implementing them in practice. These areas are the basis for creating innovations that determine the competitiveness of the state in the conditions of globalization. According to studies by the European Commission, countries with a high level of STEM and STEAM education and science demonstrate a significant contribution to the development of high-tech industries, such as information technology, biomedicine and clean energy [7, 8].

In many countries of the world, STEM and STEAM education and science have become a strategic priority. For example, in the USA, more than $3 billion is allocated annually for STEM programs aimed at preparing young people for work in innovative sectors of the economy. In Singapore, the Smart Nation program has been implemented since 2015, which includes the active introduction of STEM education into the school system to prepare for the digital economy [7, 9]. In Finland, STEM education is integrated into the general educational process, and the emphasis on an interdisciplinary approach contributes to the formation of practical skills and critical thinking.

Statistical data confirm the direct relationship between the level of STEM and STEAM education and science and innovative development. For example, in OECD countries, where more than 2% of GDP is spent on STEM and STEAM education and science [10], the level of patent activity is significantly higher than in countries with less funding. The table compares the expenditure on STEM and STEAM education and science and the number of patent applications filed per million population in selected countries. The specific expenditure on STEM and STEAM education and science and patent activity (2022) are presented in Table 1.

The data in Table 2 confirm that countries that invest heavily in education and science receive greater returns in the form of technological innovations. In the USA, Germany and Singapore, intensive funding contributes to high patent activity, while in Ukraine, low spending on STEM and STEAM education and science is reflected in modest results. Insufficient investment in STEM and STEAM education and science in Ukraine is one of the key reasons for the low level of innovative development.

Table 1 Expenditures on STEM and STEAM education and science and patent activity (2022)

Country	Expenditures, % of GDP	Patent applications per 1 million population
USA	2.5	862
Germany	2.3	832
Finland	2.1	794
Ukraine	0.4	47

Among other challenges, it is worth highlighting the imperfection of the material and technical base, insufficient integration of education with scientific research and weak cooperation with business. To overcome these problems, it is necessary to introduce new approaches to financing and managing education, focusing on the experience of developed countries.

Thus, STEM education plays a crucial role in shaping an innovative economy, which is confirmed by statistical data and successful experience of other countries. For Ukraine, the development of STEM areas is a strategic task that requires a significant increase in funding, modernization of educational programs and strengthening the connection between education, science and business.

Table 2 Elements of the university's innovation ecosystem and their contribution to achieving the Sustainable Development Goals

Sustainable Development Goals (SDG)	Initiatives of NTU KhPI	Structural divisions of NTU "KhPI"
SDG 4. Quality education	Startup school for schoolchildren	Startup Center "Spark", Research Department
	All-Ukrainian Innovation Festival	
	Scientific and educational laboratories	
SDG 9. Industry, innovation and infrastructure	Submission of innovative projects to the CIG R&LAB competition	Research Department
	Commercialization of innovations (cooperation of startups with business, implementation of ideas under the order of business)	LLC "Science Park NTU "KhPI"
SDG 11. Sustainable development of cities and communities	Submission of works by young scientists to state innovation funding programs	Research Department
SDG 17. Partnership for the Goals	Involvement of foreign partners from China to participate in competitions as investors	Startup Center "Spark", Research Department, LLC "Science Park NTU "KhPI"
	Involvement of business partners as mentors, investors, employers of startups	
	Involvement of government structures in mentoring, employment, and support of startups	

4 Justification of the Role of the University Innovation Ecosystem in the Context of Ensuring the Sustainable Development of the Country

The university innovation ecosystem is a key element that contributes to the training of specialists capable of meeting the challenges of the modern world and ensuring the sustainable development of the country. Universities play an important role as centers of integration of education, science and innovation, which allows creating new knowledge and technologies. In the previous section, it was noted that STEM and STEAM education and science are the foundation of innovative development, but its effectiveness largely depends on the presence of an innovation ecosystem that is able to combine theoretical training with practical research and entrepreneurial activity (Fig. 2).

The experience of developed countries shows that universities with strong innovation ecosystems are becoming drivers of sustainable development. For example, in the USA, universities such as the Massachusetts Institute of Technology (MIT) actively cooperate with business, creating startups that introduce advanced technologies into industry and the social sphere. In Finland, universities are centers of regional development, integrating local needs into their research programs. In Ukraine, the development of the innovation ecosystem is still at an early stage, but there are examples that demonstrate the prospects of this direction, such as the activities of LLC "Science Park NTU "KhPI" & Startup Center "Spark".

Fig. 2 All-Ukrainian innovation festival

Table 3.2 shows the main components of the university ecosystem, their functions and contribution to the implementation of the Sustainable Development Goals, such as quality education (SDG 4), industry, innovation and infrastructure (SDG 9) and partnerships for sustainable development (SDG 17).

NTU "KhPI" implements a number of initiatives aimed at creating a sustainable innovation ecosystem that supports the Sustainable Development Goals, in particular in the areas of education, innovation and sustainable community development.

SDG 4 Quality Education—initiatives that contribute to the development of high-quality education, preparing young people for real market conditions, creating startup schools for schoolchildren and holding an innovation festival that provides students with practical skills and access to the latest technologies within the NTU KhPI innovation ecosystem.

SDG 9 Industry, innovation and infrastructure—initiatives that support startups, scientific research and innovation, as well as contributing to the creation of infrastructure for cooperation with business and commercialization of ideas, such as the CIG R&D LAB educational and scholarship program from the international investment program Chernovetskyi Investment Group, which are important elements of the university's innovation ecosystem.

SDG 11 Sustainable development of cities and communities—initiatives that stimulate the development of innovations aimed at supporting state programs, involving young scientists in projects that contribute to the sustainable development of communities, and integrating these initiatives into the sustainable future of regions through the university's innovation ecosystem. For example, the Kharkiv Energy Efficient Project, which includes NTU "KhPI", has become an important stage in the sustainable development of the city through the implementation of innovative solutions in the field of energy saving and increasing energy efficiency. Young scientists, students and postgraduates of the university are actively involved in the development of technologies that contribute to reducing energy costs in the city and at enterprises. Special attention is paid to smart energy supply networks, modernization of heating and the use of renewable energy sources. NTU "KhPI" cooperates with state and international programs that finance these initiatives. This allows the city to become an example for other regions and increase competitiveness at the international level.

SDG 17 Partnership for Sustainable Development—initiatives that promote the development of international cooperation, the involvement of business partners, investors and government agencies in supporting startups and innovative projects that form the basis for the sustainable development of the innovation ecosystem of NTU "KhPI" and facilitate interaction with the global environment.

The developed innovation ecosystem of the university contributes to the transformation of science into applied technologies that help solve current socio-economic problems. For example, within the framework of an international project under **the Horizon Europe program**, NTU "KhPI" conducts research on new technologies for the effective use of solar and wind energy in Ukraine, as well as for the integration of these energy sources into the energy systems of the region. Such work allows not only to improve energy efficiency, but also to reduce CO2 emissions, contributing to sustainable development. In addition, university startups actively contribute to the development of the local economy, attracting investments and creating new jobs, which is important for the socio-economic growth of the region.

An important component of success is close cooperation between universities, business and government, which allows you to direct resources to solving specific problems. In developed countries, the share of financing of innovative university projects by business reaches 30–50%, while in Ukraine this figure remains much lower. This limits the opportunities for implementing new developments in practice and increasing the country's competitiveness. For Ukraine, the development of such ecosystems should become one of the priorities of state policy, as this will create the prerequisites for technological breakthroughs and economic stability.

To compare the sustainable innovation ecosystem of NTU "KhPI" with other universities [11–13] that have developed innovation ecosystems, it is important to pay attention to several main aspects: the structure of the innovation ecosystem, the level of cooperation with business, the presence of mentoring programs, and the role of international cooperation in the development of startups and innovations. One of the most famous and developed innovation ecosystems is the Stanford University ecosystem. This ecosystem is characterized by the close integration of the university with local businesses and technology startups, which allows for the implementation of many innovative projects [13]. It includes startup accelerators, mentoring programs from successful entrepreneurs, and close ties with investors. Stanford also actively supports international cooperation, which allows the university to attract investments and create global partnerships.

NTU "KhPI", in turn, creates an innovation ecosystem that includes startup schools for schoolchildren, a technology park, an accelerator for startups, as well as programs to support young scientists through grants and state funding. This allows the university not only to support innovation in Ukraine, but also to form a strong connection between science and business [11]. Universities such as the Massachusetts Institute of Technology (MIT) are known for their high concentration of investments in startups and innovation, in particular through the use of large infrastructures for research projects and the commercialization of ideas. These universities also have strong partnerships with government agencies and international organizations to ensure the sustainable development of innovations [12]. Universities such as MIT also have a high level of interdisciplinary cooperation, which allows them to create innovations at the intersection of different fields of knowledge. A comparison of the ecosystems of NTU "KhPI" with other universities is presented in Table 3.

The innovation ecosystem of NTU "KhPI" demonstrates significant progress in supporting startups and innovation among young people through its infrastructure focused on integration with business and scientific programs. However, compared to universities such as Stanford or MIT, there is potential for improving cooperation with international investors and developing more structured programs for young entrepreneurs. For this university, it is worth expanding its partnerships with global innovation ecosystems and investors, as well as continuing to integrate the latest technologies into the educational process and research activities.

To improve the sustainable innovation ecosystem of NTU "KhPI", aimed at supporting innovation among young people and researchers, this system aims to

Table 3 Comparison of ecosystems of NTU "KhPI" with other universities

Criteria	NTU "KhPI"	Stanford University	Massachusetts Institute of Technology (MIT)
Innovative infrastructure	Startup Center "Spark", Research Department, LLC "Science Park NTU "KhPI"	Incubators, startup accelerators	Research laboratories, technology parks
Collaboration with business	Collaboration with business through startups and innovative projects	Strong ties with local businesses and investors	Close collaboration with technology companies and investors
International cooperation	Partnerships with Chinese investors and international partners	International cooperation to attract investments and new technologies	Global partnerships, attracting international investors
Support for young researchers	Mentoring programs, grants for young scientists	Mentoring and startup investment programs	Innovation funding, grants and mentoring programs
Startup development	Support for startups through internal events (All-Ukrainian Innovation Festival, Startup School), mentoring, participation in external events (CIG R&LAB competition)	Support for startups through investments and mentoring	Startups through incubators and close cooperation with business

create conditions that encourage creativity and entrepreneurship. It is important to ensure that students and young scientists have access to resources to implement their ideas and projects. The university is creating innovation laboratories and startup accelerators, where young people can work on their ideas, receive mentoring support and learn from experienced entrepreneurs and scientists. An important step is the active cooperation of NTU "KhPI" with business, government institutions and international partners, which allows integrating innovative projects into the university environment and providing students with real opportunities to implement their ideas.

The integration of innovative technologies into the educational process is a key aspect of the development of the university ecosystem. Students and researchers of NTU "KhPI" are aimed at the opportunity to work with advanced technologies that allow not only to develop ideas, but also to implement them in the real world. This includes participation in international competitions, conferences and hackathons, where they can present their developments, exchange experience and find new partners for further cooperation. In addition, the interdisciplinary approach to learning allows you to combine different fields of knowledge, which contributes to the search for new solutions to global problems and the development of innovations [14].

An important part of a sustainable innovation ecosystem is a system of incentives for young researchers and students engaged in innovative activities. The university supports talented youth with grants, scholarships and competitions for the best innovative projects, which will contribute to the development of creative initiatives. In addition, NTU "KhPI" organizes trainings, seminars, and other events that will help students and researchers acquire entrepreneurial skills to successfully implement their ideas in practice.

Finally, it is important to maintain a culture of openness and collaboration within the university ecosystem. Creating platforms for the exchange of ideas between students, teachers, researchers and entrepreneurs allows for effective communication and stimulates the generation of new ideas. This contributes to the integration of innovations into the socio-economic environment, as well as supporting the sustainable development of the university and its partners.

5 Empirical Study of the Impact of the University's Innovation Ecosystem on the Formation of Skills and Creative Thinking of Participants in University Programs

The university's innovation ecosystem plays an important role in training specialists at various levels of education. It helps to involve not only students, but also schoolchildren and college students in innovation activities, forming in them critically important skills and thinking. As an example, we can cite startup school projects for schoolchildren, which allow students to get acquainted with the basics of entrepreneurship and technology through practical experience and, most importantly, are an example of the implementation of STEM and STEAM education ideas for training innovators. For university students (bachelor's, master's, postgraduate students), national-level events are held, such as the All-Ukrainian Innovation Festival, where participants present their ideas, develop public speaking skills and improve projects in collaboration with mentors. In order to assess the effectiveness of such educational initiatives, a survey was conducted among participants of startup schools for schoolchildren. The main hypothesis of the study is the assumption that participation in startup schools and innovation festivals contributes to the development of entrepreneurial, technological, and soft skills in students, and also allows them to master all six levels of Bloom's taxonomy in a short time compared to traditional educational approaches.

Bloom's Taxonomy is the basis for assessing learning objectives, allowing you to track progress from basic memorization of knowledge to creating a unique product. Startup School implements all stages of the taxonomy: memorizing basic concepts through lectures, understanding through discussions with mentors, applying them in developing a business model, analyzing competitors, evaluating an idea, and creating an innovative product. Thus, the learning methodology at Startup School integrates all levels of thinking, allowing participants to gain a deeper understanding of the learning material.

If you follow Bloom's taxonomy (classification), then student knowledge is only the first, simplest level of classification. This is followed by five more levels of learning objectives (or outcomes), with the first three—*knowledge, understanding, application*—being lower-order objectives (low-level thinking). And the next three—*analysis, synthesis, evaluation*—being higher-order (high-level thinking).

Startup school allows students to consistently go through all stages of learning according to Bloom's taxonomy—from mastering basic knowledge to creating their own innovative product, while developing entrepreneurial, technological and creative skills (Table 4).

A survey was conducted among participants of startup schools for schoolchildren organized by universities to investigate the impact of this form of learning on the acquisition of the stages of Bloom's Taxonomy. The survey included questions aimed at assessing the skills acquired during participation, the levels of knowledge according to Bloom's Taxonomy, as well as the challenges faced by participants (Fig. 3).

The survey results are expected to confirm that the interactive format of startup schools contributes to the development of both intellectual and practical skills of participants. In particular, the key advantages of startup schools are the formation of soft skills (communication, leadership, critical thinking) and entrepreneurial competencies (business planning, financial literacy). Support from mentors and teamwork further enhance the effectiveness of learning.

Table 4 Interpretation of the stages of student learning in the Startup School

Bloom's taxonomy stage	Startup school tasks
1. Memorization	Listening to lectures on the basics of a startup: idea generation, stages of working on a startup, presentation, pitching. Students memorize basic concepts and instructions
2. Understanding	Discussion of lecture material with mentors. Teamwork on understanding the tasks and ideas of the startup. Formation of a clear idea of market analysis, competitors, business structure
3. Application	Carrying out practical tasks: market analysis, business model development, creation of a startup concept. Application of knowledge in teamwork during the preparation of interim results
4. Analysis	Detailed analysis of competitors, strengths and weaknesses of the startup idea. Teams develop a SWOT analysis and a strategic plan for the project development
5. Evaluation	Discussing results with mentors. Evaluating your own idea and business model based on market data and feedback. Improving the project before the final pitch
6. Creation	Developing the final startup presentation, creating a unique product or solution. Preparing for pitching: rehearsals, designing the presentation. Presenting the idea to the jury for feedback and further improvement

Fig. 3 Startup school for schoolchildren of NTU "KhPI"

The next stage of the study is the analysis of the survey results and the assessment of how participation in startup schools affects the formation of skills and thinking of participants. The collected data will allow us to offer recommendations for improving innovative educational programs.

To assess the results of the participation of schoolchildren and college students in the project "Startup School for Schoolchildren" of NTU "KhPI", a survey was conducted among 100 participants. The survey took place after the completion of the curriculum and the final assessment, which lasted 1.5 months. During this time, the participants studied twice a week, 4 h each lesson in an offline format. The online survey allowed us to collect feedback and assessments from students, which allows us to obtain important information for analyzing the effectiveness of the program and determining its impact on the participants. Survey results.

1. Distribution of answers to the question about interest in the event: "Yes, very interesting"—64.71%; "It was interesting"—35.29%.
2. Distribution of skills that were most interesting:
   - soft skills (communication, leadership, critical thinking)—52.94%;
   - entrepreneurial (business planning, financial literacy)—29.41%;
   - all of the above options—5.88%;
   - technological (programming, working with innovative technologies)—5.88%.
3. Distribution of participation in future events: "Not yet, but I'm thinking about it"—52.94%; "Yes"—47.06%.

4. Average event rating: average score—4.65. Standard deviation (spread of scores)—0.79. The arithmetic mean is 4.65.

Most respondents consider the event very interesting (over 64%). This indicates a high interest of the audience. The most valuable were "soft skills"—over 52% of respondents noted them. This indicates the importance of communication and leadership competencies. Regarding future participation in similar events, more than half have not yet decided, although almost 47% are already ready to participate. The high average rating (4.65 out of 5) indicates a positive perception of the event, and a small standard deviation (0.79) shows that the assessments were quite even. Let's consider the results of the survey on the mastery of cognitive stages of training. The questionnaire asked the question: "How do you assess your level of mastery of cognitive stages during participation in the startup school?" (1—very low, 5—very high). The results of the survey on the mastery of cognitive stages by schoolchildren are presented in Table 5.

Analysis of data on the assimilation of cognitive stages by schoolchildren.

1. General level of assimilation of cognitive stages

The distribution of responses shows that the highest level of assimilation is observed at the "Creation" stage, where 41% of participants rated it at "5 (very high)", and another 40% at "4". This indicates that most respondents believe that they have acquired the skills to create new products or ideas during the startup school.

The "Memorization" stage also received high marks: 40% at "5" and 40% at "4". This means that the participants have well assimilated the basic information provided as part of the training.

The least developed stage was "Analysis", where only 29% rated this stage at "5" and 38% at "4". This may indicate the need for greater attention to critical analysis of information and its structuring.

2. Average scores by stage

For a more in-depth analysis, we calculate the average score for each stage of cognition using the formula:

$$\overline{x} = \frac{\sum_{i=1}^{n} (x_i \bullet f_i)}{N}$$

Table 5 Results of the survey on the mastery of cognitive stages by schoolchildren

Stage of cognition	1 (low) (%)	2 (%)	3 (%)	4 (%)	5 (high) (%)
Memorizing	2	5	13	40	40
Understanding	3	4	12	45	36
Application	4	6	15	42	33
Analyzing	5	8	20	38	29
Evaluating	7	9	18	35	31
Creating	3	6	10	40	41

Table 6 Results of average values of mastery of cognitive stages by schoolchildren

Stage of cognition	Average score
Memorizing	4.11
Understanding	4.07
Application	3.94
Analyzing	3.78
Evaluating	3.81
Creating	4.13

where:

x_i– assessment level (from 1 to 5);

f_i– number of respondents who chose the level x_i;

N—total number of respondents.

The results of the average values of the assimilation of cognitive stages by schoolchildren are presented in Table 6. The following conclusions can be drawn from the survey results.

1. General results. Startup school participants, on average, highly rated their achievements in all stages of cognition, with the greatest success observed in the "Creation" and "Memorization" stages, which indicates the successful implementation of the curriculum.
2. Weaknesses. The lowest average scores were received by the "Analysis" and "Evaluation" stages, which indicates the need to improve educational approaches in these aspects. Perhaps it is worth adding more practical tasks that contribute to the development of critical thinking.
3. Suggestions for improvement. Including special classes aimed at developing analytical and evaluation skills (for example, case studies or group analysis of real situations). Providing detailed criteria for evaluating completed tasks to improve the "Evaluation" stage.
4. Positive aspects. The highest scores for "Creation" confirm that participants felt involved in the creative process. This indicates a high level of motivation and the effectiveness of team work methods.
5. Recommendations for future courses. Add more interdisciplinary projects to better integrate all stages of learning. Focus on practical application of knowledge, as the "Application" score received only a 3.94 on average.

Thus, the startup school achieved high results in training participants, especially in the aspects of creativity and practical memorization. At the same time, it is worth paying attention to improving analytical and evaluation skills, which will improve the overall learning experience. Participation in startup schools really contributes to the development of key skills in schoolchildren, including entrepreneurial, technological and soft skills. The survey confirmed that the program participants not only deepened their knowledge in these areas, but also gained practical experience, which significantly improved their ability to think critically, work in a team and solve problems.

Regarding the confirmation of the survey hypothesis, the results clearly indicate that the startup school contributes to the development of all six levels of Bloom's taxonomy. During the program, participants not only memorized theoretical materials, but also actively applied the knowledge gained in practice, analyzed situations, evaluated solution options and even created their own projects. This confirms that the startup school allows you to master the higher levels of Bloom's taxonomy in a short time, significantly reducing the time to achieve deep understanding and creative application of knowledge compared to traditional educational approaches.

6 Conclusions

Analysis of the current state of funding and staffing of science in Ukraine has demonstrated significant shortcomings, in particular the critically low level of knowledge intensity of GDP (0.4%), which is many times inferior to OECD countries. This significantly hinders innovative development, limiting the possibilities of implementing scientific research and technological developments.

It has been established that STEM and STEAM education play a key role in forming human resources capable of creating and implementing innovations. For Ukraine, it is important to strengthen funding for STEM areas, which will ensure increased innovative activity and competitiveness of the country.

The innovative ecosystem of the university is critically important in achieving the UN Sustainable Development Goals. The example of the National Technical University "Kharkiv Polytechnic Institute" shows that an effective combination of educational, scientific and innovative processes contributes to the formation of skills necessary for the implementation of sustainable development strategies.

Empirical research has confirmed the high effectiveness of interactive educational programs (startup schools and innovation festivals) in developing both basic and creative skills among young people. Program participants demonstrate significant progress at all levels of Bloom's taxonomy, especially in creating innovative products and solutions.

For the further development of university innovation ecosystems, it is recommended to expand international cooperation, implement structured mentoring programs, and intensify business involvement in financing innovation projects. This will create the prerequisites for technological breakthroughs, strengthening scientific and technical potential, and achieving sustainable development of Ukraine.

References

1. Eurostat: Research and development expenditure and personnel (2023). Rezhym dostupu: https://ec.europa.eu/eurostat
2. Derzhavna sluzhba statystyky Ukrainy: Naukova ta naukovo-tekhnichna diialnist v Ukraini (2023). Rezhym dostupu: http://www.ukrstat.gov.ua
3. IERPC: Shchomisiachne opytuvannia promyslovykh pidpryiemstv. Instytut ekonomichnykh doslidzhen ta politychnykh konsultatsii (2023). Rezhym dostupu: https://www.ier.com.ua.
4. Derzhavnyi biudzhet Ukrainy na 2024 rik. Zakon Ukrainy. Rezhym dostupu: https://zakon.rada.gov.ua/laws/show/
5. Natsionalna akademiia nauk Ukrainy: Shchorichnyi zvit pro stan nauky v Ukraini (2023). Rezhym dostupu: https://www.nas.gov.ua
6. Shulkin, D.: Finansuvannia nauky—shliakh do peremohy. Rezhym dostupu: https://svit.kpi.ua
7. Pozdniakova, A.M.: Vprovadzhennia kontseptsii rozumnykh stalykh mist v Ukraini: osoblyvosti ta rekomendatsii. Problemy systemnoho pidkhodu v ekonomitsi, Vyp. **2**(2), S. 49–57 (2019). Rezhym dostupu: http://nbuv.gov.ua/UJRN/PSPE_print_2019_2%282%29_10
8. Postova, S., Mysiuk, O., Cherniak, Yu.: STEM-osvita yak kliuch do rozvytku lideriv maibutnoho: formuvannia kliuchovykh navychok. Naukovi innovatsii ta peredovi tekhnolohii (2024). Rezhym dostupu: https://doi.org/10.52058/2786-5274-2024-11(39)-1486-1498
9. Ordanovska, O., Romashchenko, K., Tsyna, V., Tsyna, A., Postova, S.: Implementation of stem education in general education institutions. Conhecimento & Diversidade **15**, 119–140 (2023), https://doi.org/10.18316/rcd.v15i40.11272
10. Stohlmann, M., Moore, T., Roehrig, G.: Considerations for Teaching Integrated STEM Education. J. Pre-College Eng. Educ. Res. **2**,(2012). https://doi.org/10.5703/1288284314653
11. Pererva, P., Ievsieiev, A., Maslak, M., Tkachov, M., Tkachova, N.: Formation of intellectual property commercialization strategies. Eastern-Europ. J. Enterp. Technol. **1**(13)(127), 80–91 (2024). https://doi.org/10.15587/1729-4061.2024.296836
12. Lee, K., Lim, S., Cho, K.: Innovation ecosystems at MIT: a comparative study. J. Technol. Transfer **45**(4), 876–895 (2020). https://doi.org/10.1007/s10961-020-09742-5
13. Sweeney, M.: Stanford university: a model for innovation ecosystems. Innov. Entrepreneurship J. **22**(1), 55–71 (2019). https://doi.org/10.1080/08985626.2019.1582643
14. Sribna, Y., Shmatko, N., Kustrich, L., Parhomenko, L., Neboha, T., Lytvynenko, A., Lytvynenko, E.: Environmental and social responsibility management in international entrepreneurship in EU countries. E3S Web of Conferences. Volume 408. International Conference on Sustainable, Circular Management and Environmental Engineering (ISCMEE 2023). Izmir, Turkey, July 12. P.15 (2023), https://doi.org/10.1051/e3sconf/202340801024

Challenges and the Latest Trends of Application of Transport ICEs as a Long-Term Prospect Energy Generating Means

Volodymyr O. Pylyov, Oleh Y. Linkov, and Vyacheslav V. Pylyov

Abstract Solving energy and environmental problems is one of the main challenges facing modern society. Transport and mechanical engineering should be involved in its implementation to the maximum extent possible. The results of scientific discussions and management decisions regarding the future of internal combustion engines determine the conceptual provisions, directions of design improvement, and the degree and quality of decarbonisation of the energy sector as a whole. The article uses a systematic approach to examine the trends in the use of internal combustion engines as a necessary part of climate-neutral transport. These problems are considered from the following aspects: the need to transition from the use of fossil energy sources to the large-scale use of RES; the objective contradiction between the use of RES and the universal use of electric vehicles as electricity consumers; the introduction of ICEs that leave a carbon footprint into technological cycles to support a climate-neutral transport industry based on RES; ways to implement the concept of a circular carbon economy that meets the requirements of climate-neutral energy while maintaining the processes with a carbon footprint. The long-term prospects for the use of transport internal combustion engines are built on this basis.

V. O. Pylyov · O. Y. Linkov
Department of Engines and Hybrid Powerplants, Educational and Scientific Institute of Power Engineering, Electronics and Electromechanics, National Technical University, "Kharkiv Polytechnic Institute", Kharkiv, Kharkiv, Ukraine
e-mail: volodymyr.pylov@khpi.edu.ua

O. Y. Linkov
e-mail: oleh.linkov@khpi.edu.ua

V. V. Pylyov (✉)
Department of Heat Engineering and Energy Efficient Technologies, Educational and Scientific Institute of Power Engineering, Electronics and Electromechanics, National Technical University "Kharkiv Polytechnic Institute", Kharkiv, Ukraine
e-mail: viacheslav.pylov@khpi.edu.ua

© The Author(s), under exclusive license to Springer Nature Switzerland AG 2025
A. Marchenko and Z. Koruba (eds.), *Decarbonization of Transport Energy Installations in the Context of Sustainable Development Strategies*, Studies in Systems, Decision and Control 635, https://doi.org/10.1007/978-3-032-05884-3_4

1 Introduction

There has been an ongoing debate lately at the state level in the world's leading countries, as well as within the global scientific community and leading engine manufacturing companies, regarding the possible phase-out of vehicles with internal combustion engines (ICE) due to their CO_2 emissions. Relevant practical scientific and technical actions are aimed at implementing the concept of decarbonization of the economy as a whole with a particular focus on road transport. First of all, this is understood, considered and applied as directions for the adoption of electric vehicles and hydrogen as indisputable components of the advisable energy framework [1, 2]. At the same time, leading experts in the industry assert that ICEs should and will retain a substantial presence in the consumer market in the foreseeable future [3–5]. Therefore, scientific discussions on this problem and scientific interpretation of promising trends in the use of transport power trains with ICE are very relevant.

A scientifically grounded interpretation of ICEs as long-term energy-generating means in modern conditions allows for continuation and intensification of efforts in the areas of utilization of alternative fuels, improving engine performance, increasing the environmental friendliness of design, reliability, and creating energy-efficient hybrid power systems and corresponding control systems.

It should be noted that for researchers in the National Technical University "Kharkiv Polytechnic Institute" the indicated scientific areas were in past and remain priorities, as evidenced by publications and reports at numerous scientific conferences. In particular, the first publications on the problems of alternative fuels [6–8] and hybrid solutions [9, 10] received further development and led to scientific dissertations. From long before the formation of strategies for industries 4.0 and 5.0 to present, significant attention is being paid to computer technologies for analysis of in-engine processes [11–13], the stress–strain state of ICE parts [14–16], scientific and technical development of the improved ICE designs [17–19] as well as ICE environmental performance. Such scientific research, along with research by ICE specialists in different countries of the world, has created a steady trend of reducing fuel consumption per unit of work and significantly reducing harmful emissions with exhaust gases, including CO_2 emissions. At the same time, the strategies of Industry 4.0 and 5.0, alongside the UN Sustainable Development Programs, encourage the formation of a general concept for the development of ICE with the main priorities of decarbonization and the use of green hydrogen. The authors' previous research experience allows conducting an analysis of individual problems associated with the stated overarching concept and formulating possible ways to solve them, which form the core focus of this article.

2 Problems and Contradictions of Ensuring a Comfortable Human Life

During the mankind development, the methods found by it for converting stored energy and using energy resources in the needed quantity or excessively have always been aimed at increasing the comfort of life. Notably, the central place in the modern energy complex is given to heat engines, including ICE. However, over the recent historical period, progressive society has realized that it is the constantly growing use of thermal energy that leads to dangerous consequences, which manifest themselves to one degree or another through energy, environmental, climate, and even food crises. As a matter of fact, the period of fighting crisis phenomena on a planetary scale has begun.

Of the entire set of requirements for a heat engine, today special attention is paid to noise levels, emissions of toxic and harmful substances, thermal environmental pollution, and the level of greenhouse gas emissions, CO_2 in particular. It is important that the number of criteria considered, as well as the magnitude and importance of each of them, are rethought and undergoes changes over time. This reassessment occur with the development of new ideas regarding the damage caused to society as a whole, scientific and technical capabilities, creation of new improved designs, and expectations of future legislative demands.

In recent years, scientists around the world have been speaking with increasing concern about the increase in CO_2 concentrations in the atmosphere. Analysts predict that the amount of emissions into the atmosphere by 2030 is going to excide twice the amount needed to limit global warming to 1.5 °C above pre-industrial levels [20]. So, a critical contradiction between the demands of humanity's use of energy and the consequences of that use arises. The negative consequences are directly caused by the technologies of obtaining this energy. Therefore, at the legislative level, the world leading countries are developing and ratifying certain strategies and programs for the transition from the modern carbon economy to the so-called zero-carbon economy.

Let us briefly consider the development of energy production methods. According to the traditional approach [21], they can be divided into separate stages.

Stage I—the use of human and animal muscular energy, open fire (biomass). It is believed that the period of its exclusive use stretched from tens of thousands of years BCE to the seventh century CE.

Stage II—use of wind and moving water energy (the period of predominant use was the seventh to ninth centuries CE).

Stage III (eighteenth century—early twentieth century CE) corresponds to the use of chemical energy from fuels accumulated in historically distant times of the existence of planet Earth: coal, oil, gas, oil shale, etc. Let us note that, e.g., the formation of coal is identified in time from 40 to 350 million years ago. A characteristic feature of this stage is that society began to use mechanical energy obtained from heat engines—it moved to the industrial stage of existence.

Stage IV (from the beginning of the twentieth century CE) is characterized by the emergence and widespread use of electric machines, in which mechanical energy is converted into electrical energy and vice versa. The energy of falling water and the energy of heat engines begin to serve as a source of electrical energy. Thus, energy sources of stages II and III are used at a new technological level.

Stage V (from the middle of the twentieth century CE) is the stage of the creation and development of nuclear power. The main feature is that society has begun to use the conversion of thermal energy through mechanical into electrical one, similar to stage IV, but the source of heat is based on the fission of the nuclei of heavy chemical elements (uranium, plutonium, and in the future, thorium). Thus, the stage corresponds to the use of the energy of atoms accumulated in historically distant times of the existence of the Universe.

The transformation of energy stored in nature by humanity into heat and work is schematically shown in Fig. 1. It should be noted that the depicted negative time scale does not characterize the duration of the stages (all of which are used and are going to continue to be used to one degree or another, including at a new technological levels), but corresponds to the how remote in time the formation of the energy source corresponding to a certain characteristic stage is. E.g., the time of energy formation in stage I is slightly longer than the time of energy appearance in stage II. This is explained by the fact that heat generation through the combustion of biomass and muscle energy is ultimately achieved through the influence of solar energy on the processes of biomass formation with the subsequent use or consumption of the later.

Thereby, until the latest historical stage VI (also shown in Fig. 1), the development of civilization was characterized by developing new technical means and using progressively more energy-intensive sources that were formed in increasingly distant time in the past.

An assessment of thermal environmental pollution in the modern system of energy sources was carried out in [5]. The impact of thermal pollution caused by ICEs is considered in relation to other main sources of energy conversion—nuclear power plants (NPP) and thermal power plants (TPP). Based on the analysis of the values of net efficiency of the specified power plants, it is shown that diesel engines have the best indicators in terms of direct thermal pollution of the environment, followed by

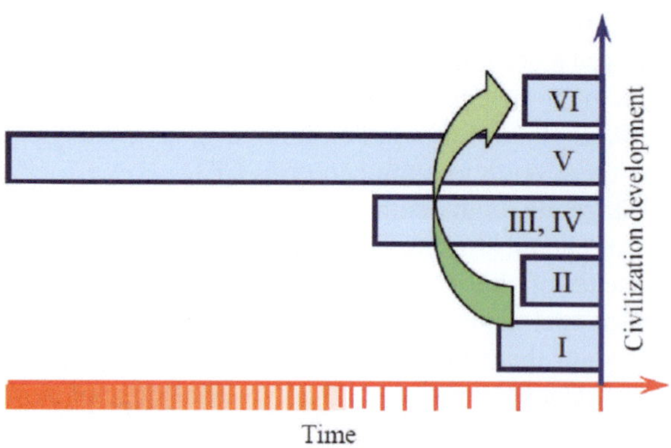

Fig. 1 The conversion of energy stored in nature into heat and work in stages of mastering: I—biomass energy; II—wind and water; III, IV—coal, oil, gas; V—nuclear energy; VI—renewable energy

gasoline internal combustion engines, then TPP and NPP. Of course, this fact should be taken into account when considering the long-term prospects for the use of ICE.

At the same time, recently, in order to radically reduce direct thermal environmental pollution and greenhouse gas emissions, it has been proposed to make an unprecedented transition to the large-scale use of renewable energy sources (RES). First of all, this applies to the direct conversion of solar energy and wind energy into electricity, and the use of biomass energy. At stage VI, a transition is made to the use of energy sources of stage II and partially of stage I—humanity becomes a witness and participant in the efforts to develop the energy industry on a new turn of the scientific and technological spiral. It is on this basis that the approach to eliminating the contradiction between the requirements of humanity's energy consumption and the consequences of this consumption is being implemented today.

In view of that, EU member states have committed to creating the first climate-neutral continent by 2050 [1]. Regarding road transport, the EU Green Deal roadmap provides the following. Road transport must be covered by emissions trading starting from 2026. The reduction in emissions from cars should reach 50% by 2030. At the same time, this agreement declares the achievement of zero emissions from new cars by 2035, which may have an ambiguous interpretation of the ways to achieve the result. Decarbonization of the economy also involves certain measures regarding air and sea transport, rational use of bioenergy resources, application of hydrogen energy technologies, etc.

3 General Aspects of Decarbonization of the Economy

Let's examine the important aspects of the creating a zero-carbon economy challenge based on the distribution of CO_2 emissions.

The first aspect is the distribution of fossil fuel emissions by regions of the world.

Figure. 2 shows the relevant data for the period starting from the beginning of stage V of energy production [22]. These data demonstrate emissions resulting from the combustion of coal, oil, gas, flaring, as well as from metallurgical and other industrial processes.

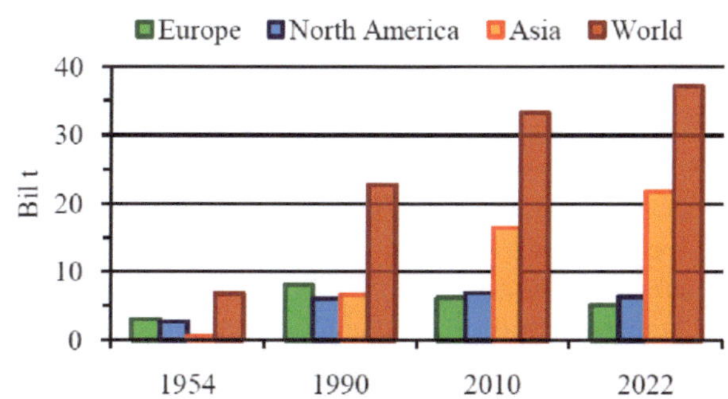

Fig. 2 Annual CO_2 emissions by regions of the world

Emissions increased more than fivefold across all regions during the reviewed period. At the same time, analysis of the data in [22] shows that since 1990 there has been a significant reduction in emissions in Europe. This decrease reached 23% in period from 1990 to 2010, and gained another 16% from 2010 to 2022. These data indicate the presence of a stable positive progress. A comparison of data for 2010 and 2022 also shows a reduction in CO_2 emissions by 8.5% in North America (in the USA by 10.9%), by 3% in Australia, as well as by 2.8% from intercontinental transportation. But emissions continue to rise in other regions and individual countries. Specifically, growth in South America from 2010 to 2022 constituted 1.9%, in Africa—16%, and in Asia—32%. At the same time, the growth in India was 68%, in China—32%, in the Middle East countries—25%, while emissions in Japan decreased by 10%.

Overall, as of 2022, contributions to total CO_2 emissions correspond to the following levels: Europe—14%; Asia—58% (China—31%, India—7%, Middle East countries—7%, Japan—3%); North America—17% (USA—13.5%); Africa—4%; South America—3%; intercontinental transportation—3%, Australia—1%.

Therefore, efforts to make Europe a climate-neutral continent would allow reducing global CO_2 emissions (in today's numbers) by up to 14%. We can agree that this value is significant. It is important to note that total emissions in Europe and the USA account for about a third of global emissions. At the same time, according to the above data, the need to introduce precautions and mechanisms to avoid unfair competition from other regions of the world becomes clear. However, the European Green Deal states that Europe has already set a positive example and encourages major international partners to set their own target dates for achieving climate neutrality [1].

The second aspect is the distribution of CO_2 emissions by economic sectors.

It is known that the main global source of CO_2 emissions is industry. The energy production consumed by industry alone accounts for over 24% of the emission total. In order of decreasing impact on the greenhouse effect, industry is followed by agriculture, forestry and land use, the shares of CO_2 emissions from which are over 18%, the municipal sector with a share of emissions close to 18% and the transport sector, the share of CO_2 emissions from which exceeds 16% [23]. However, of more importance is the analysis of the data on individual sectors of the economy, which differ in technological processes of mechanical, electrical and thermal energy consumption and, accordingly, should differ in methods of their decarbonization. The greatest greenhouse effect impact is caused by road transport (12%), followed by the municipal sector (11%), burning of the undistributed share of fuel, e.g., flaring (8%), metallurgy (7%), uncontrolled emissions from energy production (6%), technological emissions from industry, e.g., from chemical processes (5%). It attests that the use of climate-neutral transport is a significantly important factor in the process of decarbonizing the economy.

The third aspect is the distribution of energy production by resource sources.

The distribution of annual CO_2 emissions is most often considered in opposition of information on the production of mechanical energy by vehicles and the total production of electrical and thermal energy at stationary energy facilities E_m/E_{e+h}.

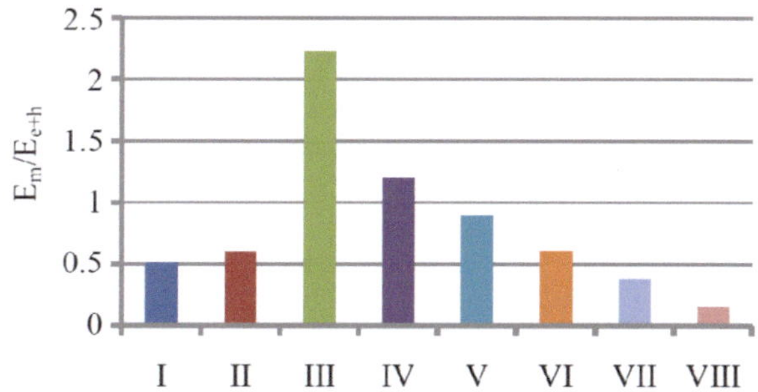

Fig. 3 Ratio of emissions from mechanical energy production (E_m) to total electricity and heat energy production (E_{e+h}): I—world; II—Europe; III—France; IV—Great Britain; V—USA; VI—Germany; VII—Japan; VIII—China

Such data for the main global economic entities based on [23] and are presented in Fig. 3.

The provided data are estimates of the possible increase in CO_2 emissions with a complete transition to the use of so-called zero-emission vehicles and, accordingly, accounting on the increase in the load on traditional stationary energy sources. The actual increase in emissions may be greater due to the higher net efficiency of ICEs.

It should be noted that it is advisable to use data not for individual countries, but for certain regions. E.g, a significantly high E_m/E_{e+h} indicator for France does indicate not higher levels of motorization and (or) energy efficiency of electricity and heat consumption, but rather insufficient electricity generation for internal needs and presence of electricity imports.

It leads to the conclusion that, when transitioning to the use of vehicles with zero emissions, the use of renewable energy as the primary energy source becomes mandatory.

Let's consider the distribution of energy production by primary resource sources in 2020 [24], i.e. before the start of the EU Green Deal, the last russian energy war with Europe, and the hot phase of the war in Ukraine. The assessment of the share of primary energy resources attributable to ICEs was carried out according to the comparison of the data [4, 24] and is shown in Fig. 4. It allows us to conclude that in case of the comprehensive transitioning to zero-emission vehicles, a significant 20–25% increase in the load on stationary power generation should be expected. These data are shown by the group I of columns.

At the same time, the replacement of traditional sources of primary resources with RES becomes a significant problem. The further development of modern economies and the growth of electricity generation needs should be taken into account. E.g., electricity consumption by data centers, artificial intelligence, and the cryptocurrency sector could double by 2026 and reach more than 1,000 TWh. This demand is approximately equivalent to the electricity consumption in Japan [25].

As for Ukraine, Fig. 4 shows that a significant disadvantage of the existing energy balance is a high share of energy generation based on the use of coal. This indicator is almost three times higher than that of developed countries, which is a significant problem for the decarbonization of the national economy. On the other hand, the consumption of primary resources to meet transport needs is half the world level.

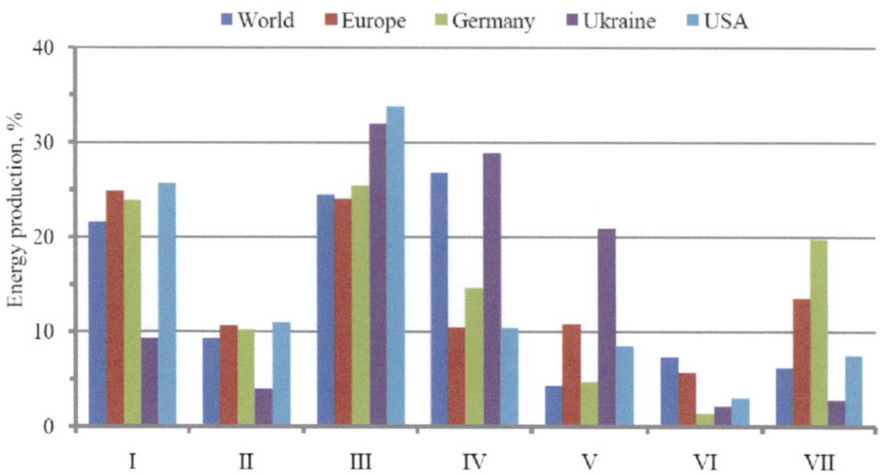

Fig. 4 Distribution of energy production by sources of primary resource: I—ICE fuels; II—oil, except for ICE fuels; III—natural gas; IV—coal; V—nuclear energy; VI—hydropower; VII—RES

This correlates well with the country's level of motorization. At first glance, this is a sufficient reason for the wider use of electric vehicles and hybrid powertrains in transport. But focusing on the entirety of the data presented in the article shows that the issue of decarbonization of the transport sector should be comprehensive, taking into account the consequences of such decarbonization for the energy sector as a whole.

The fourth aspect is the distribution of energy production by energy generating means.

It is well known that humanity's needs for energy are of a complex cyclical nature. The load on the energy system is variable depending on the season and time of day. This load is determined by the specific demand for energy consumption. When the level of consumption changes the energy generation of certain power plants must change accordingly.

The simplest and most dynamic way to achieve it is presented by the ability to change the level of mechanical energy generation using ICEs. The processes of changing the level of electrical energy generation by stationary power plants are more inertial.

Let us consider this issue in the case of a transition to the general use of electric vehicles and RES based on data on daily generation and consumption of electricity using the example of Ukraine [26] tracing the trends in the redistribution of electrical energy generation between energy generating facilities on a specific date (September 10) in 2018, 2020, 2021 with fairly close maximum energy production, $\approx$15,000 $\pm$ 1,000 MW. Let's call these dates A, B, and C.

The general nature of daily generation is the same: in stage I, from approximately 11 PM to 7 AM, there is minimal generation, in stage II, from approximately 9 AM to 4 PM, there is the first increase in generation, and in stage III, from 4 to 11 PM, the maximum peak in generation occurs. Notably, energy production at nuclear power plants and combined heat and power plants is not significantly variable. Therefore, we will be analyzing the change in the load on TPPs, hydro- and hydro-accumulating power plants (HAP and HAPP), as well as at the expense of RES and cross-border flows.

The load on the TPP at stage III (the actual energy production at the specified stage on the selected dates was: A—5,150 MW, B—4,750 MW, C—3,350 MW). The relative generation of electrical energy by selected energy generating means, assuming these values for each of the considered dates to be 1, is presented in Table 1.

It should be noted that date A actually corresponds to the case of using traditional generation systems with minimal use of renewable energy. There, at stage I, when the relative generation at the TPPs is reduced to 0.58, the expedient approach lies in reversing water supply at HAPP. In this case, the share of HAPP consumption is 0.24 of TPP generation. This made it possible to decrease the needed lowering of energy generation at the TPP. But, even when using the reverse water supply mode, this lowering remained significant. On the other hand, the mode of energy consumption by HAPP in stage I allowed to accumulate additional volumes of water for their use in stages II and III with the total share of generation of HPPs and HAPPs up to 0.38.

Now, let us consider the impact of the contribution of increased RES capacities on the redistribution of energy production among energy generating facilities. Maximum electricity generation from RES on the exemined dates in Ukraine increased by more than 9 times, from 400 MW on date A to 3,770 MW on date C and reached 25%

Table 1 The Relative Distribution of Electricity Production in Ukraine by Power-Generating Means with Variable Loading

The period of the day	TPP	RES	HPP and HAPP	Cross-border flows
Date A (10 Sept. 2018)				
I	0.58		−0.24	
II	1.00	0.08	0.10	
III	1		0.38	
Date B (10 Sept. 2020)				
I	0.71		−0.07	
II	0.76	0.70	−0.14	−0.12
III	1	0.14	0.54	
Date C (10 Sept. 2021)				
I	0.61	0.13	−0.09	
II	0.62	1.12	−0.21	−0.07
III	1	0.04	0.84	
Date D (12 Feb. 2022)				
I	0.72	0.11	−0.24	
II	0.96	0.59	0.24	
III	1	0.31	0.70	0.10

of total consumption. This led to significant changes in the regulation of traditional generation systems.

From Table 1, it can be seen that at stage II on date B, the share of RES was 0.7, and on date C had already exceeded the generation of TPP and amounted to 1.12. On dates B and C, reverse water supply to the HAPP became more suitable at stage II (with consumption shares of 0.14 and 0.21). It was at this stage the electricity was exported using cross-border flows (with transmission shares of 0.12 and 0.07). Overall, at stage II, generation exceeded demand by almost a quarter relative to the TPP load at stage III. At the same time, energy generation at the TPP at stage II actually corresponded to the generation level at stage I (0.76 and 0.62). Therefore, a further increase in the share of RES at the given capabilities of HAPPs and cross-border flows is going to lead to an increase in the operating time of TPPs at partial load modes and may lead to the need in operation at loads lower than acceptable. At the same time, in stage III, when there is a significant reduction in the share of RES contribution, the maximum need for energy consumption occurs with the highest load at TPP (with shares of 1.0), HAP and HAPP (in total with shares of 0.54 and 0.84).

The presented data shows that, whenever possible, hydropower capacity should be increased, which is a significant source of regulation for daily generation. E.g., in Germany during 2019 the share of wind energy increased by 16%, solar energy by 2%, and hydro energy by 21%. But, at the same time, the total share of hydro energy reached only 4% [27]. Clearly, the possibilities of increasing hydropower capacities are limited.

Let us examine the data on the operation of the generation system in the early autumn and winter periods by comparing date C (September 10, 2021) and date D (February 12, 2022). On date D, total electricity consumption increased by 1.38 times and reached 19,250 MW. Generation at TPPs increased by 1.59 times (to 3,300 MW), and RES capacity decreased by 1.88 times (to 2,000 MW). The relative generation of electricity by chosen energy generating means on date D is also presented in Table 1.

It can be seen that at stage II, from the costs of water return to the HAPP on date C with a share of 0.21, a transition to energy production at the HPP and HAPP took place with a share of 0.24 on date D. Also, due to cross-border flows from energy exports on date C at stage II with a share of -0.07, a transition to imports happened on date D at stage III with a share of 0.1. At the same time, at stage III, the share of energy production at HPP and HAPPs increased to 0.7.

At the point, it is obvious that daily and annual fluctuations in renewable energy production currently require appropriate regulation of energy production from other possible sources. And with a further increase in the share of renewable energy sources, the comprehensive use of electric vehicles, and a reduction in the share of fossil energy sources, the problem of regulating electricity generation will grow significantly.

It is also necessary to take into account the instability of solar and wind energy production, as it depends on weather conditions. In particular, this instability can be a negative factor even if production exceeds the average expected value. This related to the possibility of regulating the consumption balance here leading to an unacceptable reduction in the capacity of other generation sources.

E.g., on April 24, 2023, the wholesale price of electricity in the Netherlands reached negative values—the wholesale price of electricity on the Amsterdam Epex Spot exchange dropped to –739.96 Euros/(MW·h) and was negative all day [28]. The EU power markets experienced record periods of negative prices in 2024. 468 h of negative prices, 60% more than a year earlier, was recorded in Germany according to data from Epex Spot; in France, below-zero hours more than doubled, reaching 356; Spain experiences negative prices for the first time—247 h [29]. The reason was excessive generation in the RES sector—the amount of energy produced at wind farms exceeded total demand. Another example is Ukraine's sale of surplus electricity to Europe at the end of February 2024, even with the cessation of renewable electricity generation [30].

From the above, we can conclude that among the obstacles to the expansion of RES remain the challenges of connecting new capacities to existing electricity grids and ensuring an acceptable balance between energy generating facilities. Therefore, new technologies for cyclical, not fully predictable, energy production that meet the requirements of ensuring the decarbonization of the economy require new corresponding technologies for electricity consumption. The use of electric vehicles as vehicles with zero emissions by themselves does not contribute to solving the problem above.

4 The Stance of the Industry's Leading Experts on the Future of Internal Combustion Engines

Among a significant number of published works in world scientific journals, in our opinion, it is sufficient to highlight three that, in a concentrated form, present convincing viewpoint on the need to maintain the ICE segment in the consumer market and, accordingly, the further prospects for the use of transport power trains with ICE.

As early as 2017, 27 scientists under the auspices of the Scientific Society for Automotive and Engine Technology (Wissenschaftliche Gesellschaft für Kraftfahrzeug- und Motorentechnik—WKM) published a policy document "The Future of the Combustion Engine / Assessment of the Diesel Engine Situation" [3], which presented, among other things, the following theses:

- WKM is extremely concerned about the trend of discussing a ban on the combustion engine. A ban on combustion engines e.g. from 2030, would have negative effects on fighting climate change.
- There will be more combustion engines worldwide in 2030 than today. Continued intense research and development and the preservation and holding of technology leadership are appropriate. WKM predicts a long-lasting need for combustion engines and, in particular, also for diesel engines.

Another significantly expanded program publication released in 2019 is the editorial paper in the International Journal of Engine Research "IJER editorial: The future of the internal combustion engine" [4]. Its weighty theses state:

- There are still no real alternatives that can compete with the IC engine over the entire range of applications that they cover.
- For electrification, electricity has to be produced, either by the IC engine (in the case of a hybrid vehicle), or from a power generating station.
- Perhaps there is a political advantage to drawing power from the grid in that unwanted emissions are "not in my back yard". Successful technologies must be market competitive in cost, user requirements, lifecycle emissions and lifecycle efficiency.
- It is very likely that highly efficient "fully flexible" engines with hybridized solutions will be a big part of sought-after efficiency improvements, as well as emission/GHG reductions.

Similar and no less categorical are the conclusions of other scientists, e.g., formulated in 2022 [5]:

- Compared to other traditional and most common energy sources, ICEs have the least negative impact on the environment. This applies both to general thermal pollution and to the formation of greenhouse gas CO_2;
- the leading role of ICEs in the energy systems of road vehicles, today and in the future, is to expand the applications of hybrid systems, and the concept of improving all types of ICEs is going to have the strategic direction of ensuring an increase in operational efficiency, economic and environmental efficiency in high boosting modes of operation, concurrently with high operational reliability of the design.

So, theory and practice show today that transport ICEs are considered by the scientific community as energy-generating means of a long-term perspective.

The set of many-sided assertions laid out in the previous sections of the article by and large add to the above the following:

1. Humanity faced the fact of unquestionable elimination of the contradiction between the requirements of energy consumption and the consequences of this consumption. An analysis of the literature shows that one of the established positions corresponds to the following statement: the global contradiction is proposed to be eliminated based on the decarbonization of the energy sector, meaning creation of zero-carbon and climate-neutral economies. Examples of the energy industries in a number of leading countries demonstrate the real possibilities for reducing CO_2 emissions significantly.
2. Vehicles are a major factor in reducing CO_2 emissions, so their use must be carried out on a new conceptual basis. When transitioning to the use of vehicles with zero emissions by themselves, climate-neutral transport, the use of renewable energy as the primary source of energy becomes mandatory.

3. The use of electric vehicles as zero-emission ones by themselves is inconsistent with the need for cyclical, not fully predictable, energy consumption from RES, in particular with solar and wind power generation.

It is important that when referring the particulars in provisions 1–3 the essential difference between the terms *zero-carbon* and *climate-neutral* economy should be noted, both of which involve achieving the effect of decarbonization of the energy sector. Accordingly, a distinction should be made between the terms zero-emission transport and climate-neutral transport.

The first of these terms implies the absence of a carbon footprint from the operation on the base of a certain energy source or conduction of a certain technological process. The second one means implementation of a fundamentally new technological cycle using traditional and renewable energy sources, the combined use of which, regardless of the presence of CO_2 emissions and the cyclical, not fully predicted operation of RES, ultimately does not lead to an increase in CO_2 in the atmosphere.

Still, in our opinion, it is desirable today to have a discussion not so much about the essence of the terms, but about the implementation of the specified terms through in-depth scientific, technical and technological substantiation of ways to answer the challenge of decarbonization. It comes out of this that the scientifically based further use of ICEs, which undoubtedly leave a carbon footprint, should in the future be included in the technological cycles of supporting climate-neutral transport.

In general, promising technological processes to support a climate-neutral economy will allow in the future to minimize and neutralize the carbon footprint of other individual and aggregate critical sectors of the economy as a whole. One confirmation of it is that the development of a new generation of climate-neutral energy, particularly in the USA, relies among other approaches on the use of fossil fuels [31].

5 The Latest Conceptual Provisions on the Use of ICE in Power Trains of Climate-Neutral Transport

Based on the information provided in [1, 4, 31–34] and other sources, a set of main ways to ensure climate-neutral economy is identified. They are based on six main processes in the global energy conversion system. Let us lay out these processes.

1. Widespread use of natural processes of CO_2 binding based on photosynthesis. The processes of this type have two directions of application.

The first of them involves restoring the flora and planting greenery at the naturally barren territories, which is considered a quick and cheap solution for absorbing and storing carbon. According to [1], restoring forests, peatlands and other lands is a realistic way to create a more climate-resilient environment.

The second way is the intensive use of biofuels, through the cultivation of energy crops in particular. According to estimates [4, 34, 35], bioenergy effectively contributes to the phase-out of fossil fuels and is a quick-payback practice. Recently woody crops are beginning to be preferred over herbaceous ones. E.g., 9 Europe countries are using technologies for growing Swedish energy willow by now.

At the same time, it is important that the application of natural CO_2 sequestration technologies should be accompanied by strict new criteria for cutting down forests and by protection of areas with high biodiversity value [1].

2. Artificial Technological Processes for Binding Atmospheric CO_2.

Today, such processes have already entered the pilot testing stage. They rely on using RES, such as a wind turbine, to generate electricity. Subsequently, hydrogen and oxygen atoms in the water are separated using electrolysis. Hydrogen is combined with carbon dioxide from the air to produce synthetic carbon–neutral methanol. A company in Chile is already operating using this technology, and production of the synthetic fuel is planned in Spain, Saudi Arabia, and the United States.

The next step is envisioned to involve creating synthetic versions of gasoline and diesel fuel for road transport and kerosene for aviation. Importantly, synthetic fuel (also known as e-fuel) is being developed with the aim of producing it in quantities sufficient to power most cars worldwide [4, 32, 33, 36].

3. Synthesis of Carbon-Free Fuels.

First of all, it involves the use of so-called green hydrogen. In recent years, the scientific community and governments of leading countries in the world have been developing concepts for hydrogen energy, which include, along with solving other problems, the need to transition ICEs from traditional fuels to hydrogen ones.

A promising option for the use of ICEs today is considered to be the combustion of two-component fuels, which include such additives to hydrocarbon fuels as hydrogen, water, oxygen components, and even innovative chemical components, e.g., NH_3 [4, 5]. For this porpose, unlike the traditional ammonia synthesis technology, which involves notably significant CO_2 emissions, low-tech, low-energy consuming ammonia synthesis technology without a carbon footprint has already been developed and piloted [37, 38].

Projects for H_2 and NH_3 production are actively progressing in Saudi Arabia, Europe, the USA, Japan, South Korea, and Australia. Ammonia shipments from Saudi Arabia have already been sent to Japan and South Korea. Japan and Australia have announced plans to increasingly wide use NH_3 as a fuel for electricity generation and marine engines.

4. Carbon capture to reduce the amount of greenhouse gases in the atmosphere is also considered a fairly promising direction for ensuring the decarbonization of the energy industry, especially providing fuels containing fossil carbon are going to continue being used for both ICEs and TPPs in the foreseeable future. Examples of such projects include Saudi Arabia and the United States [33, 39]. E.g., one of the world's largest integrated energy and chemical companies, Saudi Aramco, is implementing a project to capture and re-inject (sequester) CO_2 into the oil reservoir at the Ghawar oil field. This technology also has an effect in increasing oil production.

Saudi Aramco is also working on developing mobile carbon capture and storage technologies for cars and trucks. It has achieved the capture and storage of up to 40% of CO_2 from vehicle exhaust. The carbon dioxide is supposed to be unloaded at gas stations for the later sequestration.

5. Use of natural carbon-based fuels.

These technologies currently undergo a decrease in their use. This applies most to the use of coal and oil. Said concept is generally accepted. E.g., the Energy Strategy of Ukraine provides for a 2.45-times reduction in coal use and a 1.6- times reduction in petroleum products use by 2035 compared to 2015. Obviously, this reduction must be agreed with other technologies to ensure energy balance.

At the same time, the use of dual-fuel ICE is also gaining appeal to researchers, among other uses as a means of ensuring the flexibility of RES. This is a cheaper and simpler solution compared to using electricity batteries. The use of gas fuel leads to a reduction in CO_2 emissions by 20–30% [40]. Converting existing diesel engines to the dual-fuel process is a relatively simple technology.

6. The main natural carbon-free fuel today is uranium, with its use in NPPs. Along with the above-mentioned technologies for the use of renewable energy according to processes 1–4, nuclear energy is considered a necessary direction for the complete satisfaction of energy needs.

Thus, Germany postponed the plans to close three NPPs [41]; at the initiative of France, 11 European countries formed an alliance to develop joint projects in the field of nuclear energy based on the use of small reactors [42]; Ukrainian Energoatom and the US company Holtec International signed an agreement to build up to 20 power units with small reactors in Ukraine, which will increase the total installed capacity of NPP in the country by 25% [43], as part of the implementation of the agreement, companies such as Mitsubishi, Hyundai, etc. are going to be involved in the project.

It is important that modular reactors are considered as a replacement for thermal power generation. In addition, and no less important, these plants are brought to peak power and execute load shedding in less than 30 min. Therefore, they are considered as a necessary source of maneuvering power in the energy system containing RES.

In general, the need for a new step forward in the development of science and technology based on a new conceptual vision regarding the integration of the use of existing and promising technologies stands out as extremely necessary. In concert, they are instrumental to meet the requirements for climate-neutral energy in the actual presence of processes with a carbon footprint. This conceptual idea has already been called the circular carbon economy [33, 39].

The full application of the circular carbon economy concerns ICEs in transport. It is expected that by 2030 there will be close to two billion cars in the world, of which only 8% are battery electric vehicles. This highlights Germany's recent actions in an unprecedented blocking of an EU agreement to ban new sales of fossil fuel cars starting from 2035. Italy also supports this decision, and support is expected from Poland and Bulgaria [44].

According to the International Energy Agency, as a result of the russian military aggression in Ukraine, global demand for coal in 2022 reached a historic high, with investment in the coal industry increasing almost six times above the level of one planned for climate goals by 2030. Investments in oil and gas production have also increased. However, only $1 trillion of investment went into fossil carbon-containing fuels, while $1.7 trillion went into developing a set of low-carbon technologies, primarily RES [45]. This surge in investment in climate-neutral energy indicates apparent changes bound to have positive consequences for energy markets and climate change.

Scientific, technical, technological, and managerial prospective solutions, as well as the state of investment projects, by now determine and significantly expand the prospects for the use of ICEs in transport. All these solutions have to be based on a comprehensive approach, which, of course, involves the integration of ICE into the circular carbon economy. Thus, research in increasing the efficiency of ICE in the broad sense of the term is essential. It reveals prospects for a wider application of ICE in the future.

6 General Conclusions

Understanding the significant impact of greenhouse gas emissions on the further development of modern society has led to a sharp debate about the content of strategies, programs, scientific, technical and managerial decisions regarding the phase-out of road transport with ICE.

The article considers the main provisions regarding the prospects for further development and use of ICE as a necessary and integral part of the energy complex of the future, which is already coming. A systematic review of the above-mentioned issues shows:

1. It is clearly proposed to eliminate the contradiction between the requirements of humanity's use of energy and the consequences of this use on the basis of an unprecedented transition from fossil energy sources to the large-scale use of RES. The main goal of this transition is to radically reduce greenhouse gas emissions into the atmosphere, primarily CO_2. One of the directions for achieving this goal is to reduce emissions from cars, and in the long term, to remove such emissions.

Solving this global problem requires coherence of the consumption needs and RES capabilities based on the implementation of innovative approaches to capacity balance, the use of new fuels, materials, designs, and technological solutions in an extensive system of energy conversion from its traditional and new sources to end consumers.

2. The issue of decarbonization of the global economy in general and transport in particular is considered from various aspects, which shows the following:

- since the development of nuclear power, greenhouse gas emissions from fossil fuels have increased more than fivefold. These emissions are unevenly distributed across regions and countries. In particular, today emissions in Europe and the United States account for about a third of global emissions;
- the analysis of the distribution of CO_2 emissions was conducted not by sectors of the world economy, but by their components, which should differ in the methods of their decarbonization. From this view, it is clear that the greatest impact on the greenhouse effect is caused by road transport—12%, followed by housing construction (11%), burning of the unallocated share of fuel (8%), metallurgy (7%). Thus, the use of climate-neutral transport is undoubtedly a significant factor in the process of decarbonization of the economy;
- it is shown that the ratio of CO_2 emissions from the production of mechanical energy by vehicles to the total production of electrical and thermal energy in the world is 0.5:1, for Europe—0.6:1, for the USA—0.9:1. These data indicate that when transitioning to the use of climate-neutral transport, the use of renewable energy as a primary source has significant prospects;
- the distribution of energy production by primary resource sources is presented. It is shown that the transition to the comprehensive use of zero-emission vehicles requires an additional increase in the load on stationary energy industry, taking into account the high net efficiency of ICEs, by at least 20–25%. Thus, the numerical value of the minimum additional RES capacities is estimated;
- it is important that RES depend on unstable energy flows present in nature. Therefore, fluctuations in RES production today significantly complicate the balance of energy production from other possible sources. This proves that new technologies for the not fully predictable use of RES, which meet the requirements of ensuring the decarbonization of the economy, require new fitting technologies for electricity consumption.

Overall, the above demonstrates that there is an objective contradiction between the use of renewable energy and the comprehensive use of electric vehicles as consumers of electrical energy.

3. The position of leading experts in the engine industry regarding the future of internal combustion engines is unambiguous—transport ICEs are energy-generating means of the long-term perspective. In addition, it is necessary to distinguish the fundamentally different terms zero-emission transport and climate-neutral transport, both of which involve achieving a decarbonization effect. This allows concluding that the continued use of ICEs, which leave a carbon footprint, should be incorporated in technological cycles to support a climate-neutral transport sector based on renewable energy.
4. The world is developing the concept of a circular carbon economy these days, which meets the requirements of global climate-neutral energy cycle with the actual presence of processes with a carbon footprint. This progressive approach involves the use of conceptual approaches based on: the use of natural and artificial processes of binding atmospheric CO_2 to obtain carbon fuels; the synthesis of

carbon-free fuels; the capture of CO_2 from burned carbon fuels with sequestration in the oil reservoirs.

All of the mentioned approaches are at least at the pilot stage. And experts involved in various areas of renewable energy application claim that each of the presented approaches contends to become mainstream.

The analysis presented in the article outline the base for long-term prospects for the use of transport ICEs in the future and our principled commitment to the further intensive development of automotive, ship, aviation, and stationary ICEs.

References

1. The European Commission: Factsheet. Delivering the European Green Deal (2021). https://commission.europa.eu/publications/delivering-european-green-deal_en. Last updated 9 Oct 2023
2. The European Commission: Communication COM/2020/301: A hydrogen strategy for a climate-neutral Europe (2020). https://knowledge4policy.ec.europa.eu/publication/communication-com2020301-hydrogen-strategy-climate-neutral-europe_en. Accessed 11 Mar 2024
3. WKM: The future of the combustion engine/Assessment of the diesel engine situation (2017). Available via https://www.wkm-ev.de/images/20170708-englisch-Die-Zukunft-des-Verbrennungsmotors.pdf. Accessed 11 Mar 2024
4. Reitz, R.D., Ogawa, H., Payri, R., et al.: IJER editorial: the future of the internal combustion engine. Int. J. Engine Res. **21**(1), 3–10 (2020). https://doi.org/10.1177/1468087419877990
5. Marchenko, A.P., Parsadanov, I.V., Strokov, O.P.: Dvyhuny vnutrishnoho zghoriannia i navkolyshnie seredovyshche (Internal combustion engines and environment). Dvyhuny vnutrishnoho zhoriannia **2**, 3–12 (2022). https://doi.org/10.20998/0419-8719.2022.2.01
6. Marchenko, A.P., Semenov, V.G.: Alternative biofuel from rape oil derivatives. Chem. Technol. Fuels Oils **37**, 183–185 (2001). https://doi.org/10.1023/A:1017976722355
7. Marchenko, A.P., Semenov, V.G.: Alternative biofuel based on rapeseed oil derivatives. Khim. Tekhnol. Topl. Masel **3**, 31–32 (2001)
8. Marchenko, A., Osetrov, A., Linkov, O. et al.: Simulation of biofuels combustion in diesel engines. In: Fang, Z. (ed) Biodiesel-Feedstocks, Production and Applications, Intech, pp. 407–432 (2012). https://doi.org/10.5772/52333
9. Parsadanov, Y.V., Teplytskyi, A.A., Solodovnykov, V.V., et al.: Prymenenye vodotoplyvnoi emulsyy v avtotraktornom dyzele (chast I). Dvyhately vnutrenneho shoranyia **1**, 118–121 (2011)
10. Marchenko, A.P., Parsadanov, I.V., Prokhorenko, A.O. et al.:Osoblyvosti protsesu zghoriannia v dyzeli pry roboti na vodopalyvnii emulsii (Features combustion process in diesel engines when working at water-fuel emulsion). Dvyhately vnutrenneho shoranyia **1**, 3–10 (2016). https://doi.org/10.20998/0419-8719.2016.1.01
11. Parsadanov, I., Marchenko, A., Tkachuk, M. et al.:Complex assessment of fuel efficiency and diesel exhaust toxicity. SAE Int. Technical Paper 2020-01-2182 (2020). https://doi.org/10.4271/2020-01-2182
12. Belousov, E., Marchenko, A., Gritsuk, I. et al.: Research of the gas fuel supply process on the compression stroke in ship's low-speed gas-diesel engines. SAE Int.. Technical Paper 2020-01-2107 (2020). https://doi.org/10.4271/2020-01-2107
13. Parsadanov, I.V., Lal, A.G.: Kontseptsyia orhanyzatsyy protsessa shoranyia v oppozytnom dvukhtaktnom dyzele pry vysokom urovne forsyrovanyia (The concept of combustion process organisation in a boxer two-stroke diesel engine at a high level of boosting) Dvyhuny vnutrishnoho zhoriannia **1**, 45–51 (2021). https://doi.org/10.20998/0419-8719.2021.1.06

14. Marchenko, A., Tkachuk, M.A., Kravchenko, S. et al.: Experimental tests of discrete strengthened elements of machine-building structures. In: Tonkonogyi, V. et al. (eds.), InterPartner 2019: Grabchenko's international conference on advanced manufacturing processes, Odessa, September 2019. Lecture notes in mechanical engineering. Advanced manufacturing processes, Springer, Cham, pp. 559–569 (2020). https://doi.org/10.1007/978-3-030-40724-7_57
15. Pylyov, V.O., Linkov, O., Samoilenko, D. et al.: The influence of load modes on the resource reliability of engine parts of agricultural machinery. In: Transport Means 2020: Proceedings of 24th international scientific conference, Kaunas, September 2020, Part I, pp. 107–113 (2020)
16. Marchenko, A., Pylyov, V.V., Linkov, O.: Estimation of strength of the combustion chamber of the ICE piston with a TBC layer. In: Nechyporuk, M., Pavlikov, V., Kritskiy, D. (eds.), ICTM 2020: Integrated computer technologies in mechanical engineering—2020. Lecture notes in networks and systems, vol 188. Springer, Cham, pp. 415–426 (2021). https://doi.org/10.1007/978-3-030-66717-7_35
17. Samoilenko, D., Marchenko, A., Cho, H.M.: Improvement of torque and power characteristics of V-type diesel engine applying new design of Variable geometry turbocharger (VGT). J. Mech. Sci. Technol. **31**(10), 5021–5027 (2017). https://doi.org/10.1007/s12206-017-0950-2
18. Marchenko, A., Grabovskiy, A., Tkachuk, M.M. et al.: Detuning of a supercharger rotor from critical rotational velocities. In: Ivanov, V. et al. (eds.), DSMIE 2021: Advances in design, simulation and manufacturing IV, Lviv, June 2021. Lecture notes in mechanical engineering, vol 1. Springer, Cham, pp. 137–145 (2021). https://doi.org/10.1007/978-3-030-77823-1_14
19. Linkov, O., Marchenko, A., Pylyov, V.V. et al.: Assessment of the parametric reliability of the side surface of the diesel engine piston during its design. In: Transport Means 2023: Proceedings of 27th international scientific conference, Palanga, October 2020. Part I, pp. 340–346 (2023). https://doi.org/10.5755/e01.2351-7034.2023.P1
20. Harvey, F.: Cop26: world on track for disastrous heating of more than 2.4C, says key report. In: The Guardian, 9th November 2021 (in press) (2021)
21. Atkins, P.W.: The Second Law. Scientific American Books, W.H, Freeman & Co, New York (1994)
22. Ritchie, H., Rosado, P., Roser, M.: CO_2 and Greenhouse Gas Emissions. Annual CO_2 emissions by world region. In: OurWorldInData. CO_2 and Greenhouse Gas Emissions (2023). Available via https://ourworldindata.org/grapher/annual-co-emissions-by-region#explore-the-data. Accessed 11 Mar 2024
23. Ritchie, H., Rosado, P., Roser, M.: Emissions by sector: where do greenhouse gases come from? In: OurWorldInData (2024). Available via https://ourworldindata.org/emissions-by-sector#citation. Accessed 11 Mar 2024
24. bp Statistical review of world energy. 71st edition (2022) BP. Available via https://www.bp.com. Accessed 11 Mar 2024
25. IEA (2024) Electricity 2024. Analysis and forecast to 2026. Available via https://www.iea.org/reports/electricity-2024. Published January 2024
26. National Power Company Ukrenergo (2022) Transmission and dispatching. https://ua.energy/en/. Accessed 12 Feb 2022
27. Ruter, G., Ya, B.: Nimechchyna: bilshe zelenoi enerhetyky, menshe vuhilnoi (Germany: more green energy, less coal). In: DW, 15 January 2020 (in press) (2020)
28. Morison, R., Fokuhl, J.: Dutch Power Prices Turn Negative as Green Power Floods Grid. In: Bloomberg, 19 April 2023 (in press) (2023)
29. Brendel, E., Farhat, E.: German Power Slips Below Zero as Negative-Price Phenomenon Grows. In: Bloomberg, 02 January 2025 (in press) (2025)
30. National Power Company Ukrenergo (2024) Stan enerhosystemy (State of the energy system). https://t.me/Ukrenergo/2671. Accessed 29 Feb 2024
31. Lefebvre, B.: How American energy helped Europe best Putin. In: Politico, 28 March 2023 (in press) (2023)
32. Vyhranenko, Y.: Formula-1 prahne staty vuhletsevo-neitralnoiu do 2030 roku (Formula 1 aims to become carbon neutral by 2030). In: Garagepeppers, 08 July 2022 (2022)

33. Saudi Arabian Oil Co: Managing our footprint. Circular carbon economy (2024). https://www.aramco.com. Accessed 11 Mar 2024
34. Volk, T.A., Berguson, B., Daly, C., et al.: Poplar and shrub willow energy crops in the United States: field trial results from the multiyear regional feedstock partnership and yield potential maps based on the PRISM−ELM model. Global Change Biol. Bioenergy **10**(10), 735–751 (2018). https://doi.org/10.1111/gcbb.12498
35. Bioenergy Association of Ukraine: Energy Crops (2024). https://uabio.org/en/energy-crops. Accessed 11 Mar 2024
36. Ravikumar, D., Keoleian, G., Miller, S.: The environmental opportunity cost of using renewable energy for carbon capture and utilization for methanol production. Appl. Energy **279**(115770), 1–12 (2020). https://doi.org/10.1016/j.apenergy.2020.115770
37. Saudi Arabian Oil Co: First accredited low-carbon ammonia shipment for power generation dispatched from Saudi Arabia to Japan (2023). https://www.aramco.com. Accessed 11 Mar 2024
38. Li, J., Xia, Y., Song, X., et al.: Continuous ammonia synthesis from water and nitrogen via contact electrification. Proc. Natl. Acad. Sci. **121**(4), 1–5 (2024). https://doi.org/10.1073/pnas.2318408121
39. Sick, V., Armstrong, K., Cooney, G. et al.: The need for and path to harmonized life cycle assessment and techno-economic assessment for carbon dioxide capture and utilization. Energy Technol. 1901034. **8**(11) (2019). https://doi.org/10.1002/ente.201901034
40. Karimi, M., Wang, X., Hamilton, J., et al.: Status, challenges and opportunities of dual fuel hybrid approaches-a review. Int. J. Hydrogen Energy **46**(70), 34924–34957 (2021). https://doi.org/10.1016/j.ijhydene.2021.08.008
41. Lindner, C.: Germany says it is no longer reliant on Russian energy. In: BBC news, 19 January 2023 (in press) (2023)
42. Simon, F.: Eleven EU countries launch alliance for nuclear power in Europe. In: Euractiv, 01March 2023 (in press) (2023)
43. Interfax-Ukraine News Agency: Energoatom, Holtec International agree to build up to 20 SMR reactors in Ukraine (2023). https://en.interfax.com.ua. Published 24 April 2023
44. UNN: Nimechchyna rozliutyla YeS pislia toho, yak zablokuvala uhodu pro zaboronu prodazhiv avtomobiliv na vykopnomu palyvi (Germany angers EU after blocking deal to ban fossil fuel car sales) (2023). https://unn.ua. Published 07 March 2023
45. Ambrose, J.: Invasion of Ukraine 'has fuelled funding boom for clean energy. In: The Guardian, 25 May 2023 (in press) (2023)

Studying the Energy Systems State in Modern Megacities Provided Total Transition to Electric Vehicles

Wolter Martin, Meshkov Denys, and Mischenko Mykyta

Abstract This study examines the impact of transitioning to electric vehicles (EVs) on energy systems in modern megacities. Using the National Transport Strategy of Ukraine for 2030 as a framework, the paper discusses global trends in reducing greenhouse gas emissions through the adoption of EVs and Industry 4.0/5.0 technologies. It highlights challenges for power generation systems, focusing on energy efficiency and capacity requirements to replace internal combustion engine vehicles with EVs. The findings reveal significant opportunities for emissions reduction, aligning with EU policies for sustainable urban mobility and supporting the integration of renewable energy solutions in metropolitan areas.

1 Introduction

1.1 Relevance

According to the National Transport Strategy of Ukraine for the period up to 2030 [1] "...global trends in the development of transport systems indicate the need for rapid unification of transport technologies and regional mobility projects. Transport has become more and more energy-saving and "green", safe and friendly to the consumer and the natural environment. Many countries around the world plan to replace the

W. Martin
Institute of Electric Power Systems, Otto-von-Guericke University Magdeburg, Magdeburg, Germany
e-mail: martin.wolter@ovgu.de

M. Denys (✉) · M. Mykyta
Department of Engines and Hybrid Powerplants, Educational and Scientific Institute of Power Engineering, Electronics and Electromechanics, National Technical University "Kharkiv Polytechnic Institute", Kharkiv, Ukraine
e-mail: denys.meshkov@khpi.edu.ua

M. Mykyta
e-mail: mykyta.mishchenko@ieee.khpi.edu.ua

© The Author(s), under exclusive license to Springer Nature Switzerland AG 2025

A. Marchenko and Z. Koruba (eds.), *Decarbonization of Transport Energy Installations in the Context of Sustainable Development Strategies*, Studies in Systems, Decision and Control 635, https://doi.org/10.1007/978-3-032-05884-3_5

majority of cars with internal combustion engines by electric vehicles by 2030.Road transport accounts for 90 to 95% of emissions into the air. The EU transport policy has set a goal of halving the use of conventional fuel cars in cities by 2030 and completely replace them by 2050. Among the expected results is "…reduction of the total volume of the air pollutants emissions, conditionally reduced to carbon monoxide taking into account the relative aggressiveness of the main pollutants to 70% (from 2015 level)".

According to the White Paper of the European Commission—Plan for the development of a single European transport space—on the way to a competitive and resource-efficient transport system [2] «According to the White Paper of the European Commission [2], the EU aims to drastically reduce greenhouse gas emissions, targeting an 80–95% reduction below 1990 levels by 2050, with at least a 60% reduction in the transport sector.»

It is also important to note that the improvement of the design and operational processes of bulk power plants, together with the wide implementation of Industry 4.0 and 5.0 technologies—primarily IT technologies—has ensured a significant reduction in specific fuel consumption, which, in turn, leads to reduced CO_2 emissions and lower harmful exhaust emissions [3].

In this connection, an urgent question arises regarding the efficiency assessment of the existing electric energy generation system and its capacity to support the transition of modern megacities from traditional internal combustion engine vehicles to hybrid and fully electric vehicles. This includes evaluating the shift from large centralized power plants (e.g., coal or nuclear) to smaller, more flexible power generation units such as gas-fired plants and cogeneration systems, which can better adapt to the changing demands of urban electrification.

The purpose of the study is to investigate the effectiveness of the use of environmentally friendly road transport in terms of CO_2 emissions into the environment from existing sources of electricity production provided that the car park of modern megacities is completely converted to electric vehicles. The transition to electric vehicles is a key aspect of achieving climate goals, especially in urban environments like Berlin, Warsaw, and Kyiv, which present unique challenges and opportunities.

1.2 Research Analysis

The works of many scientists all over the world are devoted to the problem of the efficient use of electric energy in the transportation sector. The study [3] examines the impact of switching to hydrogen (H_2) as a fuel for Dubai buses with varying degrees of penetration on the reduction of carbon dioxide (CO_2) emissions. The publication [3a] is dedicated to the study of the effective use of natural gas in Ship's Low-Speed Gas-Diesel Engines, which, as is known, provides a significant reduction in CO_2 emissions compared to traditional petroleum-based fuels. The paper [4] analyses four key European and German transport policies and assesses the possibility of their improvement in order to increase their potential for reducing greenhouse gas emissions in road transport by using the ecological potential of electric vehicles.

The scientific work [5] examines the influence of factors affecting CO_2 emissions, namely population activity, electricity intensity, production structure, energy efficiency and the influence of the fuel emission factor.

The focus on researching the impact of daily electricity production from nuclear and renewable energy on achieving the goals of carbon neutrality in three of the largest nuclear energy countries (i.e. the USA, China and France) is presented in the scientific work [6]. This methodology for estimating and comparing marginal impacts of carbon emissions and prices (also known as side effects) of generation in a country to other countries is presented in [7]. The methodology is applied to analysed possible decarbonization paths from 2018 to 2030 and 2040 for EU countries considering the expected expansion of power generation and transmission capacities as well as the gradual abandonment of nuclear and fossil fuels.

A comparative assessment on the effectiveness of using vegetable-based fuel in a diesel engine regarding the environmental friendliness of its use is presented in papers [8, 8a,]. In work [9], it is noted that the use of biofuel, namely ethyl alcohol, leads to the improvement of environmental characteristics in internal combustion engines. It is indicated that with an increase in the content of ethanol in the fuel the concentration of toxic substances (CO, NO_2, SO_2, CH_2O) in the exhaust gases of internal combustion engines decreases.

In the study [10], the transition to a low-emission society requires a radical reduction of emissions in the electricity production sector. The study [11] quantitatively evaluates the factors of carbon emissions from electricity production in the EU during two sub-periods, 2000 to 2007 and 2007 to 2015. The work [12] examines the driving factors of carbon dioxide emissions from electricity production in the European Union for the years 2000 to 2018, namely: the effects of economic activity, population number, electricity intensity, electricity trade, energy intensity, production structure and the emission factor.

Thus, the work of many scientists is devoted to the issues of greening electricity production, the transition of road transport to ecological types of fuel, the reduction of pollutant emissions of the production sources of electric energy and road transport.

Along with this, the problem of increasing indirect (made in the process of production and transmission of electric energy) air pollutants emissions from electric vehicles during their operation remains outside the attention of scientists.

1.3 Sources of Electricity Production in Megacities

Analysis of the main sources of electricity production in Berlin, Warsaw and Kyiv, presented in Fig. 1, shows that each of these cities uses a variety of energy sources to generate electricity.

Coal and gas-fired power plants account for about 65% of total electricity generation in Berlin according to data of 2021 [13]. Also, important sources of energy

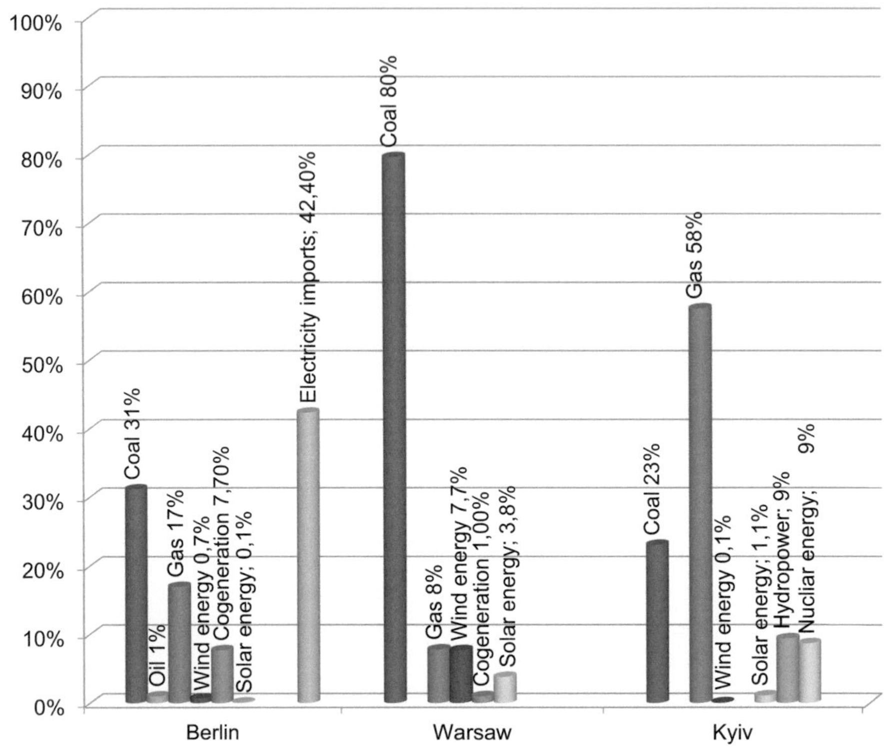

Fig. 1 Analysis of the main sources of electricity production in Berlin, Warsaw and Kyiv

are nuclear and wind power plants, which account for about 22% of the total electricity generation in Berlin. In the near future, it is planned to reduce the use of coal and increase the share of renewable energy sources. Taking into account the world‘s desire to reduce greenhouse gas emissions and increase the level of renewable energy, it is planned to increase the share of electricity generation at wind and solar power plants in the future [14]. It is also worth noting that Berlin has plans to increase electricity production from renewable sources to 65% by 2030 [15] meaning a significant reduction in energy production from conventional sources such as coal and gas.

The largest share of electricity production in Warsaw is taken up by coal-fired power plants, which account for about 80% of the total electricity generation in the city. The rest is produced at gas power plants and hydroelectric power plants, while nuclear energy is imported. In the near future, it is planned to reduce the use of coal and increase the share of renewable energy sources, such as solar and wind energy.

The largest share of electricity production in Kyiv is taken up by thermal power plants that operate on natural gas and coal. They make up about 75% of the total electricity generation in the city. The rest is produced by hydroelectric power plants, nuclear power plants and solar power plants. In the near future, it is planned to

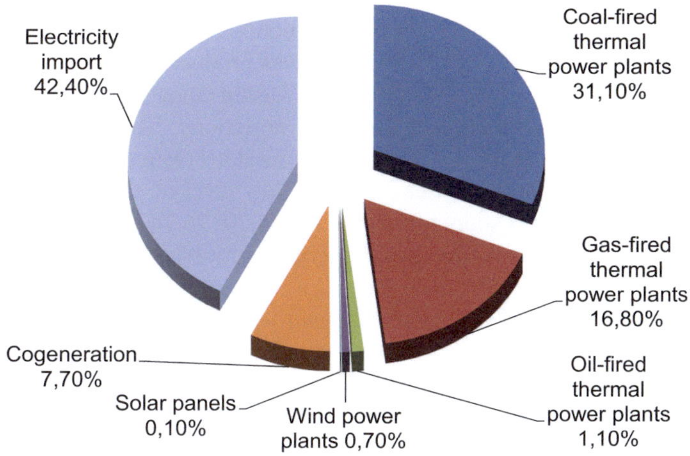

Fig. 2 The main sources of electricity production in Berlin and their percentage contribution to the total amount of electricity in the city according to data for 2021

increase the share of electricity generation from renewable sources, such as solar and wind energy as well as the development of small hydroelectric plants.

1.4 Analysis of the Structure of Electricity Generation Sources

The main sources of electricity production in Berlin and their percentage contribution to the total amount of electricity in the city according to data from 2021 are presented in Fig. 2.

Coal-fired power plants in Berlin are one of the main sources of greenhouse gas emissions. According to data of 2021, coal-fired power plants in Berlin produced about 7.4 million tons of CO_2. Gas-fired power plants emitted about 3.5 million tons of CO_2. The use of renewable sources of electricity production, including wind and solar power plants, can help reduce the share of greenhouse gas emissions from the total amount of emissions caused by the operation of other sources of electricity production.

Regarding electricity generation in Berlin, it can be noted that according to the German transmission network operator 50Hertz, in 2021 coal-fired thermal power plants produced 15.8 TW · h of electricity, while gas-fired thermal power plants—7.9 TW · h of electricity [16].

Data on greenhouse gas emissions were published in the Statistischer Bericht E IV 5—j/21 of 2021, which was prepared by the Amt für Statistik Berlin-Brandenburg.

This report contains information on the amount of carbon dioxide, nitrogen oxides and sulfur emissions caused by electricity production in Berlin.

In 2021, around 32.7TW · h of electricity was produced in Berlin. The preliminary energy and CO_2 balance of 2021 for Berlin shows a slight increase in energy consumption and energy-related CO_2 emissions. The nationwide coefficient (110,288 kg CO_2/TJ) for carbon dioxide emissions from electricity production increased in the 2021 reporting year due to the low level of electricity fed into the grid from renewable energy sources compared to 2020 (96,787 kg CO_2/GJ). This led to an increase in emissions from electricity consumption both in households (15.7) and in other sectors [17].

According to the results of the study, the main sources of electricity production in Warsaw were gas and coal-fired thermal power plants, as well as waste incineration. A total of 27 sources of electricity production were identified. Overall, the study found that electricity generation in Warsaw remains a strong source of greenhouse gas emissions, with gas and coal-fired thermal power plants being the main sources of emissions. The average statistical volume of electricity production in Warsaw during 2021 is about 23.6TWh. In Warsaw in 2021, electricity was produced using the following sources presented in Fig. 3.

CO_2 emissions from electricity production in Warsaw in 2021 by types of power plants are:

- Coal-fired power plants—9.689 million tons of CO_2.
- Gas power plants—1.738 million tons of CO_2.
- Wind power plants—0.035 million tons of CO_2.
- Solar power plants—0.011 million tons of CO_2.

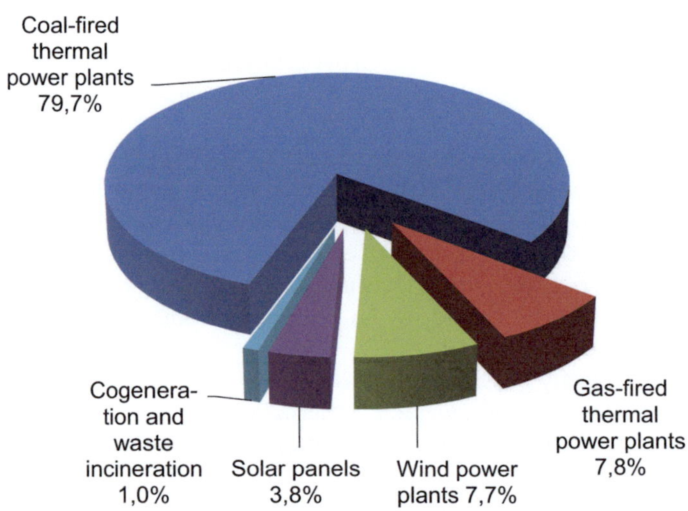

Fig. 3 The main sources of electricity production in Warsaw and their percentage contribution to the total amount of electricity in the city according to data for 2021

- Cogeneration and waste incineration—1.868 million tons of CO_2.

The total amount of greenhouse gas emissions from electricity production in Warsaw for a year is about 13.341 million tons of CO_2.

In Kyiv, as in many other cities, there are several sources of electricity production. The main sources of electricity production in Kyiv are: TPPs (thermal power plants) on natural gas and coal, and HPPs (hydroelectric plants) on the Dnieper River and reservoirs located on the territory of Kyiv and the region. There are also several distribution networks operating in Kyiv that ensure the delivery of electricity to consumers.

The percentage contribution of each source of electricity production to the total volume of production in Kyiv may vary depending on various factors, such as demand for electricity, technical features of production sources, and the state's energy policy. However, over the past few years, the largest source of electricity production in Kyiv has been natural gas and coal-fired thermal power plants [18]. Thermal power plants have been the largest source of electricity production in Kyiv, the main of which are TPP-6, TPP-5, Darnytska TPP, and Kyiv Hydroelectric Power Plant and Kyiv Hydroelectric Power Plant have also made an important contribution to the generation of electricity. A component of nuclear energy is the use of the power transmission line (LEP) Rivne NPP—Substation "Kyivska"—a power transmission line between the Rivne NPP and the substation "Kyivska" with a voltage of 750 kV, which supplies the city of Kyiv and the Central and Northern regions. The introduction of transmission lines made it possible to change the balance of electricity generation and increase the share of nuclear generation, the capacity of which was limited by networks.

According to the State Statistics Service of Ukraine [19], in 2021, 10.3TWh of electricity was produced in Kyiv. This is about 6.2% of the total amount of electricity produced in Ukraine in the same year.

The main producers of electricity in Kyiv are Kyiv CHPP-5—3.5 TWh, Kyiv CHPP-6—2.8TWh, Darnytska CHPP—2.0 TWh, Energy CHP—1.0 TWh. In Kyiv in 2021, electricity was produced using the sources presented in Fig. 4.

According to statistical data [20] emissions of pollutants into the atmospheric air from stationary sources of pollution (the total amount of substances or materials of physical, chemical or biological origin that entered the air pool from stationary sources of pollution as after passing through dust and gas treatment plants as a result of incomplete capture and purification on organized sources of pollution, and without cleaning from organized and unorganized sources of pollution) in 2021 amounted to 38.8 thousand tons. Of these, emissions of individual pollutants into the air from stationary sources of pollution in the city of Kyiv for 2021 are: carbon monoxide (CO)—1.6 thousand tons, methane—18.1 thousand tons, nitrogen dioxide (NO_2)—6.0 thousand tons, sulfur dioxide (SO_2)—6.8 thousand tons, non-methane light organic compounds—2.9 thousand tons, carbon dioxide (CO_2)—3.9 million tons. Along with this, the emissions into the atmospheric air of certain pollutants by categories of stationary sources of pollution, namely from the energy sector in 2021, according to the data of the Main Department of Statistics in the Kyiv region [21],

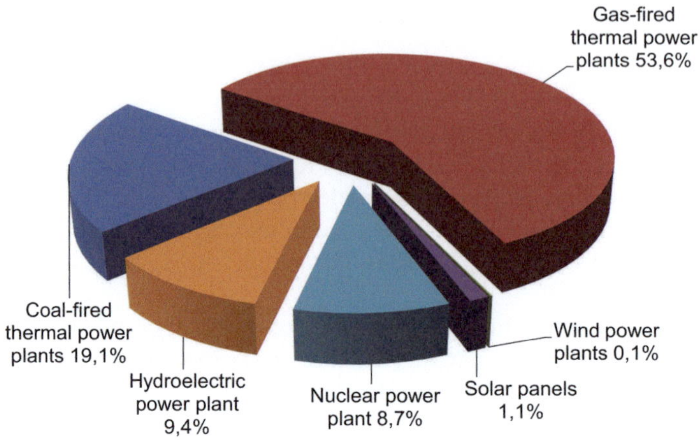

Fig. 4 The main sources of electricity production in Kyiv and their percentage contribution to the total volume of electricity in the city according to data for 2021

are: nitrogen dioxide (NO_2)—4168.3 tons, nitrogen dioxide sulfur (SO_2)—24,991.1 tons, carbon monoxide (CO)—1248.6 tons, carbon dioxide (CO_2)—3.3 million tons.

At present, each of the analysed cities—Berlin, Warsaw, and Kyiv—utilizes a mix of energy sources for electricity production, combining traditional sources such as coal and gas with renewable sources like wind and solar energy. In response to global climate challenges and the aim to reduce greenhouse gas emissions, all three cities are planning to increase the share of renewable energy in their energy mix in the coming years.

In addition to electricity generation, road transport represents another major environmental challenge due to its significant contribution to air pollution and greenhouse gas emissions. Key environmental impacts of road transport include:

- **Emissions of hydrocarbons and nitrogen oxides (NOx)**: Internal combustion engines commonly used in vehicles emit substantial amounts of hydrocarbons and NOx, which contribute to smog formation, acid rain, and pose health risks to humans.
- **Greenhouse gas emissions**: Road transport accounts for about 20% of total CO_2 emissions in the European Union, with passenger cars emitting an average of 150 g of CO_2 per kilometres travelled. This contributes significantly to climate change and global warming.
- **Particulate air pollution**: Motor vehicles emit particulate matter such as sulphur dioxide and soot particles, which are especially harmful in densely populated urban areas with high traffic volumes.

When compared to aviation, road transport produces a lower per-kilometres CO_2 emission rate per passenger (150 g for passenger cars versus approximately 257 g for domestic flights). However, the widespread use of road transport results in its overall emissions being much higher, emphasizing the need for measures to decarbonize both sectors.

2 Analysis of Urban Car Fleets and Their Environmental Impact

According to the Federal Environment Agency, the shares of gasoline and diesel engines in the German car fleet are expected to be roughly equal by 2030, at around 30% and 35%, respectively [22]. The share of electric and hybrid cars is expected to increase and reach about 25% by 2030.

As for the average mileage for each vehicle type, the figures vary depending on factors such as vehicle model, year of manufacture, driving style, and operating conditions. It is important to note that the average mileage does not depend directly on the propulsion method but rather on the usage patterns of vehicle owners. For example, individuals who purchase electric vehicles or hybrids often tend to drive shorter distances, making these types of cars more suitable for urban or suburban environments. In general, the average mileage for gasoline and diesel cars can reach around 15,000 km per year, while for electric and hybrid vehicles, the annual mileage is often lower due to their frequent use for shorter trips.

According to official statistics published on the official website of the city of Berlin [23], at the beginning of 2021, Berlin's car fleet consisted of 1.79 million registered cars. The breakdown of engine types in the Berlin fleet for 2021 was as follows:

- Gasoline cars: 1.02 million (57.2%).
- Diesel cars: 0.51 million (28.9%).
- Hybrid cars: 0.1 million (5.9%).
- Electric cars: 0.027 million (1.5%).
- Gas-powered cars: 0.034 million (1.9%).
- Other types of engines: 0.083 million (4.6%), which may include experimental or niche technologies such as hydrogen-powered vehicles, though their presence in Germany is minimal.

Thus, at the beginning of 2021, the majority of cars in the Berlin fleet continued to run on gasoline or diesel. However, the share of hybrid and electric cars is still increasing, which indicates a trend towards more environmentally friendly cars.

Despite the growing popularity of environmentally friendly cars, petrol and diesel cars continue to be popular among local car owners in Berlin.

According to the latest available data (2021), the number of registered cars in Warsaw is about 0.75 million units [24]. The distribution of engine types in the Warsaw fleet for 2021 was as follows:

- Gasoline engines: 0.45 million (61%).
- Diesel engines: 0.24 million (32%).

- Hybrid cars: 0.045 million (6%).
- Electric cars: 0.007 million (1%).

Warsaw also has the highest level of registration of hybrid and electric cars in Poland.

According to the State Automobile Inspectorate of Ukraine, the fleet of passenger cars in Kyiv (in 2021) was about 1.2 million units. In percentage terms, the distribution of engine types in the Kyiv fleet was as follows:

- Gasoline engines: 0.8 million (67%).
- Diesel engines: 0.36 million (30%).
- Hybrid cars: 0.024 million (2%).
- Electric cars: 0.012 million (1%).

The number of registered cars in Kyiv in 2021 is 1.12 million. The maximum daily number of active cars in Kyiv as of January 2022 is 1.13 million units. In 2021, there were 17% more cars in Kyiv than in 2020.

Taking into account the above-mentioned indicators of the volumes of electricity production in Berlin, Warsaw and Kyiv, and the CO_2 emissions caused by them, as well as taking into account the analysis of the car park of these cities, it is appropriate to study the effectiveness of the use of hybrid cars and electric cars by analyzing the indicators of power, environmental friendliness and calculation of CO_2 emissions into the environment.

3 Comparative Analysis of Vehicle Drivetrains and Their Environmental Impact

For comparative analysis, we will consider four types of Volkswagen cars that are popular in these European cities and correspond to different types of drivetrains or engines: internal combustion (gasoline and diesel), hybrid, and fully electric.

To conduct a comparative analysis for all three European cities, it is proposed to use a single vehicle model, which will allow for a comparison of results regardless of design differences. For this purpose, it seems reasonable to select the Volkswagen Golf as the reference vehicle. This car is officially available and popular in the markets of Germany, Poland, and Ukraine. The manufacturer continuously modernizes it, making it one of the most advanced and environmentally friendly vehicles, incorporating nearly all modern technologies aimed at reducing fuel consumption and harmful emissions.

A detailed analysis of the comparison of the Volkswagen Golf with a 1.4 TSI engine with a capacity of 150 hp [25], CRLB 2.0 TDI diesel engine [26], Volkswagen Golf GTE [27] with a hybrid drivetrain based on a 1.4 TSI 150 hp engine and fully electric e-Golf [28] in terms of power and environmental performance:

- by power—all three cars have the same engine power—150 hp. However, the Volkswagen Golf GTE additionally has an electric motor that adds another 102 p to the total output, making it significantly faster and more dynamic than the regular Volkswagen Golf with a 1.4 TSI engine or a 2.0 TDI diesel engine;
- in terms of environmental friendliness—the Volkswagen Golf with a 1.4 TSI engine has CO_2 emissions of about 126 g/km, the diesel has a slightly higher figure of 132 g/km, while the Volkswagen Golf GTE with a hybrid power plant has CO_2 emissions of 40 g/km, which are much smaller. Also, the Volkswagen Golf GTE can drive in fully electric mode for up to 50 km, which makes it possible to make short trips without using gasoline;
- fuel consumption—the Volkswagen Golf with the 1.4 TSI engine has an average fuel consumption of approximately 6.0 l/100 km in urban conditions and 4.8 l/100 km on the highway, the version with a 2.0 TDI diesel engine has an average fuel consumption in urban conditions 5.2 l/100 km and 4.0 l/100 km on the highway. In the Volkswagen Golf GTE, the average fuel consumption, depending on the driving mode, can be from 1.4 to 2.0 l/100 km, which is several times less than that of a regular Volkswagen Golf;
- cost and running costs—the Volkswagen Golf GTE is a more expensive car than the Volkswagen Golf with a 1.4 TSI or 2.0 TDI engine, as hybrid technology requires a more complex and expensive power plant. However, thanks to the hybrid technology, the Golf GTE can be significantly more economical to operate, as it has a significantly lower level of fuel consumption and helps to reduce the level of carbon dioxide emissions;
- according to the level of harmful emissions of other polluting substances—in addition to CO_2, it is also important to consider the level of emissions of other harmful substances. According to the manufacturer, the Volkswagen Golf GTE complies with the European emission standard EURO 6, which regulates the level of emissions of nitrogen oxides (NOx) and other pollutants. Thus, the Golf GTE is a greener option in terms of emissions of other pollutants.

The energy that powers the e-Golf electric motor is stored in lithium batteries with a capacity of 35 kWh, which is enough for 231 km. Accordingly, the hatchback's energy consumption per 100 km is 14.1 kWh, which is slightly more than, for example, the BMW i3 electric car.

To perform an analysis of the environmental friendliness of each type of drivetrain of the compared cars, we will take into account the passport data of the CO_2 indicators declared by the manufacturer. While the electric version of the e-Golf has no CO_2 emissions during operation, its charging relies on electricity, the production of which from sources such as coal and gas power plants does emit CO_2. Although nuclear power plants do not produce direct CO_2 emissions, their share in the overall energy mix should be considered when assessing the environmental impact of electric vehicles. The analysis uses official statistical data from national agencies and indicators for urban car fleets. CO_2 calculations are based on EU-standard methodologies.

From the analysis of emissions from sources of electricity production, it is known that 32.7 TWh of electricity is produced in Berlin per year, which accounts for emissions of 10.9 million tons of CO_2. According to statistical sources, Warsaw generates 23.6 TWh of electricity per year and emits 13.341 million tons of CO_2. The situation is a little worse in Kyiv, where 12.69 TWh of electricity is produced annually, which accounts for CO_2 emissions in the amount of 3.3 million tons.

Considering these data, it is possible to calculate the emissions from electric vehicles, which are the average annual mileage of 15.000 km and the amount of CO_2 emissions per kilometer of mileage. To perform this calculation, it is necessary to determine the mass of CO_2 emissions equivalent to the production of 1 kW of electricity.

The calculation of g_W (CO_2), kg/kW—the quantitative indicator of CO_2 emissions caused by the production of 1 kW of electricity was performed according to formula 1, the results are shown in Fig. 5 and Table 1.

$$g_W(CO_2) = \frac{E_{CO_2}}{W}, \tag{1}$$

where E_{CO2}—gross emission of carbon dioxide, million tons per year;

W—volume of generated electricity, TWh/a.

We perform the analysis of the environmental friendliness of each type of car on the condition that the average mileage of each type is 10.000 km per year. Therefore, the emissions of gasoline, diesel and hybrid cars will remain unchanged for each

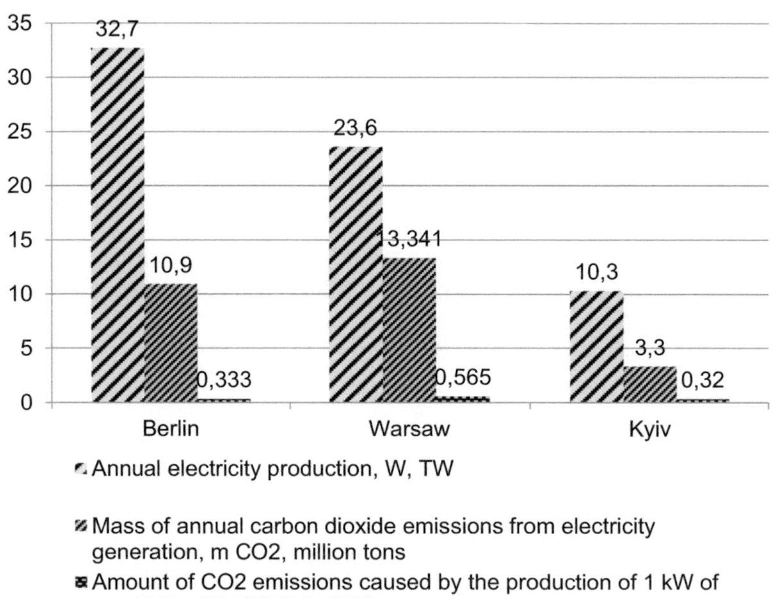

Fig. 5 Quantitative indicators of energy efficiency

Table 1 Quantitative indicator of CO_2 emissions from sources of electricity production

City	E_{CO2}, million tons	W, TWh/a	g_W (CO_2), kg/kWh
Berlin	10.9	32.7	0.333
Warsaw	13.341	23.6	0.565
Kyiv	3.9	12.69	0.320

city. However, emissions from an electric car will differ, this is due to the difference in the mass of CO_2 emissions per kilowatt of electricity in each city.

The indicator of CO_2 emissions from an electric vehicle per kilometer of the driving cycle is calculated according to formula (2), the results are presented in Fig. 6 and Table 2.

$$k_{CO_2} = G_{energy} \cdot g_W(CO_2) \quad (2)$$

where g_w (CO_2)—CO_2 emissions, g/km;

G_{energy}—energy consumption (for e-Golf 14.1 kWh/100 km).

The results shown in Table 2 and Fig. 6 indicate that the lowest rate of CO_2 emissions from land transport belongs to a car with a hybrid power plant, which, according to the results of calculations, is inferior to an electric car and emits carbon dioxide into the air due to the production of electrical energy for its operation.

The rapid growth of electric vehicles plays a big role. Despite the fact that manufacturers of electric vehicles indicate the environmental cleanliness of such vehicles,

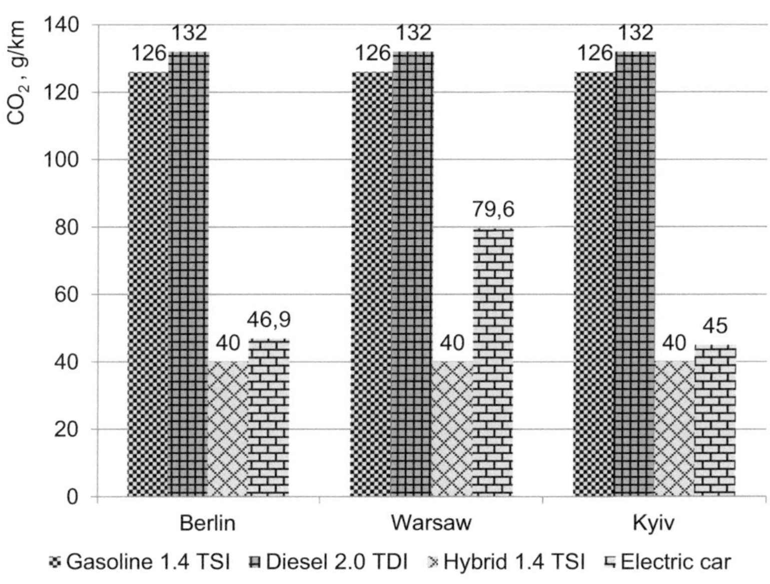

Fig. 6 CO_2 Emission indicators from ground transport

Table 2 Carbon dioxide emissions from land transport

Type of power plant	Annual mileage, km	CO_2 emission rate according to passport data, g/km	CO_2 Emission indicators, g/km			Annual indicators of CO_2 emissions, kg/year		
			Berlin	Warsaw	Kyiv	Berlin	Warsaw	Kyiv
	10,000	*According to passport data*						
Gasoline 1.4 TSI		126	126	126	126	1260	1260	1260
Diesel 2.0 TDI		132	132	132	132	1320	1320	1320
Hybrid 1.4 TSI		40	40	40	40	400	400	400
		According to the results of the calculations						
Electric car		–	46,9	79,6	45,0	469	796	450

this is a false statement. Let's look at this issue from the point of view of the energy sustainability of electric vehicles—the use of electricity produced by the energy complex for the operation of the vehicle and the pollution of the environment by carbon dioxide emissions from its production. The strategic task outlined in the Plan for the Development of the Single European Transport Space clearly defines a 50% reduction in the use of cars on "traditional types of fuel" in urban transport by 2030, and the gradual abandonment of such vehicles and the achievement of CO_2-free space in the main urban centers by 2050 year. Solving the task set and accepted in the European community only by switching to electric vehicles, as the most ecological type of vehicles, is not an exceptionally effective method.

We simulate the situation based on the indicators of electricity production, the composition of the car fleet and the indicators of Table 1 for Berlin, as a city with more than a million inhabitants in 2021, while we take as the studied model the fact that the entire car fleet of the city consists only of Volkswagen e-Golf electric cars.

Let's calculate and compare for real indicators and the studied model:

- the amount of electricity consumed by all electric vehicles per year, based on an annual mileage of 15.000 km, energy consumption for the e-Golf is 14.1 kWh per 100 km of mileage;
- emissions of carbon dioxide from the production of electricity consumed by electric vehicles per year.

Table 3 shows the projected CO_2 emissions under different scenarios, highlighting the impact of renewable energy integration.

In the modeled scenario, Berlin's car fleet is assumed to consist entirely of electric vehicles, amounting to 1.79 million units. However, this figure requires further explanation. Germany's national goal is to reach 15 million electric vehicles by 2030, representing approximately 18.75% of the total vehicle fleet. For Berlin, which accounts

Table 3 Comparative characteristics of energy consumption and CO_2 emissions from electric vehicles in Berlin

№	Indicators	Actual condition (2021)	Modeled condition	Adjusted modeled condition
1	Annual volume of electricity production (TWh)	32.7	32.7	32.7
2	Number of electric cars (million units)	0.027	1.79	0.5
3	Electricity consumed by electric vehicles per year (TWh)	0.076	5.048	1.41
4	Share of electricity consumed by electric vehicles from total production (%)	0.23	15.43	4.31
5	CO_2 emissions from electricity consumed by electric vehicles per year (tons)	25,308	1,680,980	126,363

for about 5% of the German population, this translates to an estimated 750,000 electric vehicles under equal distribution. However, regional disparities must be considered: wealthier southern regions are likely to have a higher adoption rate of electric vehicles, while urban residents in cities like Berlin often rely on public transport rather than personal cars. Adjusting for these factors, a more realistic estimate for Berlin's electric vehicle fleet in 2030 might range between 300,000 and 500,000 vehicles, significantly lower than the modeled 1.79 million.

The analysis of the obtained results shows, that the share of electricity consumed by electric vehicles per year from the total volume of electricity production in the studied model exceeds the result obtained according to the real indicators of 2021 in Berlin by 67.08 times. In fact, this indicator of the studied model is 15.43%, which is 1.81 times higher than the cumulative percentage of electricity production (from the total volume of electricity production in 2021) by all renewable (ecological) sources of electricity (Fig. 2), which was—8.5% including cogeneration. This gives reason to emphasize not only the need to increase the production of electric energy in Berlin, but also the aggravation of the issue of the development of renewable energy, since there is a direct dependence of **"ecological transport—non-ecological energy sources—export of electric energy"**.

The results of the calculations of the studied model indicate that CO_2 emissions from the production of electricity consumed by electric vehicles per year, which amount to 1.680,980 tons, exceed CO_2 emissions from the production of electricity consumed by electric vehicles per year by 66.4 times in the real conditions of Berlin in 2021.

Every year there is an increase in the number of cars in large cities, which adds a significant influence on the results obtained in the simulated situation. **The total transition to electric cars will cause the need to increase the production of electrical energy, and taking into account the rejection of traditional energy sources, will lead to the need to increase the capacity of renewable "green" energy.** While the shift to electric vehicles increases electricity demand, their overall carbon footprint is significantly lower than that of conventional vehicles, particularly as renewable energy sources grow in share.

It should also be emphasized that the results were obtained taking into account only CO_2 emissions into the environment from motor vehicles (electric vehicles) during their operation. It does not take into account emissions caused by stationary sources such as automobile factories that manufacture cars and enterprises that concentrate the production of individual components of these vehicles. It also should be noted that the disposal of used parts of electric cars is not an ecological process. In addition, each component: production, operation and disposal require energy costs, which additionally affects the quantitative indicators of environmental pollution by various types of emissions, in particular CO_2.

In view of the obtained results, the policy of full transition to electric vehicles will aggravate the problem of rapid introduction of "green energy" in order to eliminate the increase in demand for electricity, prevent the growth of its export, and as a result increase the level of environmental pollution from its production.

4 Conclusions

1. Studies show that while the transition to electric vehicles is not a completely CO_2-neutral solution under the current energy mix, it significantly improves the carbon footprint compared to internal combustion engine (ICE) vehicles. Electric vehicles produce no direct emissions during operation, and even when considering the indirect emissions from electricity production, their overall CO_2 balance is typically better than that of conventional cars. However, the growing fleet of electric vehicles will inevitably increase electricity demand, requiring additional power generation capacity. To maximize the environmental benefits of electric vehicles, this increased demand must be met with renewable energy sources, which will further reduce CO_2 emissions and contribute to achieving climate goals.
2. According to the results of the study, it was the hybrid drive that showed the best environmental performance, which may be due to the optimal operating modes of the internal combustion engine paired with an electric motor. An alternative to the "electrification" of road transport on the way to reducing carbon dioxide emissions into the environment is also the use of alternative fuel in diesel engines, namely biodiesel—biofuel for diesel engines, which is made from vegetable oils or animal fats and has similar characteristics to diesel fuel, so it can be used in such engines both in mixtures with ordinary diesel fuel and in its pure form.

However, the main technical problem with the widespread use of biofuel diesel is the need to modify existing diesel engines, since it has some differences in physical properties compared to conventional diesel fuel. The advantages of biofuel compared to diesel fuel include the renewability of the raw material base and significantly lower emissions of carbon dioxide into the atmosphere.

3. Along with this, one of the most promising alternative fuels for gasoline diesel engines is hydrogen. It is characterized by high energy indicators, unique kinetic characteristics, environmental cleanliness, unlimited raw material reserves and the possibility of obtaining it in large volumes, which determines the perspective of its use. However, the presence of still unresolved technological issues related to storage, transportation, refueling and use in cars prevent the widespread use of hydrogen technology in transport.

References

1. Кабінет Міністрів України. Розпорядження від 30 травня 2018 р. № 430-р Київ. Про схвалення Національної транспортної стратегії України на період до 2030 року. Available at: https://zakon.rada.gov.ua/laws/show/430-2018-%D1%80#Text
2. (2011) Європейська Комісія. Біла книга—План розвитку єдиного європейського транспортного простору—на шляху до конкурентоспроможної та ресурсоефективної транспортної системи. Luxembourg: EU Publishing Center, 28 p. Available at: https://brdo.com.ua/....
3. Awad, A., Bouabid, A., Almansoori, A., Al Hajaj, A.: Analysis of CO_2 emissions reduction on the future hydrogen supply chain network for Dubai buses. Int. J. Hydrogen Energy **54**, 256–266 (2024). https://doi.org/10.1016/j.ijhydene.2023.04.121
4. Peiseler, L., Cabrera Serrenho, A.: How can current German and EU policies be improved to enhance the reduction of CO_2 emissions of road transport? Revising policies on electric vehicles informed by stakeholder and technical assessments. Energy Policy **168**, 113124 (2022). https://doi.org/10.1016/j.enpol.2022.113124
5. Li, L., Raza, M.Y., Cucculelli, M.: Electricity generation and CO_2 emissions in China using index decomposition and decoupling approach. Energ. Strat. Rev. **51**, 101304 (2024). https://doi.org/10.1016/j.esr.2024.101304
6. Kartal, M.T., Pata, U.K., Depren, Ö., Erdogan, S.: Effectiveness of nuclear and renewable electricity generation on CO_2 emissions: Daily-based analysis for the major nuclear power generating countries. J. Clean. Prod. **426**, 139121 (2023). https://doi.org/10.1016/j.jclepro.2023.139121
7. Romano, E., Mutschler, R., Hollmuller, P. et al.: Spatial carbon and price spillovers among EU countries on their pathway toward net-zero electricity supply. Energy Econ. **131**, 107349 (2024). https://doi.org/10.1016/j.eneco.2024.107349
8. Марченко, А.П., Прохоренко, А.А., Осетров, А.А., Смайлис, В., Сенчила, В.: Сравнительная оценка эффективности применения растительных топлив в дизельном двигателе. Двигатели внутреннего сгорания, №1, С.46–51 (2004). Available at: https://www.kpi.kharkov.ua/archive/%D0%9D%D0%B0%D1%83%D0%BA%D0%BE%D0%B2%D0%B0_%D0%BF%D0%B5%D1%80%D1%96%D0%BE%D0%B4%D0%B8%D0%BA%D0%B0/Dvs/2004_1/13.PDF
9. Гомонай, В.І., Богоста, А.С., Лобко, В.Ю., Тацькар, А.Р.: «Етиловий спирт як екологічне паливо для двигунів внутрішнього згорання», Науковий вісник Ужгородського

університету, Сер.: Хімія., №1, С. 82–87 (2011). Available at: https://nbuv.gov.ua/UJRN/Nvuuchem_2011_1_21
10. Gaure, S., Golombek, R.: True or not true: CO_2-free electricity generation is possible. Energy **259**, 124998 (2022). https://doi.org/10.1016/j.energy.2022.124998
11. Rodrigues, J.F.D., Wang, J., Behrens, P., de Boer, P.: Drivers of CO_2 emissions from electricity generation in the European Union 2000–2015. Renew. Sustain. Energy Rev. **133**, 110104 (2020). https://doi.org/10.1016/j.rser.2020.110104
12. Karmellos, M., Kosmadakis, V., Dimas, P., et al.: A decomposition and decoupling analysis of carbon dioxide emissions from electricity generation: evidence from the EU-27 and the UK. Energy **231**, 120861 (2021). https://doi.org/10.1016/j.energy.2021.120861
13. Berlin Energy Agency. EEAP Berlin EN. Available at: https://www.buendnis-klimaschutz.de/fileadmin/_migrated/content_uploads/EEAP-Berlin_EN_02.pdf
14. Berlin Energy Transition Dialogue 2021. Multiple Challenges Renewable Energy Takes Center Stage. Available at: https://www.unendlich-viel-energie.de/en/details/news/berlin-energy-transition-dialogue-2021-multiple-challenges-renewable-energy-takes-center-stage
15. The State of Berlin's Energy and Climate Protection Program 2030. Available at: https://www.berlin.de/senuvk/umwelt/klimaschutz/programm_klima_energie/files/2030_klimaschutzprogramm_en.pdf
16. Amprion. Responsibility and Climate Protection: Renewable Energies in Numbers. Available at: https://www.amprion.net/verantwortung/klimaschutz/erneuerbare-energien/erneuerbare-energien-in-zahlen
17. Statistischer Bericht E IV 5—j/21. Amt für Statistik Berlin-Brandenburg. Available at: https://www.statistik-berlin-brandenburg.de/e-iv-5-j
18. (2020) Паливно-енергетичні ресурси України. Статистичний збірник. Київ, Available at: https://ukrstat.gov.ua/druk/publicat/kat_u/2020/zb/12/Zb_per.pdf
19. (2022) Державна служба статистики України. Статистичний збірник «Регіони України» Частина I. Київ, 2022. Available at: https://ukrstat.gov.ua/druk/publicat/kat_u/publ2_u.htm
20. Державна служба статистики України. Головне управління статистики у Київській області. Available at: http://kyivobl.ukrstat.gov.ua/p.php3?c=1630&lang=1
21. Umweltbundesamt. Clean Air Strategy 2030 for German Cities. Available at: https://www.umweltbundesamt.de/en/press/pressinformation/clean-air-strategy-2030-for-german-cities
22. Statistik Berlin-Brandenburg. Economy and Environment. Available at: https://www.statistik-berlin-brandenburg.de/wirtschaft/umwelt
23. Warsaw Car Statistics. Available at: https://warszawa.naszemiasto.pl/ile-jest-samochodow-w-warszawie-ze-spisu-wynika-az-1-7/ar/c1-9194555
24. Volkswagen Golf Models. Available at: https://www.autocar.co.uk/car-review/volkswagen/golf-2009-2013
25. e-Golf Specifications. Available at: https://hevcars.com.ua/volkswagen/e-golf-36kwh-2017/

Sustainable Development of Transport Objects Through the Implementation of Industry 4.0 and Industry 5.0 Technologies

Mykola M. Tkachuk, Grabovskiy Andriy, Alokhin Vitalii, Mykola A. Tkachuk, and Vasiliev Anton

Abstract This article reviews sustainable design and production strategies for transport elements, emphasizing the life cycle assessment (LCA) of internal combustion engine parts. Industry 4.0 and 5.0 technologies offer opportunities to reduce environmental impact while enhancing resource efficiency. The study critically analyzes existing approaches to digital integration and highlights the need for active human-system collaboration. Practical examples include developing diesel locomotive engines with optimized performance and environmental characteristics. The findings support strategic advancements in sustainability-focused manufacturing and operation, aligning with modern industry goals.

1 Introduction (General Provisions and Principles of Sustainable Manufacturing)

Let us consider some general issues using the example of transport objects. Sustainable manufacturing, particularly of internal combustion engine (ICE) components, is becoming an increasingly important topic in the context of modern industry. Given global environmental challenges, reducing carbon emissions and natural resource

M. M. Tkachuk (✉) · G. Andriy · M. A. Tkachuk · V. Anton
Department of "Theory and Computer-Aided Design of Mechanisms and Machines", National Technical University, "Kharkiv Polytechnic Institute", Kharkiv, Ukraine
e-mail: m.tkachuk@tmm-sapr.org

M. A. Tkachuk
e-mail: tma@tmm-sapr.org

V. Anton
e-mail: AVasiliev@tmm-sapr.org

A. Vitalii
Department of Foundry Production, National Technical University, "Kharkiv Polytechnic Institute", Kharkiv, Ukraine

© The Author(s), under exclusive license to Springer Nature Switzerland AG 2025
A. Marchenko and Z. Koruba (eds.), *Decarbonization of Transport Energy Installations in the Context of Sustainable Development Strategies*, Studies in Systems, Decision and Control 635, https://doi.org/10.1007/978-3-032-05884-3_6

consumption are critical tasks for all manufacturing industries, including transport infrastructure and transport engineering. Internal combustion engines remain the foundation of the modern automotive industry, which continues to play an important role in the global economy. Despite the trend toward transport electrification, particularly the growing popularity of electric vehicles, ICEs remain widely used, especially in commercial transport, heavy-duty machinery, military applications, and other specialized vehicles. However, increasing attention to environmental challenges, particularly climate change, necessitates a reevaluation of approaches to the production of these engines.

The production of internal combustion engine components has traditionally been associated with high energy consumption, extensive use of natural resources, and significant greenhouse gas emissions and waste generation. Changing these approaches to a more sustainable method is a necessary condition for ensuring the long-term development of the industry, combining economic efficiency with social responsibility and environmental sustainability. Sustainable manufacturing encompasses a set of measures aimed at minimizing environmental impact and reducing resource consumption while maintaining high productivity and product quality.

In the context of the production of internal combustion engine components, this means optimizing processes, implementing the latest technologies and materials, and transitioning to cyclic manufacturing models that reduce waste. Process optimization is a key aspect of sustainable manufacturing. Reducing the consumption of energy, water, raw materials, and other resources is an important step toward reducing the environmental footprint of production. This can be achieved through energy efficiency [1].

Energy efficiency is the foundation of any sustainable manufacturing strategy. Manufacturing enterprises can implement advanced technologies to reduce energy consumption. For example, using automated production management systems allows for energy optimization, reducing its use during idle times or periods of low activity. The advantages of energy-efficient technologies include not only reducing production costs but also decreasing greenhouse gas emissions. Modern factories can also integrate renewable energy sources, such as solar panels or wind turbines, to cover part of their energy needs. This is particularly important in the context of combating global warming.

A crucial issue is also decarbonization, that is, reducing carbon dioxide emissions into the atmosphere from internal combustion engines during operation.

Waste reduction. Reducing waste in production is crucial for achieving sustainable development. Manufacturing enterprises can implement waste management systems that reduce waste volume and minimize negative environmental impact. One approach is the introduction of closed-loop production cycles, where the waste from one process becomes raw material for another. For example, chips and other metal residues generated during the machining of internal combustion engine components can be collected, remelted, and reused. This not only reduces the amount of waste but also promotes more efficient resource utilization, decreasing dependence on the extraction of new raw materials.

Implementation of new materials. Innovations in materials science play a vital role in the sustainable manufacturing of ICE components. The use of modern materials allows for weight reduction, increased strength, lower production costs, and minimized environmental impact.

Composites and lightweight alloys. Lightweight materials, such as aluminum alloys and carbon fiber-based composites, are becoming increasingly popular in the automotive industry. They enable weight reduction of components, positively impacting vehicle fuel efficiency. For example, a lightweight aluminum engine block can reduce the total vehicle weight by several dozen kilograms, leading to significant fuel savings and a corresponding reduction in CO_2 emissions. Additionally, these materials often exhibit superior corrosion resistance and durability, extending component lifespan and reducing the need for frequent replacements and repairs.

Circular economy in ICE component manufacturing. The circular economy is a new paradigm focused on maximizing resource utilization and minimizing waste. It involves creating closed cycles where products and materials are used as efficiently as possible, and waste is transformed into new resources.

Design for recycling or reuse. This design approach is a key principle of the circular economy. It involves creating products that can be easily disassembled and recycled after their life cycle ends or reused in newly manufactured assemblies.

Material unification. The use of identical or compatible materials in different ICE components also simplifies recycling processes. This reduces the number of different materials that need to be processed and simplifies waste sorting and handling. For example, if all plastic engine parts are made from the same type of plastic, they are much easier to recycle together than if they were made using different types of plastics.

Recycling waste and old components and secondary metal processing. Recycling waste and used components is an important aspect of the circular economy, allowing for reduced consumption of primary resources and minimizing environmental impact. Most metals used in the production of ICE components can be reused after recycling. For example, old engines or their components can be disassembled, metal parts recast, and the recovered metal used for manufacturing new components. This reduces the need for extracting new resources and lowers energy consumption in production.

Sustainable manufacturing of cast ICE components is a crucial step toward reducing the automotive industry's environmental impact. Implementing innovative technologies, optimizing manufacturing processes, and transitioning to a circular economy help decrease negative environmental effects, enhance resource efficiency, and ensure sustainable industry development. However, achieving these goals requires coordinated efforts from manufacturers, governments, and scientific institutions.

This paper focuses on the development and implementation of sustainable development approaches based on Industry 4.0 and Industry 5.0 technologies using internal combustion engines as an example.

This goal involves attention to the issues discussed and described further.

2 Targeted Development and Lifecycle Management Within the Framework of Sustainable Manufacturing of Internal Combustion Engine Components

Participants in the complex supply chains of automotive industry components, regulators, and investors seeking to invest in sustainable manufacturing methods are increasingly interested in considering the carbon footprint of vehicles throughout their lifecycle—from material extraction and production to manufacturing, use, and disposal. Lifecycle assessment of technical products is becoming common practice in various industries, but in the automotive sector, methodologies for determining carbon emissions from cradle to grave have so far been diverse and therefore incomparable. This is expected to change with the adoption by the United Nations Economic Commission for Europe (UNECE) Working Group on Pollution and Energy of a technical mandate for an informal working group on vehicle lifecycle assessment, both as a whole and separately for its components. The goal of this new working group, led by the governments of Japan and the Republic of Korea, is to develop an internationally harmonized procedure for determining the carbon footprint of vehicles, including energy consumption, regardless of technology (internal combustion engine, electric, hybrid) used, as well as for comparison between different models employing the same technology [2].

Ensuring both the satisfaction of end-user needs and minimal environmental impact is a highly complex task. In the entire process, it is commonly said that the design stage of internal combustion engine components determines 80% of their future environmental impact [1]. One way to control this impact at an acceptable level is to manage the end-of-life phase of products through design activities.

There is a need to improve the environmental orientation of cast component design, and managing their end-of-life phase is one approach to achieving this. Many studies suggest that this path can be initiated by various stakeholders, such as:

- Customers who demand more environmentally friendly engine manufacturing technologies, both for complete engines and spare parts;
- Companies seeking to reduce their product's environmental impact while increasing revenues;
- Regulations favoring low-environmental-impact engines and requiring manufacturers to incorporate end-of-life considerations from the design stage [3].

Currently, the primary focus is on end-of-life strategies that maximize component value, known as reuse strategies. These strategies have key characteristics that engineers must consider facilitating their initial implementation.

These end-of-life strategies and their implications for ICE component design are now well understood and widely adopted among companies. However, discussions remain open to propose new strategies that retain increasing levels of residual value in used engine components with the goal of sustainable manufacturing of innovative products. Repurposing, meaning that components that have reached the end of their service life can be reprocessed for applications different from their original use, is one such approach to extending their lifecycle.

High-Residual-Value End-Of-Life Strategies

The need to define an end-of-life strategy for components arises when they are classified as "waste." The European Commission defines waste as: "any substance or object that the holder discards or intends or is required to discard." Depending on its type, characteristics, and operating conditions, a rejected engine component may follow one strategy or another. ISO proposes a classification of end-of-life strategies through ISO 14062 [4], ranked according to potential environmental benefits: prevention, reuse, recycling, energy recovery, and disposal.

Given that energy recovery and landfill disposal are not sustainable strategies, as they do not recover any component elements, both residual value and material are lost. These strategies are grouped under the category of "waste" (see point 1 in → Fig. 1). Recycling (see point 2 in → Fig. 1) involves recovering materials from used components to avoid the extraction of new raw materials, thereby limiting environmental impact and supply chain issues. The recycling strategy eliminates the final residual value and instead restores only the materials.

Strategies that recover material while retaining residual value are known as reuse strategies. These can be divided into two different sub-strategies: direct reuse and remanufacturing. Direct reuse (see point 3 in → Fig. 1) is a process in which the quality and market demand allow continued use of the same component by another buyer. The remanufacturing strategy (see point 4 in → Fig. 1) applies to components that must undergo a new manufacturing process before being reintroduced to the market. Indeed, direct reuse and remanufacturing are aimed at ensuring that, like newly manufactured ICE cast components, they provide at least the same guarantees and performance as new products and for the same application.

The direct reuse strategy can be defined as "any operation in which components or parts that are not considered waste are used again for the same purpose for which they were originally designed" [5]. This statement is valid provided that the performance of the directly reused product is as high as that of a new one to achieve the same functionality. In any case, the product must be in sufficient working condition for immediate reuse. Components that are reused directly are often considered second-hand products, and their parts are used for repairing other technical objects rather than becoming products in themselves. From a manufacturing process perspective, the direct reuse strategy involves collecting already used components from the waste stream, cleaning, sorting, and testing them.

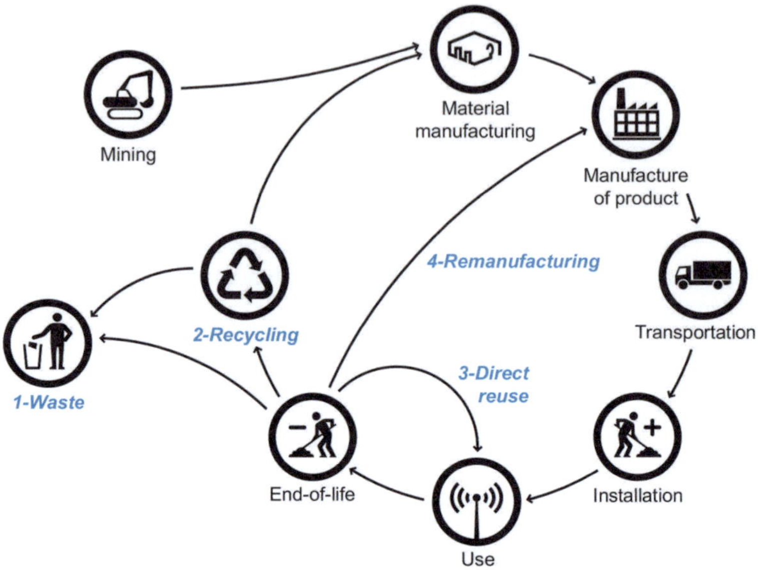

Fig. 1 Four end-of-life strategies for technological products [1]

The remanufacturing strategy has been extensively studied over the past decades. Some early works [6] provided the first definition of remanufacturing, formulating it as "an industrial process in which worn-out components are restored to a condition similar to new." This definition was later adapted by the European Commission [7], which describes remanufacturing as "a series of manufacturing steps applied to a component or product that has reached the end of its life, with the goal of restoring it to new or improved characteristics with an appropriate warranty." In other words, the remanufacturing process seeks to extract as much residual value as possible from the initial production cycle.

The main drawback of the mentioned strategies lies in the efficiency of reused components. Direct reuse and remanufacturing, even if the components are as good as new, may be less efficient than entirely new parts due to technological advancements.

Design for direct reuse and remanufacturing, which includes end-of-life strategies with the highest residual value retention from used components, has already become a reality in some companies and is increasingly demanded by society, aligned with sustainable development ambitions. While direct reuse is mainly a matter of logistics and quality control, remanufacturing aims to restore components to their original specifications. These two strategies have been comprehensively examined in recent research, and their key characteristics have been categorized into three dimensions: product size, manufacturing processes, and business model characteristics.

Most research on remanufacturing impacts has focused on qualitative analysis, with very few studies [8] providing a detailed literature review highlighting that most research examines the economic effects, while only a few studies quantify the environmental contributions using transparent inventory data sourced from primary sources. As previously mentioned, the earlier sustainability requirements are integrated into the design and planning of an ICE component, the greater the impact on its future characteristics.

On average, compared to manufacturing new components, the use of remanufactured parts results in the following savings: According to the Ellen MacArthur Foundation [9], there is an 80% reduction in energy demand, 88% in water consumption, 92% in chemicals usage, and 70% in waste generation avoidance. This aligns with research [10], which, based on expert consultations in China, reported a 46–90% reduction in materials use for remanufactured products compared to new products and a 68% to 82% reduction in energy consumption, alongside a 30–70% cost reduction. Other sources indicate an annual material savings of 14 million tons due to remanufacturing in the automotive sector, as reported by BORG Automotive [11].

Similarly, research [12] reports an 88% reduction in materials use for remanufactured ICEs and estimates that producing a new diesel engine requires approximately 2.3 times more energy than remanufacturing one.

According to [13], in 2023, GM introduced a program that repurposes components from End-of-Life (EoL) vehicles. These engineer-approved parts are then sold to customers who need them. The GM Customer Care and Aftersales (CCA) program has become a key initiative in the circular economy, helping to reduce raw materials demand and prevent waste from ending up in landfills by enabling vehicle parts to be reused through remanufacturing.

In 2023, the CCA program sold over 700,000 units in the United States, covering thousands of unique parts, including engines, transmissions, and other components that meet engineering specifications. Alongside new service parts, remanufactured components enable vehicle repairs, extending their lifespan. The CCA team collaborates with dealers and suppliers to encourage the remanufacturing of parts that are currently not eligible for reuse.

3 Design of ICE Components for Sustainable Development

Sustainable design has been a research area that has significantly evolved over the past 30 years. Since the late 1980s, when the first definitions of sustainable development emerged, design has been recognized as a crucial area for achieving more responsible and resilient models of production and consumption.

Initially, academic and industrial discourse primarily focused on identifying environmental issues and developing different approaches to address problems caused by industrial activities. These efforts were mainly corrective actions applied at the end of the product lifecycle to mitigate environmental damage caused by manufacturing processes [14].

By the late twentieth century, a product lifecycle vision was established, recognizing that negative impacts can occur at all stages of a product's lifecycle, from material extraction to final disposal. This led to the formulation of the eco-design concept, an industrial approach aimed at reducing environmental impact at every stage of a product's lifecycle. Eco-design became one of the most widely adopted sustainable design concepts in industry and remains relevant today [15].

Eco-design was initially presented as a product-oriented approach within linear production and consumption models. However, the need for a more systemic perspective on solutions led to the development of methods for achieving greater sustainability. During the 2000s, many authors and organizations highlighted the necessity of a holistic approach to environmental and social issues, shifting research toward systemic views of sustainability [16].

In recent years, the concept of the circular economy has emerged as a key solution for fostering sustainable product-service systems (PSS) [17]. Circular economic principles integrate previous sustainability ideas into a structured framework, emphasizing regenerative and restorative industrial systems.

The most widely accepted definition of the circular economy describes it as "an industrial system that is restorative or regenerative by design." It replaces the "end-of-life" concept with restoration, shifts toward renewable energy, eliminates toxic chemicals that hinder reuse, and aims to eliminate waste through optimized materials, products, and system design, including business models [9]. This zero-waste closed-loop system approach aligns with the fundamental transformations required for sustainable development.

However, not all circular solutions inherently lead to sustainable alternatives. There are specific requirements related to how these solutions are designed, and several barriers and challenges—such as mindset shifts, achieving critical mass, consumer behavior changes, regulatory updates, and financing the transition to circularity—must be addressed.

Defining sustainable development in the design of automotive industry components requires a deep understanding of the causes and consequences of engineering decisions throughout the entire product lifecycle and the triple bottom line of sustainability. Sustainable design objectives help mitigate complexity through a formalized problem description. To this end, various decision-support mechanisms have been explored, including:

- (a) Use-case-driven breakdown into subsystems,
- (b) Systematic reduction of the solution space,
- (c) Design assistance tools ensuring alignment with sustainable design goals.

The implementation of automated design/engineering tools (CAD/CAE) and lifecycle management (PLM) systems as interfaces for interacting with engineering tools can make the information retrieval and decision-making support process more seamless.

The concept of Sustainable Product Development (SPD) has been discussed in academic research since the early 1990s, with a strong emphasis on the environmental perspective [1]. However, nearly thirty years after the initial debates on sustainable development, industrial manufacturing remains far from being truly sustainable. This realization raises the question of which specific challenges must be addressed in product design to improve the overall situation.

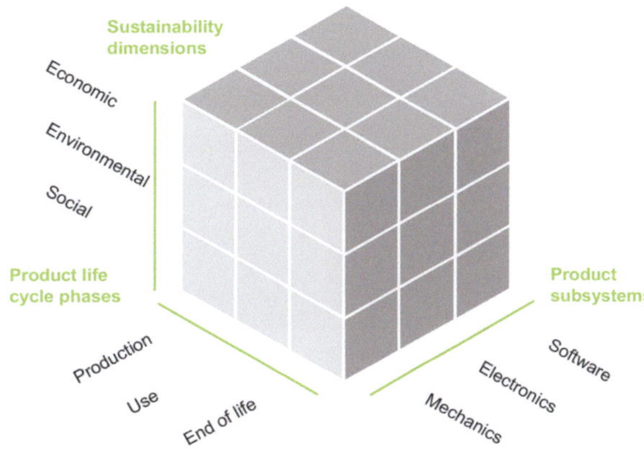

Fig. 2 Modules of Sustainable Development Tasks for Internal Combustion Engine Components [1]

From a practical perspective, the success of SPD implementation is influenced by several factors, such as:

- The motivation of market participants in the production of ICE components (e.g., incentives for integrating sustainability into product design).
- Available resources (e.g., time allocated for SPD application).
- Path dependencies (e.g., existing contracts with suppliers of non-sustainable materials).

Although these practical barriers can be overcome through appropriate managerial oversight, significant potential remains untapped in SPD research. The primary challenge in this context is finding ways to reduce the complexity associated with SPD decision-making. However, sustainable design can only be fully realized if engineers develop subsystems in accordance with their impact on the triple bottom line (economic, environmental, and social sustainability) at every stage of the product lifecycle (→ Fig. 2) [1]. One approach to addressing this non-trivial challenge is to break down the problem into smaller, more manageable subproblems—so-called "problem modules".

Such an approach may prove insufficient due to the various interdependencies between problem modules. Thus, the key task of SPD research is to enable design engineers to anticipate these dependencies through methodological guidelines, as well as by providing enhanced access to knowledge and information. In other words, instead of searching for solutions to specific problem modules, it is necessary to develop new mechanisms that increase transparency in engineering decision-making.

The complexity of cause-and-effect chains poses a serious challenge for assessing judgments in sustainable manufacturing development. One of the most effective ways to reduce complexity factors throughout the entire process is to define goals that determine the most significant impact factors on the problem, providing a foundation for the development of further decision-making models. Therefore, refining "sustainable design goals" can be formulated in a complementary manner with key technical or functional requirements.

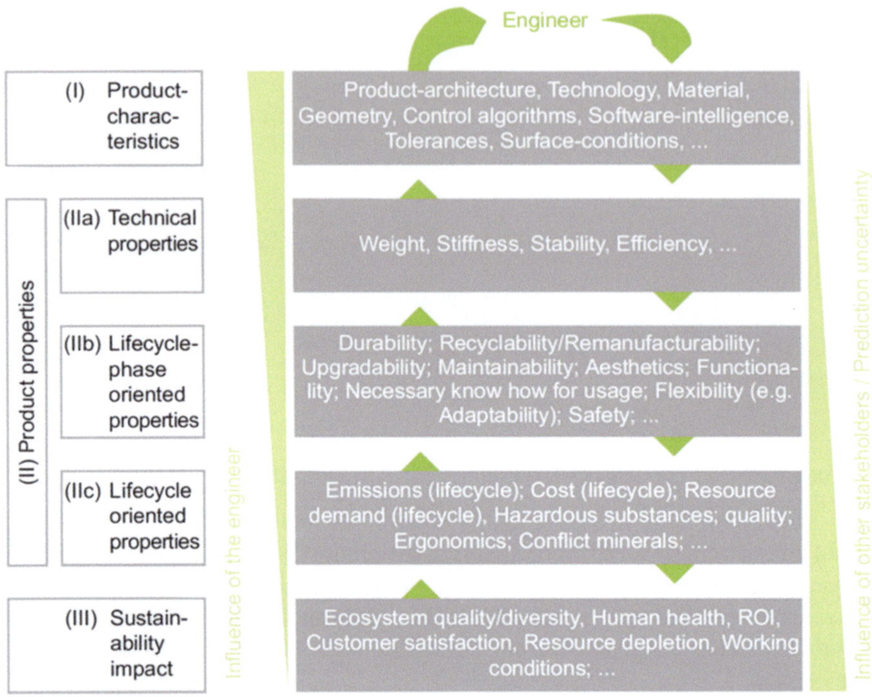

Fig. 3 Scheme of the connection between design decisions and sustainability impact [1, 18]

The starting point for solving this issue is identifying cause-and-effect relationships between engineering decisions and their impact on sustainable development. This is achieved by categorizing different types of information into specific groups in a fixed order, as illustrated in → Fig. 3 [1].

This scheme is based on the separation of product characteristics (I) and properties (II), within the framework of the "Product Development with Managed Properties" approach. Category (III) denotes the product's impact on the sustainable development of various stakeholders, such as the environment, the health aspects of employees and customers, and the financial stability of the company (III). To link the design perspective (categories I and II) with the sustainability perspective (III), the product properties of category II were divided into three subclasses, each considering the product's life cycle perspective (IIa–IIc).

Category IIa consists of technical properties that are directly determined as a result of engineering decisions based on product characteristics. The definition of the material and geometry of the characteristic determines, for example, the weight as a technical property. By combining the technical properties and supplementing them with influences from outside the system, the boundary-oriented life cycle properties (IIb) are determined.

4 Analytical Study of the Impact of Industry 4.0 and 5.0 Concepts on the Development of Environmentally Responsible Production of ICE Parts

Traditional production decision variables include cost, quality, flexibility, and time; however, recently, resource scarcity and pollution issues, as per the agreement at the UN Climate Change Conference in Paris in December 2015, are forcing a revision or integration of these standards, which are unable to account for these aspects. Furthermore, increasing competition driven by the globalized economy depicts a complex scenario for the entire industry and, in particular, for a mature technology such as ICE parts manufacturing.

One promising solution to these problems is the implementation of sustainable development alongside traditional production decision variables (→ Fig. 4) [18]. Sustainable development is defined quite broadly as meeting humanity's current needs; in more pragmatic terms, the "triple bottom line" is generally accepted, where environmental, economic, and social aspects are considered together.

Based on current trends, complex technical systems such as ICE parts and assemblies often undergo life cycle assessments in individual studies or in combination with environmental impact assessments. Manufacturing technologies have always been a prevailing factor in industrial progress. Today, along with fundamental technologies, computer-integrated technologies are also developing, which determines the success of the economy and the place in a constantly updated world.

In the fields of power engineering, automotive industry, etc., power plants or generator sets based on ICEs play a key role, which are currently used in both civil and military sectors and are of significant importance. High demands on the main components of ICEs in terms of performance characteristics, ensuring reliable operation in various operating conditions, raise questions about the development of modern technologies for manufacturing ICEs and their components, as well as the preparation of production itself. The choice of manufacturing method here plays a decisive role and affects technical and economic indicators. The limited amount of available literature indicates greater concern for the energy and environmental

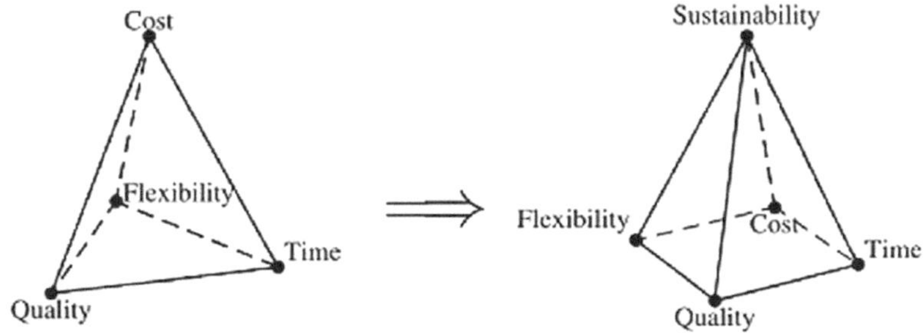

Fig. 4 Change in decision-making attributes in modern manufacturing to incorporate sustainability [18, 19]

consequences of casting technologies than for material issues of reducing sand and water consumption. Although the extraction and production of basic raw materials are an important part in determining sustainable production criteria.

Today, there is a concept that "recycled aluminum requires only 5% of the energy of primary aluminum" [19], while ignoring the energy of auxiliary processes used in the recycling stage to bring the material to a state where it can be reused.

In 2008, a research paper was published [20], which compared the embodied energies in the production of automotive components at two stages of the material life cycle—"material" (i.e., extraction and creation of materials for use at the production stage) and "use". Their findings clearly outlined that the energy involved during the "use" phase of the vehicle is much greater than during the "material" phase of the materials. Today, manufacturers in the automotive industry face three main questions:

- What is the best way to assess the environmental impact of a vehicle?
- Is reducing the mass of assemblies by reducing the density of the material always the best way to reduce the environmental impact of a vehicle?
- How should design engineers choose the right materials that have the least environmental impact throughout the entire life cycle of the materials used in vehicle production?

In [21], an analysis of the environmental impact of material substitution in passenger vehicles is presented. This process is very complex and depends on many assumptions that must be considered before making final decisions. According to this study, based on the analysis of more than one hundred primary sources and considering the selected parameters, namely a 1.6L inline 4-cylinder block, replacing cast iron parts with aluminum alloy components does not create more environmentally friendly vehicles, considering the total energy of production and the actual fuel economy achieved. In fact, to recover the difference in production energy throughout the entire material cycle, it is necessary to drive a car with an aluminum alloy cylinder block for at least 143,000 km to 840,000 km, depending on the exact production method and vehicle driving style. This is a direct result of the high primary energy content in aluminum alloys and the very small mass savings achieved by the replacement, which is less than 1% of the total vehicle mass.

For some production scenarios, the break-even distances calculated from the results of this study are close to the expected vehicle lifespan. However, for most production scenarios, break-even distances significantly exceed the vehicle lifespan.

In particular, the foundry industry is a major supplier of metal components for a wide range of industries, from automotive to aerospace and the energy sector. Despite this important key role for our society, the foundry industry must implement important future-oriented topics such as digitalization and Industry 4.0, 5.0, thereby missing the opportunity to unlock the full potential for efficiency improvements. The reasons are diverse and range from a high proportion of manual work steps to the conservative nature of the industry and the lack of know-how for new technologies. For this reason, digital solutions are needed to meet the circumstances of this industry sector, dominated by small and medium-sized enterprises, often with a high proportion of manual work steps. But, in the future, transparent registration and visualization of all relevant

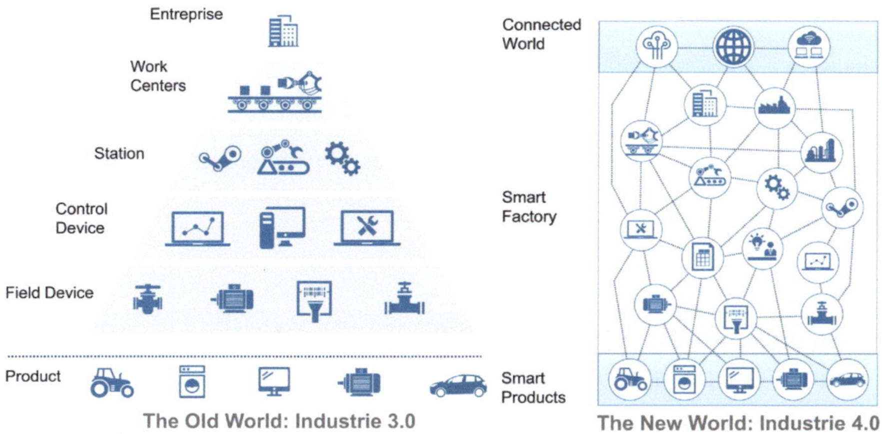

Fig. 5 Transitional and emerging technologies used in Industry 4.0 [22]

parameters of the manual casting process are needed to improve reproducibility and, thus, increase sustainability by reducing defects and energy resources. Furthermore, the implementation of new principles meets the requirements of the new European Union initiative Industry 5.0, which adheres to a human-centric approach to future production models as a complement to Industry 4.0.

From the mentioned new and transitional technologies (→ Fig. 5) [22], the following main segments can be distinguished:

- Industrial Internet of Things (IIoT)—a tool for creating real-time production and trade networks;
- Big Data and Analytics. Collection, processing, and analysis of technological, commercial, and financial data on an enterprise and trade network scale to make decisions on increasing product competitiveness;
- Autonomous Robots. Industrial robots equipped with a cyber-physical system that can independently (autonomously) perform tasks instead of human specialists. Collaborative robots (cobots), which can interact with both other robots and personnel;
- Additive Manufacturing, based on 3D printers, which enable the production of both experimental prototypes of parts and small batches of them;
- Simulation. Creating a virtual production model alongside real production allows the virtual model to be used for adjustments, with subsequent transfer of changes to real production. This will save time and finances when choosing the optimal course of action for "Foundry 4.0";
- Cloud Technology, providing the fastest possible data processing from both internal and external information sources (equipment, warehouses, suppliers, trade networks);
- Augmented Reality—creating enhanced capabilities for personnel to analyze the current situation, repair equipment, and train.

As Industry 4.0 digital innovations affect not only how enterprises produce, but also how they invest in product design and engineering technologies. Industry 4.0 principles allow for significant improvement in part design through the integration of software components (e.g., sensors, GPS) that are connected to machines or other physical objects and allow data to be collected from the field. Integrating products with sensors in the IoT opens up new opportunities at every stage of the ICE part's life cycle. Access to data and information generated during product use allows engineers to continuously monitor performance.

5 Sustainable Development of Transport Objects in the Modern World: From Problems to Solutions

Here, various aspects of the problem (in the context of transport objects) should be highlighted:

1. Societal needs → logistics schemes → transport infrastructure → computer modeling of the stages of developing requirements for the technical characteristics of vehicles → design of infrastructure and traffic flow control algorithms;
2. Initial requirements for vehicles → models of analysis and synthesis → numerical models → computer tools → technical solutions for real objects;
3. Vehicle → system approach → mathematical modeling of processes and states in individual elements → numerical modeling → patterns of influence of varied parameters and technical characteristics → recommendations on progressive technical solutions.

In addition, the role and place of the individual in the interaction with information technologies should be considered. In particular, we are talking about interaction with computer modeling tools at all stages of the product life cycle.

It is also important to compare real objects with their mathematical and computer models.

The issues of involving artificial intelligence, cloud resources, supercomputers, etc., in solving modern problems are also important.

A balance of criteria in the relationship "economy; ecology (e.g., decarbonization); comfort, etc." is also important.

All of the listed aspects are partially covered below.

6 Individual: Object, Subject, or Synergy

Sustainable development directions in the context of transport consist in meeting public demands: economy, ecology, and comfort. With the transition from Industry 4.0 to Industry 5.0, the emphasis is increasingly shifting from economic problems to environmental problems, in particular, decarbonization. The next trend is the emphasis on ensuring comfortable living, working, and leisure conditions. However, these well-known emphases and trends have a deeper structural composition and various accents in many contexts. It should be noted immediately that, in addition to the shift of attention in the transition from Industry 4.0 to Industry 5.0 to the needs of an individual in terms of meeting their needs, a new level of activity of this individual in the interaction "world—environment—individual" arises.

This process, in turn, has two sides.

First, I—is a gradual transition from simple passive consumption of goods from society and the information environment by the individual to their active role in these interactions. That is, despite the growing role of digital reality tools in recent decades and the use of the benefits they deliver, in the future, an individual will strive to orient this interaction according to the principle "for oneself". In turn, this will have two tendencies:

4. Creating "a digital world for a specific individual" instead of "a digital world for everyone";
5. The active role of an individual in shaping their "digital world", i.e., "individual adaptation".

Secondly, II—is an increase in the individual's responsibility to both society and the digital environment.

These features of modern trends are transferred from the level of an individual to the level of a work collective, corporation, and society. This also applies to the spheres of business and science.

Thus, modern trends form a multitude of digital individual "multiverses". Therefore, it is necessary to prepare for new patterns in the development of individual social spheres, including—in industry, energy, and transport.

7 The Triangle "Real Sector of Economy—Science—Information Technologies"

To determine the above-mentioned patterns and trends, one should turn to the real sector of the economy. Indeed, this sector remains the leading one in tandem with the information sector. This is due to the fact that the primary motive for the development of information technologies is the needs of the real sector of the economy.

Another important factor is the role of science in the triangle "real sector—science—information technologies". Here, it should be noted that there are also several trends.

In particular, this is a two-way intensive exchange between the components of this mentioned "triangle". Thus, science "borrows" the instrumental tools of information technologies in its activities. Indeed, scientific activity requires the intensification of both the activities of individual scientists and cooperation within scientific teams, as well as—between scientific groups. In addition, the formation of not only "databases" but also “knowledge bases” is taking place. These knowledge bases concentrate on the products of intellectual activity. These products begin to be processed by “artificial intelligence” tools, with the involvement of cloud resources and supercomputers. On the other hand, information technologies create a “language of interaction” with the real sector of the economy. That is—these are both formats and channels of information exchange between science and the real sector of the economy.

On the other hand, the development of information technologies has largely exhausted the possibilities of development on a traditional basis. At the same time, not only a “ceiling of opportunities” but also a “wall of progress” has formed. This “wall” consists in the fact that due to the unsatisfied needs of information technologies, the development of the scientific paradigm itself is stimulated.

Here, mutually competing trends arise.

The first of them is the desire for rapid development of information technologies based on the latest branches of mathematics. The other trend concerns the deepening of knowledge of the objective laws of the universe, which are the subject of study of natural sciences.

The first of these trends mainly affects the development of software and algorithmic tools for information processing. The second trend relates more to the corresponding hardware.

Here, it should be noted that a complete tool for the activities of both science and the real sector of the economy is not individually taken software or hardware tools, but balanced software and hardware complexes. Therefore, this implies the development of scientific directions that “are responsible” for both of these directions.

8 Risks, Errors, and Myths

In addition to the positive trends that are obviously manifested in the interaction “science—real sector of the economy—information technologies”, there are also certain risks, errors, and myths in the problem under consideration.

Existing risks lie in shifting the emphasis in this “triangle” to information technologies. Despite penetration into all areas of activity, these technologies are not “self-sufficient” but satisfy objective needs, in particular, the real sector of the economy and science. This is the main source and motive for development.

As for dangerous errors, they consist in giving preference to the development of information technologies to the detriment of, for example, science. This creates obstacles to the development of both science and information technologies themselves. After all, there is a separation of these technologies from the base of their development.

And, finally, there is a whole range of myths that hinder progress. For example, this is the myth of unlimited resources and capabilities of information technologies. In fact, such limitations exist. Importantly, the most important are not so much quantitative as—qualitative limitations. Thus, those that are currently difficult to formalize and translate into digital format. Accordingly, instead of exclusive information technology tools, it is necessary to use their hybrid with a human expert.

9 Balance of Criteria

If we consider, for example, vehicles as objects that exist in a chain of stages of their life cycle, it should be noted that they have certain features. Thus, these objects themselves appear as a result of public demand. This demand also forms requirements for the technical and environmental characteristics of these vehicles, considering logistics schemes and infrastructure structure. In addition, at each of the stages (including—design, technological preparation of production, manufacturing, operation, disposal, etc.), a number of parallel processes take place. These are, in particular,—processes of scientific and information support [23, 24].

It is important that when creating certain vehicles, several criteria, requirements and restrictions are considered. These are, as mentioned, economic, environmental, technological, and other criteria and restrictions. Importantly, considering certain circumstances, the specific weight of different criteria and restrictions may change. Thus, certain technical solutions will be initiated. At the same time, on the one hand, such technical solutions, which is important can only be substantiated by objective laws of science. On the other hand, considering the current development of information technologies, they serve as the predominant tool for describing the corresponding technical solution.

10 Examples of the Application of the Proposed Approaches and Principles

As an example of the application of the approaches and principles described above, one can consider, for example, locomotive engines. Such engines are created in a few countries. The very operating conditions, economic and environmental requirements that operate in a particular country or region generate certain technical solutions for such engines. Apart, in Ukraine, series of locomotive engines D-70, D-80, D-100 and others have been developed. During their development, studies of working processes in these engines, analysis of kinematics, dynamics and strength of mechanisms and power elements, as well as the development of technological methods for strengthening the most loaded parts, etc., were carried out. → Figs. 6, 7, 8 and 9 show illustrative materials of scientific support for developments at various stages of the life cycle, which are carried out using information technologies through numerical modeling with the involvement of modern software and hardware complexes.

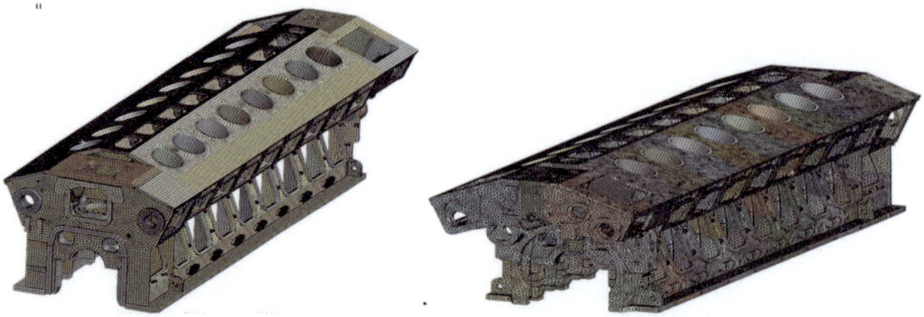

Fig. 6 Example of a technical solution variant for engine elements of the D-70, D-80, and D-100 series, which were created, manufactured, and modernized in Ukraine for main-line locomotives

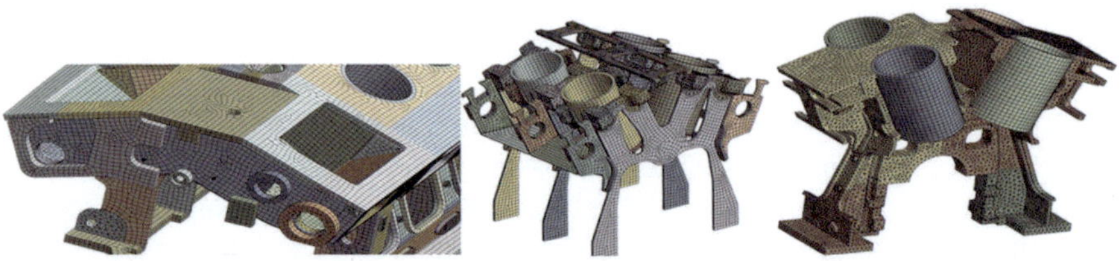

Fig. 7 Finite element models of locomotive engine elements

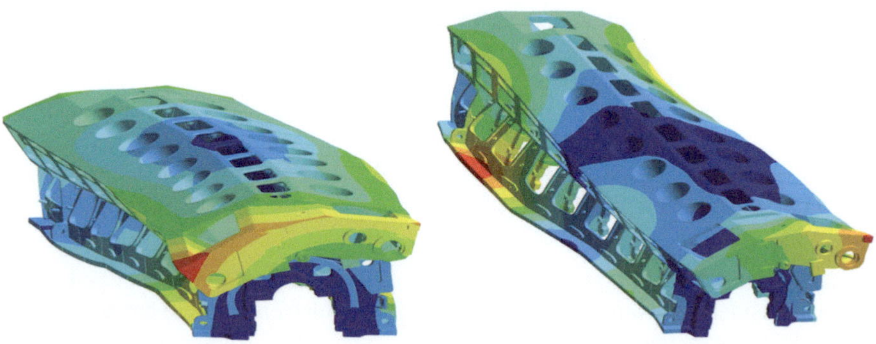

Fig. 8 Vibration modes of locomotive engine elements

As an illustration of information support for products at all stages of their life cycle, let us consider the processes of creation, research, manufacturing, and operation of locomotive engines.

Without going into details, let us note only a few points.

The operating conditions of locomotives in many regions of Eastern Europe and Asia are quite harsh. This creates stringent requirements for their technical, economic, and environmental characteristics. In combination with the general layout features of

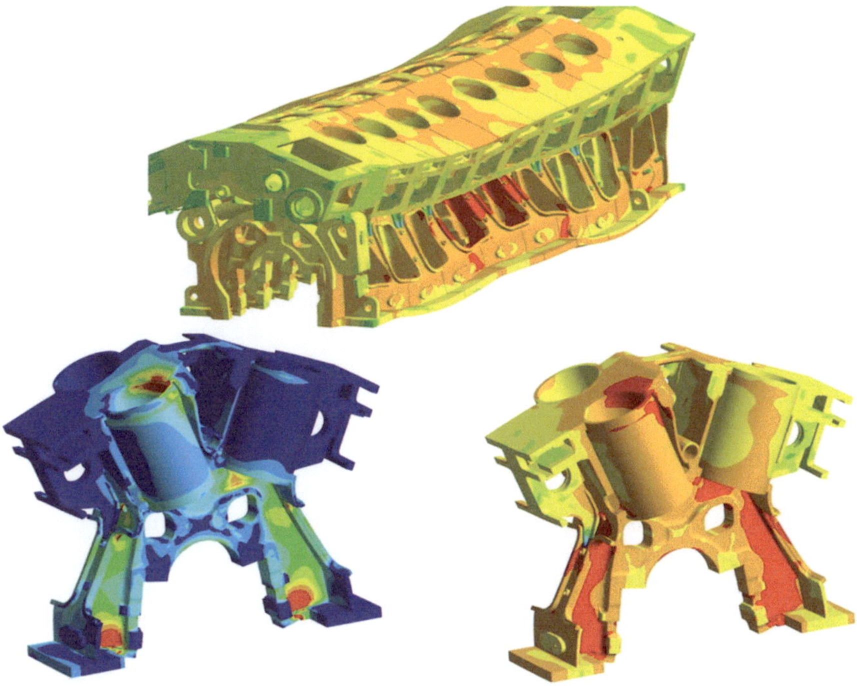

Fig. 9 Distribution of equivalent von Mises stresses in locomotive engine elements

locomotives, the structure and design of the elements of their engines were generated (→ Fig. 6).

At various stages of creating this type of highly economical, reliable, and environmentally friendly engine, various CAD/CAM/CAE systems were used (→ Figs. 7, 8 and 9). This includes modeling of working processes, structures, as well as technological processes of manufacturing and operation.

As a result, highly efficient locomotive engines have been created. They meet modern requirements for environmental friendliness, economy, manufacturability, etc. The very process of their creation is accompanied by information-computer systems, scientific research, and the creation of new technologies. This also applies to the control systems of these engines in operation, etc. Another property is the flexibility of technical solutions and the reserve for modernization. It should be noted that the needs for such engines, as well as the requirements for them, originate from public needs.

11 Conclusions

The analysis of the main trends and trends in sustainable development, particularly in the field of transport, provides grounds for the following conclusions.

1. The main directions of social development in the main spheres are the implementation of the principles of Industry 4.0 and 5.0. They are reflected in many publications, documents, and in the practice of corporations and business in general. However, there are unresolved problems. Among these problems is the uncertainty of some principles. This concerns the role of the individual in interaction with the real sector of the economy, on the one hand, and information technologies, on the other.
2. In the triangle "individual—real sector of the economy—information technologies", active two-way connections and exchanges are realized. At the heart of these relationships is active scientific support at all stages of the life cycle of products, goods, and services.
3. One of the conclusions is the danger of the dominance of information technologies in interaction with society, science, technology, and the economy. This, on the contrary, should be mutual influence and interconnection. Moreover, these relationships are in a dynamic balance, which can lead to a sequential change of "leading" and "driven" links.
4. One of the most important aspects of Industry 5.0, which distinguishes it from Industry 4.0, is the predominant emphasis on humans. Sustainable development involves a predominant focus on the needs of the individual, including—economy, ecology, comfort, etc. At the same time, it is necessary to determine an increasingly active, not passive, role of the individual in the formation of these needs, as well as in the specific weight of each of them in the system of evaluation criteria and constraints.
5. In addition to the activity of the individual, the role of their responsibility is growing. This becomes another trend in the development of information technologies.
6. An important conclusion is that information technologies are not able to solve the entire set of problems that arise only through their own capabilities. Therefore, it is advisable to entrust the part of the problems that are not formalized at the current stage to an expert. And then the synergy "expert—information system" produces both solutions to the problems of the real sector and the formation of new problems of science, as well as trends in the development of society.
7. The examples of information support for various stages of the life cycle of vehicle transport elements illustrate the possibilities of these information technologies. The interconnection of these stages and tasks of scientific support, as well as trends in the development of the real sector of the economy, is also demonstrated. In particular, for vehicles, there is a combination of multidirectional requirements: economy, manufacturability, etc. At the same time, the synthesis of new progressive technical solutions for vehicle elements involves

solving not so much optimization as compromise tasks. Along with the progress of sustainable development, the balance of compromises is shifting. In particular, now more and more attention is paid to environmental problems. In this vein, the role of decarbonization of working processes, for example, in engines and power plants, is growing. (see paragraphs 8–10).

8. Life cycle management, for example, in the sustainable production of ICE parts, provides significant progress in reducing environmental impact and increasing resource efficiency. The introduction of innovative technologies, optimization of production processes, and the transition to a circular economy reduce negative environmental impact, increase resource efficiency, and ensure sustainable industry development. The integration of LCA, the use of innovative materials and Industry 4.0 technologies create a platform for the transition to a sustainable future.
9. Sustainable design of ICE parts is based on the use of modern technologies, innovative materials, and the integration of environmental practices throughout the product life cycle. The transition to sustainable parts production not only reduces environmental impact but also increases the competitiveness of enterprises. At the design stage with the implementation of advanced practices of "design for specific goals", economic and environmental benefits can be achieved through reuse and recovery of parts.
10. Thus, to realize the potential benefits of smart manufacturing, the implementation of Industry 4.0, 5.0 principles must be accompanied by the introduction of production management systems that allow storing, sharing, and using data, and ensuring proper information management throughout the product life cycle. The integration of digital technologies, automation, and a focus on the human factor solves the problems of environmental friendliness and economic efficiency simultaneously.

However, to achieve the goals of these concepts using advanced technologies, coordinated efforts are needed from manufacturers, governments, and research institutions.

References

1. Stark, R., Seliger, G., Bonvoisin, J.: Sustainable Manufacturing. Cham Springer International Publishing (2017)
2. Unece.org.: UNECE starts regulatory work on automotive life cycle assessment UNECE (2023). [online] Available at: https://unece.org/sustainable-development/news/unece-starts-regulatory-work-automotive-life-cycle-assessment. Accessed 13 January 2025
3. Goodall, P., Rosamond, E., Harding, J.: A review of the state of the art in tools and techniques used to evaluate remanufacturing feasibility. J. Clean. Prod. **81**, 1–15 (2014). https://doi.org/10.1016/j.jclepro.2014.06.014
4. International Standards Office: ISO/TR 14062 Environmental management—Integrating environmental aspects into product design and development. Geneva: ISO (2002)
5. Directive 2008/98/EC of the European Parliament and of the Council of 19 November 2008 on Waste and Repealing Certain Directives: Official J. Europ. Union, 3–30 (2008)

6. Pigosso, D.C.A., Zanette, E.T., Filho, A.G., Ometto, A.R., Rozenfeld, H.: Ecodesign methods focused on remanufacturing. J. Clean. Prod. **18**(1), 21–31 (2010). https://doi.org/10.1016/j.jclepro.2009.09.005
7. European Commission, Directorate-General for Communication: Circular economy action plan: for a cleaner and more competitive Europe. Publications Office of the European Union (2020). https://doi.org/10.2779/05068\xEF\xB7\x92
8. Gutowski, T.G., Sahni, S., Boustani, A., Graves, S.C.: Remanufacturing and Energy Savings. Environ. Sci. Technol. **45**(10), 4540–4547 (2011). https://doi.org/10.1021/es102598b
9. SYSTEMIQ and Ellen MacArthur Foundation, Achieving 'growth within' (2017). [online] Available at: https://www.ellenmacarthurfoundation.org/achieving-growth-within. Accessed 13 January 2025
10. Liu, Z., Li, T., Jiang, Q., Zhang, H.: Life cycle assessment-based comparative evaluation of originally manufactured and remanufactured diesel engines. J. Ind. Ecol. **18**(4), 567–576 (2014). https://doi.org/10.1111/jiec.12137
11. Borgautomotive-group.com: Corporate Responsibility (2024). [online] Available at: https://borgautomotive-group.com/corporate-responsibility?L=3. Accessed 13 January 2025
12. McKenna, R., Reith, S., Cail, S., Kessler, A., Fichtner, W.: Energy savings through direct secondary reuse: an exemplary analysis of the German automotive sector. J. Clean. Prod. **52**, 103–112 (2013). https://doi.org/10.1016/j.jclepro.2013.02.032
13. General Motors: Sustainability (2023). [online] www.gm.com. Available at: https://www.gm.com/commitments/sustainability. Accessed 13 January 2025
14. Bhamra, T., Hernandez, R.J.: Thirty years of design for sustainability: an evolution of research, policy and practice. Design Sci. **7**,(2021). https://doi.org/10.1017/dsj.2021.2
15. Ceschin, F., Gaziulusoy, I.: Evolution of design for sustainability: from product design to design for system innovations and transitions. Design Studies [online] **47**(47), 118–163 (2016). https://doi.org/10.1016/j.destud.2016.09.002
16. Environment, U.N.: Design for Sustainability: A Step-by-Step Approach (2017). [online] UNEP—UN Environment Programme. Available at: https://www.unep.org/resources/report/design-sustainability-step-step-approach. Accessed 13 January 2025
17. Hernandez, R.J.: Sustainable product-service systems and circular economies. Sustainability **11**(19), 5383 (2019). https://doi.org/10.3390/su11195383
18. Pagone, E., Jolly, M., Salonitis, K.: A tool to promote sustainability in casting processes: development highlights. Smart innovation, systems and technologies, 172–184 (2017). https://doi.org/10.1007/978-3-319-57078-5_17
19. Sustainable materials—with both eyes open the future in practice the state of sustainability leadership. [online] Available at: https://www.cisl.cam.ac.uk/system/files/documents/julian-allwood-sustainable-materials-with-both-eye.pdf. Accessed 13 January 2025
20. Ashby, M., Coulter, P., Ball, N., Bream, C. (2012). Available at: https://proyectaryproducir.com.ar/public_html/Seminarios_Posgrado/Material_de_referencia/Ashby%20-%20ecoaudit.pdf. Accessed 13 January 2025
21. Wiener Motorensymposium: International Vienna Motor Symposium (2025). [online] Available at: https://wiener-motorensymposium.at/en/. Accessed 18 March 2025
22. Heidel, R., Hoffmeister, M., Hankel, M., Döbrich, U.: Industrie 4.0—The Reference Architecture Model RAMI 4.0 and the Industrie 4.0 Component (2019)
23. Golovianko, M., Terziyan, V., Branytskyi, V., Malyk, D.: Industry 4.0 vs. industry 5.0: co-existence, transition, or a hybrid. Procedia Comput. Sci. **217**, 102–113 (2023). https://doi.org/10.1016/j.procs.2022.12.206
24. Zizic, M.C., Mladineo, M., Gjeldum, N., Celent, L.: From Industry 4.0 towards Industry 5.0: a review and analysis of paradigm shift for the people, organization and technology. Energies **15**(14), 5221 (2022). https://doi.org/10.3390/en15145221

Impact of Renewable Fuels on Chemical Energy Release and Power Boosting in Vehicle Powertrains and Cogeneration Powerplants

Vítek Oldřich, Kyjovský Štěpán, Bortel Ivan, and Macek Jan

Abstract This study explores renewable fuels (e-fuels) as alternatives to conventional fuels in vehicle powertrains and cogeneration plants. Analyzed fuels include hydrogen, ammonia, biomethane, and dual-fuel mixtures, with a focus on lean combustion and turbocharging. Experimental and simulation methods evaluate ignition, heat release, and efficiency under high boost pressures. Findings highlight the potential of e-fuels to enhance grid stability and reduce dependency on fossil fuels, while addressing production and transport challenges. Results support the transition to sustainable energy systems in transportation and power generation.

1 Introduction

The proposed ban of new registration of cars with internal combustion engines (ICE) and following European regulations, as [1] and [2] will interrupt long-term continuous development of vehicles powered by them alone or in combination with electric motors and small batteries in the form of different lay-outs of hybrid electric vehicles HEV. It would be reasonable if adequate surrogate for those products exists.

Unfortunately, neither from the customers' point-of-view (total cost of ownership, TCO, durability, unlimited range and readiness to use, thermal comfort, safety,

V. Oldřich (✉) · K. Štěpán · B. Ivan · M. Jan
Department of Automotive, Combustion Engine and Railway Engineering, Czech Technical University in Prague, Prague, Czech Republic
e-mail: oldrich.vitek@fs.cvut.cz

K. Štěpán
e-mail: stepan.kyjovsky@fs.cvut.cz

B. Ivan
e-mail: ivan.bortel@fs.cvut.cz

M. Jan
e-mail: jan.macek@fs.cvut.cz

© The Author(s), under exclusive license to Springer Nature Switzerland AG 2025
A. Marchenko and Z. Koruba (eds.), *Decarbonization of Transport Energy Installations in the Context of Sustainable Development Strategies*, Studies in Systems, Decision and Control 635, https://doi.org/10.1007/978-3-032-05884-3_7

etc.) nor for holistically assessed environmental impact—[3]—the fully equivalent product of general use exists.

It can be illustrated by the lack of interest to buy such vehicles as the first car for a family if no governmental financial support is available. Moreover, there are other solutions, attracting customers with still existing potential to be environment-friendly at the same level as pure battery electric vehicles, BEV. Those solutions are not excluded out of Europe, as in USA or Japan, not speaking about emerging markets in other parts of world. This might be just the current situation, which will be overcome by future development of BEVs, but with the same probability by the following development of advanced ICEs, especially in hybrid cars, and by yet not ready hydrogen technologies in ICEs and later in fuel cell (FC) cars. The truck and bus market shows clearly the time for abandoning ICEs is still far away. Moreover, the current political tension and close-to-war situation calls for military vehicles, for which ICE—perhaps in hybrid layout—[4], as well, offers the only feasible solution.

It does not mean the ICE powertrain should use fossil fuels only: The alternative, low-carbon or even almost-zero-carbon renewable fuels are already available, although in small amounts for high price. It could be changed if the same financial resources, as, e.g., for battery development, are directed to this domain. The energy efficiency WTT of synthetic fuels is worse than for direct WTG battery charger input, but the investment costs for infrastructure change are very low and application of synthetic fuel similar to fossil hydrocarbon one may be immediate—unlike the stepwise and expensive change of all vehicles in a fleet. Additionally, the synergy between car, bus and truck development is supported by accepting this way as one of possible alternatives.

Obviously, the best mobility system change is not clearly predictable. In such a position, more ways should be investigated, waiting for better based future decision and maintaining all well-developed features of the current mobility system with new energy carriers.

Moreover, Europe as a traditional manufacturer of highly sophisticated cars and trucks, whereas the best developed battery technologies have come from Far East. The majority of material production is currently in force of China, especially due to the lack of high level environment protection there, which is reflected by low prices for selected customers. It enhances the need for keeping parallel development of more possible solutions of future car instead of banning some potentially useful ways, in which Europe has kept its leading position.

The current paper describes the Central European car-fleet situation and tools for analyzing it in a predictive way, helping to find possible scenarios and decisions. It shows examples of achieved results showing the way of reaching compromise between strictly stated environmental goals and the need to keep competitiveness of European industry in internal and external markets, which is decisive for growing of European countries GDP. The more resources, the higher environmental goals can be achieved. It briefly shows examples of renewable fuel ICE systems considering combustion, turbocharging and synergy with hydrogen FC development. The basic condition for European transition to cleaner mobility should aim at the best compromise, not at ideologically set black-and-white goals only, despite the fact it might be temporarily attractive.

2 Holistic Assessment of the Current Fleet—Example of Personal Cars

A tool for the holistic assessment of future scenario of vehicle fleet development based on WTW and LC analyses and suitable vehicle data including vehicle weight impacts has been developed—[5] and amended by optimization results, as in [6–8]. The combination of statistical data, including integral data on transport performance and fuel or electricity consumption in different domains of transport sector, is used to find and correct inconsistencies, typical for any report-based statistics. Approx. 6,600,000 M1 + N1, altogether incl. rail 8,000,000 surface transport vehicles, are considered. The examples of integral results for road passenger transport are presented in Fig. 1.

Mean road energy consumption under real driving conditions can be calculated—[5]—from the following Eq. (1), whereas 10 kWh is equal almost accurately to 1 L of diesel oil B7 blended with 7% of biodiesel.

$$E_{road} = \frac{1}{36 s_{tot}} \int_{s_{tot}} \frac{dW_{wheel}}{\eta_{trans}\eta_{PM}} = \frac{1}{36 s_{tot}} \int_{s_{tot}} \frac{F_{wheel} ds_{wheel}}{\eta_{trans}\eta_{PM}} [kWh/100\ km] \quad (1)$$

Resistive force loading powertrain mover is expressed assuming optimum gear for vehicle speed and driving resistance force at wheels, simplified to

$$F_{wheel} \doteq \left(m_{veh,curb} + m_{load}\right)(K_{roll} + A) + K_D c_{wheel}^2 \quad (2)$$

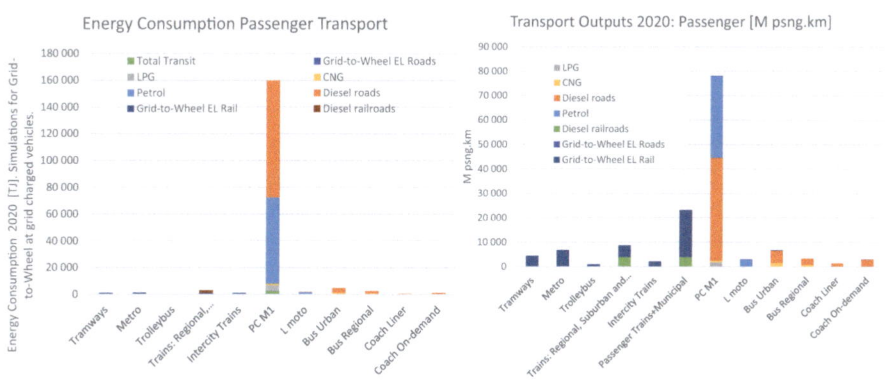

Fig. 1 Energy consumption and transport outputs of the surface transport passenger vehicles in the Czech Republic before start of Covid-19 pandemic

where acceleration & slope term $\boldsymbol{A}$ (i.e., recuperable mechanical energy source, important for hybrid and battery vehicle simulation) stays for inertia and gravity forces at inclines.

Primary mover efficiency can be expressed as a function of relative losses—[9]. They are described by primary mover normalized power and speed based on simulations of relevant sizes and types of powertrains (ICEs, e-motors, PEM FCs, hybrids after ECMS optimization, etc.). Normalized PM speed is set close to optimum efficiency band, i.e., the optimum gear is determined:

$$\eta_{PM} = \frac{P}{P+L} = \frac{1}{1 + L_{rel}(F_{wheel}, c_{wheel}, a)} \tag{3}$$

in which relative loss $\boldsymbol{L}$ is simulated by non-linear regression, fitted to results of experiments or detailed simulations. The charging and discharging loss for batteries in different operation modes (charging, driving, energy recuperation) is added together with transmission loss.

Real driving energy consumption can be simplified by representing integral in Eq. (1) for mean representative vehicle speed with mean virtual slope, if averaged energy consumption and annual odometer mileage is estimated from statistics for different types of vehicles. The additional condition is time-of-operation, or mean speed combined with odometer data. This simple representation of operation of different fleet segment energy consumption (i.e., vehicle classes, e.g., as mini, small, lower medium, …, luxury and SUV for M1 personal cars) enables extrapolation for new powertrain lay-outs with different efficiency and possibilities of recuperation using alternative fuels of better emission potential. The set of nonlinear algebraic equations for mean energy consumption and transport output, based on Eqs. (1)–(3), yields the representative speed, slope and cargo capacity utilization.

Vehicle mass, changed for other powertrains and/or storage systems is an input for such a prediction, considering rated range of operation, which is important especially for BEV, PHEV or CNG, CH2 and LNG. Moreover, the energy supply for auxiliaries is respected, if not covered by otherwise wasted energy, e.g., as typical for heating of vehicles with an ICE. Altogether 45 vehicle categories, each for 8 emission regulation levels (Euro 0–Euro 6 d) have been simulated and calibrated with 36 typical ICEs, 12 e-motors and 2 PEM FCs, using 23 sources of energy and combinations of them (hybrids, dual-fuel). The integral consumption for the whole vehicle fleet and different fuels (gasoline, diesel oil, CNG, LPG, electricity for trolleybuses, trams, metro trains or different types of railway stock) have been confronted with integral data of fuels sold in considered country or electricity consumed by rail transport, which is the standard part of annual statistical reports.

3 Impacts of New Energy Vectors on Vehicle Energy Consumption and GHG Emissions

The prediction of future fleet averaged GHG emissions based on bottom-to-up approach with large number of vehicle powertrain types was based on calibrated simulations, combining GT Suite with simplified models form in-house large engine, motor and vehicle database and experience from European research projects together with optimization algorithms elaborated in past—[5–7, 9–11]. That is why new combustion and charging concepts are developed and assessed at the CTU premises, as described in the second part of this contribution. Moreover, new powertrain layouts and their control are developed for HEV, PHEV and FCEV, which exceeds the contents of the current paper, however. The examples of weighted averages for domestic fleets (from Pre-Euro up to Euro 6d-AP) showing efficiency and normalized energy consumption parameters are presented in Fig. 2.

The model has been designed using longitudinal dynamics of a vehicle, prediction of individual vehicle powertrain maps in GT Suite and in-house optimizations of ICE, BEV and different hybrid vehicle control systems. Potential of biofuels and electrification can be assessed for different conditions of electricity production and WTT additional emissions, as available in European regulations—[1]. Some of results are presented in Fig. 3. The large difference between assumption of zero emissions for electrified cars in European regulation is clearly visible together with the fact of possible improvement using low-carbon fuels burnt in ICE.

The data on WTW energy consumption (e.g., [12, 13]) including net emission factor of electricity consumed can be used as a part of Life Cycle Analysis (LCA), which is involved in the developed assessment tool Fig. 4. CTU tool tries to be not biased against electrified vehicles, which is visible from comparison with measured GreenNCAP data—[3]. Despite rather optimistic assumption of future mean emission factor in Europe for the year 2029, the total emissions of vehicles with batteries

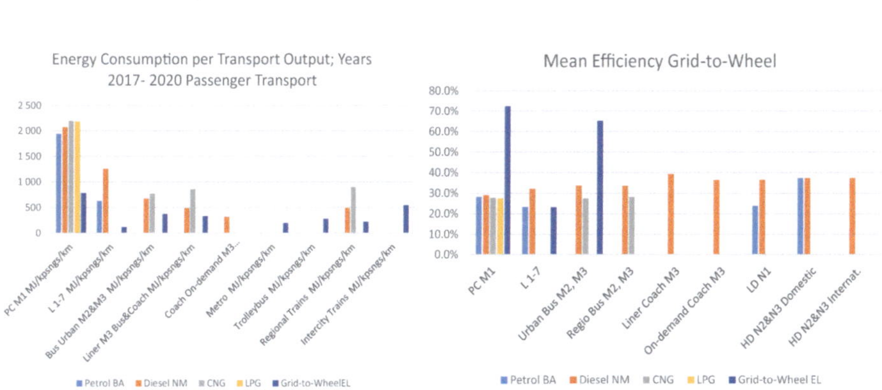

Fig. 2 Specific energy consumption per passenger transport output (millions of passenger*km) and mean grid or tank to wheel efficiency for the currently used road vehicles

Fig. 3 Personal car GHG emissions for ICE and electrified powertrains

produced in China are worse than for standard diesel car during the whole durability period of a battery.

The well-known fact about almost constant specific energy of different hydrocarbon or hydrogen fuels is roughly followed by energy density, as well, as presented in Fig. 5. After WWII, Kapica stated empirical hypothesis of limited density of energy

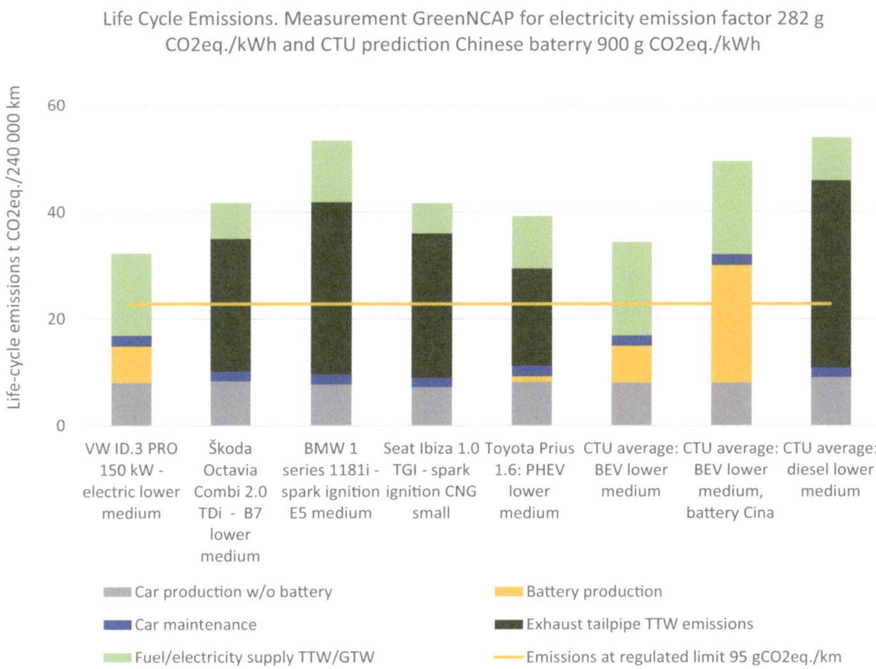

Fig. 4 Results pf CTU LCA assessment for the whole fleet sector compared to public data of GreenNCAP [3] considering measurements at specimens of vehicles

flux, generalized from Poynting's vector in W/m^2, valid for any of basic energy input for transformation—[14]. Since the main issue of the current BEVs is the mass of storage, it is interesting to compare, what are the limits of energy flux in combustion machines and electric ones.

For this purpose, it is interesting to compare the power output as brake parameters of a machine and power input of chemical or electrochemical energy, respecting energy conversion efficiency. Specific input power of any combustion engine can be expressed as the enthalpy flux in a combustible mixture of fuel of LCV $\boldsymbol{H_u}$ with stoichiometric A/F ratio $\boldsymbol{L_t}$ and air excess $\boldsymbol{\lambda}$ and related to already mentioned brake power with different, but machine size independent multipliers $\boldsymbol{K}$:

$$\frac{\dot{m}_f H_u \eta_e}{P} = \frac{H_u \eta_e}{1+\lambda L_t}\frac{\rho_{mix} c A}{P} = \frac{H_u}{1+\lambda L_t} * \left\langle \begin{array}{l} \frac{\eta_e \rho_{mix} c_{MP} A_P i_{cyl}}{p_e c_{MP} A_P i_{cyl}} K_{ICE} \\ \frac{\eta_e \rho_{mix} c_{flow} A_{flow}}{(\Delta h_T(T_{T1},\pi_T) - \Delta h_C(T_{C1},\pi_T))\rho_{mix} c_{flow} A_{flow}} K_{turbine} \end{array}\right. \quad (4)$$

whereas the averaged net operation power $\boldsymbol{P}$ has been expressed either from the brake mean effective pressure and piston speed for an ICE or from specific power of a gas turbine, which is equivalent to the difference of enthalpy heads at both turbine and compressor, dependent on pressure ratio and turbine inlet temperature. In both cases

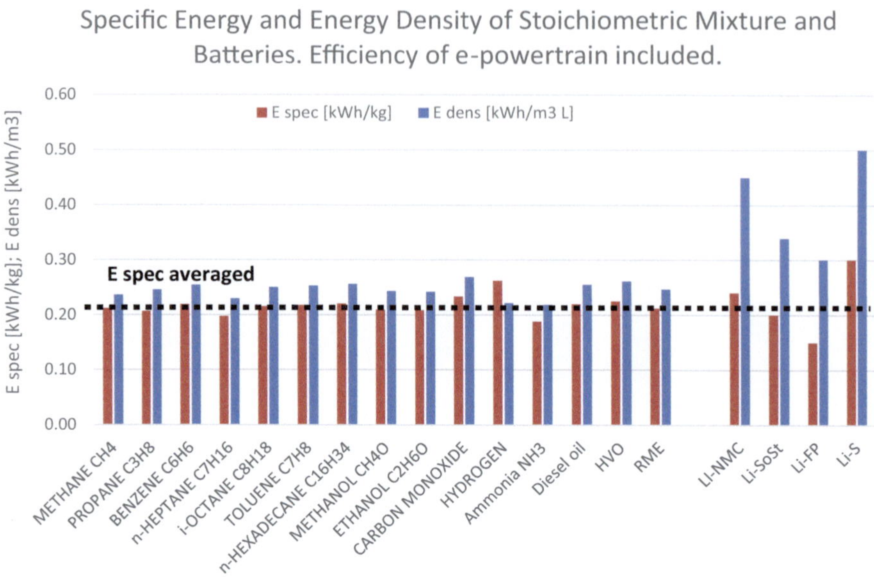

Fig. 5 Specific energy and energy density for stoichiometric mixtures of different fuels suitable for vehicles and those of accumulator batteries of different chemistries

the ratio of brake power and mass flow rate of mixture is roughly constant, machine-size-independent, with multipliers determined by engine development level, i.e., the value of brake mean effective pressure or turbine inlet temperature.

In the case of purely battery vehicles, the virtual mass flow rate can be expressed as the mass of a battery + e-motor per time of operation and power of e-motor as voltage times current times efficiency of e-motor. Then, simple result can be found, again with powertrain and storage specific, but size-independent multiplier, related to battery mass and output energy from a battery:

$$\frac{E_{spec,batt} m_{batt} \eta_{e-M}}{\Delta t_{oper} P_{e-M}} = E_{spec,batt} \frac{\frac{m_{batt}}{\Delta t_{oper}}}{U_{oper} I_{oper} \eta_{e-M}} = \frac{E_{spec,batt}}{\eta_{e-M}} K_{BEV} \tag{5}$$

In real powertrain with combustion engine or fuel cell, the specific energy of stored combustible fuel is many times greater, since the vehicle need not to transport stored oxygen. Batteries need oxygen or other oxidizing matter (e.g., sulfur) in a suitable compound, in contrary reducing the specific energy. Moreover, the processes are limited by forced diffusion fluxes under electrostatic forces. The volumetric energy including the case of a battery pack densities are different from gaseous mixtures, which is not surprising, however.

The following parts will present the way of combining experiments and advanced simulations to assess the potential of hydrogen dual-fuel engine, SI lean mixture hydrogen engine and issues of high levels of turbocharging for gas engines fueled by hydrogen, CNG or ammonia.

4 Alternative Fuels for the Compression Ignition Engine and the Potential to Decrease Carbon Footprint

4.1 Renewable Fuels for Compression Ignition Engines

Vehicles equipped with an internal combustion engine (ICE) still best meet the requirements for freedom and independence of movement due to the excellent range of vehicles, low space requirement, existing fuel infrastructure, and short refueling time. However, the progress of global warming pushes to reduce the carbon footprint of vehicles. The view of European regulators mainly assessed tail pipe emissions, nevertheless real impact should be considered in terms of well-to-wheel (WTW) or better whole life cycle analysis (LCA). The compression ignition engine is preferred due to its superior efficiency; however, the carbon footprint remains too high. Renewable fuels can substantially reduce the carbon footprint of road transport. The real impact depends on the production processes. The appropriate fuel to replace regular diesel oil (RD) is hydrotreated vegetable oil (HVO), but gaseous fuels lead to lower CO_2 WTW emissions in the optimal production case.

The potential of selected fuel was investigated in a CI engine adopted, but not fully optimized, to dual fuel (DF) operation. This means that the homogeneous mixture of gaseous fuel is ignited via injection of liquid fuel during appropriate operation modes. This approach provides high fuel flexibility and ensures the range of the vehicle during the development of alternative fuels. The preferred fuel for DF is hydrogen, because H_2 appears to be an energy carrier of the future excess of green electricity and, in addition, provides a significant reduction in tailpipe CO_2 assessed by the regulations. The parameters of the experimental single-cylinder engine are presented in Table 1.

Compression ratio is approximately optimized for hydrogen-based DF operation to avoid abnormal combustion [15, 16] and the diesel injector was supplemented by a copper cooling sleeve to reduce the thermal stress of the injector tip [17, 18].

Table 1 Engine parameters

Engine	Single cylinder AVL 5402
Displacement	510 cm^3
Bore/stroke	85/90 mm
Geometric compression ratio	14:1
Fueling-liquid fuel	BOSCH Common Rail, CP 4.1, 1800 bar
Fueling-gaseous fuel	Port fuel injection H_2/CH_4

The combustion chamber was not optimized for dual fuel combustion; however, suggestions for optimization for the used engine are discussed in [19].

4.2 HVO Substitution of Regular Diesel Oil

The complete results are presented in [20]. The impact of replacing RD with HVO was observed during the test based on the World Harmonized Stationary Cycle (WHSC). The experiments were carried out with the original compression ratio 17:1. However, the assumption is that the advantages of HVO should be highlighted with a lower compression ratio. Experiments confirmed the better ignitibility of HVO than that of RD oil, represented by an approximately 60% reduction in CO and HC emissions. The effect is pronounced under low load and cold conditions. The tailpipe emissions of CO_2 improved slightly by 3.6%. The impact on engine efficiency and NO_x is positive or negligible. It is appropriate to optimize dwell between pilot and main injections with respect to the higher cetane number of HVO. Otherwise, the pilot injection quantity can burn too fast, and consequently, the main injection is not sufficiently connected to the pilot combustion. It can increase the PN emissions and the rate of pressure rise. In general, HVO have a positive effect on most of the energetic and emission parameters without any changes in the ECU settings. The benefits of HVO open new possibilities for optimization of the ECU strategy to improve the NOx-PM trade-off.

4.3 Dual Fuel Diesel—Hydrogen Operation

4.3.1 The Initial Mapping of a Diesel-Hydrogen DF Engine

This chapter is based primarily on the results presented in [15, 16]. The original geometric compression ratio was reduced from 17:1 to 14:1. The reduced compression ratio shifts the occurrence of abnormal combustion, represented primarily by preignition, to higher loads. The basic mapping of DF combustion with H_2 in dependence on H_2 normalized energy contribution was performed at an engine speed of 2000 rpm. All measurements started at the point with the RD fueling only. For all supercharged conditions, the air excess ratio for the diesel base points was kept approximately constant at $\lambda = 2.2$. The λ value was obtained in naturally aspirated mode with 6 bar of IMEP. The combustion phasing, given by CA50, was controlled by the timing of diesel fuel injection.

Figure 6 shows the main limitations obtained during the mapping. CA50 was kept approximately constant between 8 and 10 degrees of CA after TDC for loads up to 14 bar of IMEP. At IMEP of 16 bar, the CA50 needs to be retarded to achieve the maximum cylinder pressure of 150 bar and maximum rate of pressure rise (RMAX) of 10 bar/degree CA. Furthermore, the highest attainable share of hydrogen energy is limited because of the prevention of preignition. Higher loads than 16 bar of IMEP lead to regular preignitions or backfires; therefore, 16 bar of IMEP was taken as a limit. At low loads (4 and 6 bar of IMEP) the limiting factor was extremely slow and incomplete combustion of an extremely lean mixture. Despite the advanced injection timing, the CA50 became extremely retarded. The reduced compression ratio is not optimal for an extremely lean operation at low loads due to the low temperature at compression end. Therefore, it is inappropriate to use dual fuel operation at low loads. The highest hydrogen shares were obtained at loads between 8 and 14 bar of IMEP. In this region, the maximum H_2 energy share almost reached 98%. At 16 bar of IMEP the limit hydrogen energy share is approximately 70%. Measurements, as described in this section, were performed as well for engine speeds of 1200 rpm, and the 3200 rpm and results approximately correspond to the presented results of engine speed of 2000 rpm.

The same mapping procedure was performed with NG instead of H_2. The NG mainly consists of methane in our experiments; therefore, the DF NG operation leads to the same results as DF biomethane combustion. Except for the carbon footprint, of course.

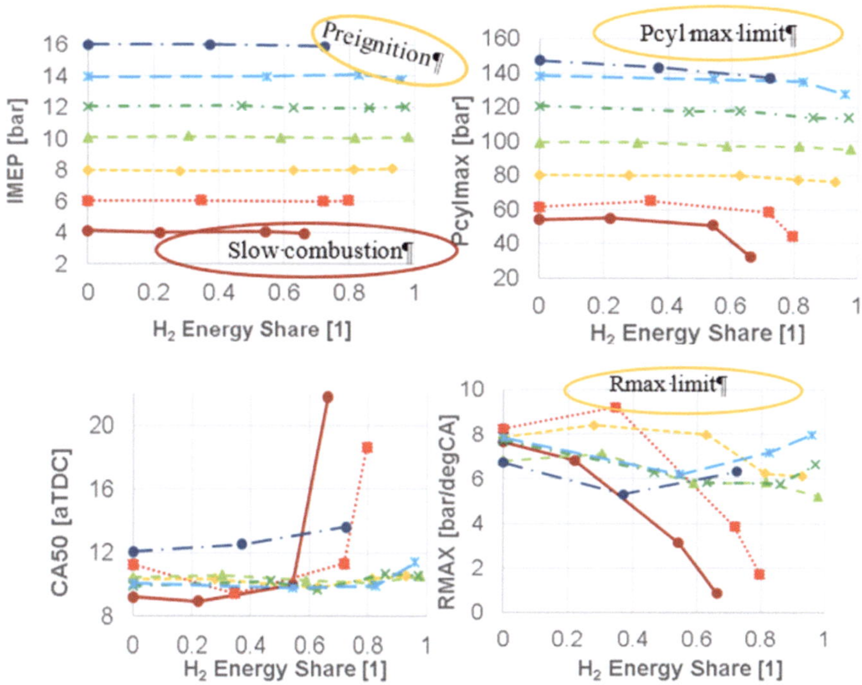

Fig. 6 Indicated mean effective pressure (IMEP),maximum cylinder pressure (Pcylmax), combustion phasing (CA50), maximum rate of pressure rise (RMAX), as a function of hydrogen energy share (engine speed 2000 rpm) [16]

Diesel - NG/H_2 eenergetic share [%]

IMEP [bar]	800	1200	2000	3200	4000	4500
16		0%	0%	0%		
14		0%	0%	0%	0%	
12		0%	0%	0%	0%	0%
10	0%	0%	0%	0%	0%	0%
8	0%	0%	0%	0%	0%	0%
6	0%	0%	0%	0%	0%	0%
4	0%	0%	0%	0%	0%	0%
2	0%	0%	0%	0%	0%	0%
1	0%	0%	0%	0%	0%	0%

speed [rpm]

H_2 energetic share [%]

IMEP [bar]	800	1200	2000	3200	4000	4500
16		74%	73%	0%		
14		98%	97%	70%	0%	
12		96%	97%	64%	42%	0%
10	95%	96%	98%	94%	91%	89%
8	97%	97%	94%	82%	74%	69%
6	86%	84%	80%	74%	69%	67%
4	65%	66%	68%	0%	0%	0%
2	0%	0%	0%	0%	0%	0%
1	0%	0%	0%	0%	0%	0%

speed [rpm]

NG energetic share [%]

IMEP [bar]	800	1200	2000	3200	4000	4500
16		90%	96%	93%		
14		95%	96%	92%	90%	
12		89%	95%	89%	90%	90%
10	87%	90%	94%	87%	84%	82%
8	85%	88%	95%	72%	65%	61%
6	82%	74%	58%	51%	29%	0%
4	81%	54%	0%	0%	0%	0%
2	0%	0%	0%	0%	0%	0%
1	0%	0%	0%	0%	0%	0%

speed [rpm]

Diesel mdot [kg/h]

IMEP [bar]	800	1200	2000	3200	4000	4500
16		20.2	32.6	50.4		
14		17.7	28.6	43.4	57.3	
12		15.2	24.7	38.8	49.3	55.8
10	8.4	13.0	21.0	32.8	41.1	46.3
8	6.7	10.7	16.7	26.6	34.2	39.0
6	5.0	8.3	12.6	19.9	25.7	29.4
4	3.4	5.0	8.5	13.4	16.8	18.9
2	1.7	2.5	4.2	6.7	8.4	9.5
1	0.8	1.3	2.1	3.4	4.2	4.7

speed [rpm]

H_2 mdot CO_2 [kg/h]

IMEP [bar]	800	1200	2000	3200	4000	4500
16		5.3	9.0	50.4		
14		0.3	0.9	13.7	57.3	
12		0.6	0.7	13.6	22.1	55.8
10	0.6	0.5	0.3	2.0	3.1	3.8
8	0.2	0.3	1.0	4.8	7.4	9.0
6	0.8	1.4	2.7	5.6	7.5	8.7
4	1.7	2.1	2.9	13.4	16.8	18.9
2	1.7	2.5	4.2	6.7	8.4	9.5
1	0.8	1.3	2.1	3.4	4.2	4.7

speed [rpm]

NG mdot CO_2 [kg/h]

IMEP [bar]	800	1200	2000	3200	4000	4500
16		16.7	26.2	39.0		
14		13.2	23.2	34.8	47.2	
12		11.3	19.8	30.7	39.9	45.6
10	5.8	9.5	17.0	26.5	34.2	39.0
8	3.9	7.5	14.7	22.3	28.9	33.0
6	2.3	5.6	12.1	17.6	22.9	29.4
4	2.1	4.3	8.5	13.4	16.8	18.9
2	1.7	2.5	4.2	6.7	8.4	9.5
1	0.8	1.3	2.1	3.4	4.2	4.7

speed [rpm]

Fig. 7 Maps of the relative amount of gaseous fuels and CO_2 for pure diesel fuel, DF hydrogen and DF-NG

4.4 Extrapolation of Experimental Data to the Full-Size Engine Characteristics

Two fueling strategies were selected from all measured data points. Pure diesel fueling, representing the state-of-the-art reference, and the DF hydrogen fueling strategy represented by the points with maximum allowable hydrogen share were selected from the systematic mapping described in the previous section. Side-by-side comparison of two fueling strategies is available. Measured data were recalculated from indicated to effective parameters, using the realistic estimate of friction losses according to [21]. To provide the input data for the simulation of the driving cycle, extrapolations were performed to increase the speed range. In this way four-cylinder engine maps were designed. The characteristics of the hydrogen (NG) energy share and the CO_2 emissions per hour are introduced in Fig. 7. Pure diesel operation was used at low loads as a result of the pure chemical efficiency and combustion stability of the DF fueling. More details about the extrapolation procedures are presented in [15, 16].

4.5 Reduction of CO_2 Emissions Assessed by Vehicle Driving Simulation

In this section, the initial assessment of the use of the impact of DF hydrogen engines in a real powertrain of the van will be outlined. For this purpose, the vehicle model was compiled in GT-Suite, assuming the vehicle parameters in Table 2. The vehicle in MPV category was assessed.

Table 2 Vehicle parameters [15]

Fueling	Diesel	Diesel-H_2	Diesel-NG
Operating vecide mass (kg)	2001	2183	2199
Diesel tank (dm^3)	50	12	12
Gaseous fuel in160 dm^3 tank (kg)	0	6.25($p_{fuel} = 700$ bar)	24($p_{fuel} = 200$ bar)
Simulated vehicle range in WLT (km)	868	455	771
Cylinders/engine volume	4/2.0444 dm^3		
Maximum torque	233–247 Nm at 1200–3200 rpm		
Maximum power	82–84 kW at 4000 rpm		

The model uses a so-called quasi-steady approach to calculate longitudinal vehicle dynamics according to the prescribed velocity profile and gear ratio. It includes all driving resistances. Fuel consumption and CO_2 emission is interpolated from the steady state engine maps for each time step selecting the proper operational point.

The results of two driving cycles (NEDC and WLTC) are presented in Fig. 8. The dark red columns represent pure diesel fueling. The CO_2 emissions of 148 g/km in WLTC correspond to diesel oil consumption of 5.76 l/100 km, which confirms that the simulation provides realistic data. The dark purple columns represent CO_2 emissions from the DF hydrogen concept. The CO_2 emissions of 44 g/km represent the significant reduction and meet the European requirements for CO_2 emissions after the year 2030. The range given by the simulation is presented in Table 2. The value of 455 km seems to be sufficient especially considering the refueling speed comparable to conventional fuels. The results of DF with natural gas (yellow columns) show that this concept do not fully exploit the potential of nature gas. The reason is the share of diesel oil and the necessity of pure diesel operation at low loads. However, the option to burn a wide range of fuels, diesel oil, diesel oil and hydrogen, and diesel oil and bio methane (or NG) allows the vehicle to remain in use while the infrastructure of renewable fuels is being developed. It should be noted that results correspond to tailpipe emissions. The overall carbon footprint depends on the technology of production of renewable fuels.

4.6 *The Extension of Opportunities of Dual Fuel Diesel-Hydrogen Engine by Hydrotreated Vegetable Oil (HVO)*

The HVO has higher cetane numbers and better ignitibility in cold conditions than regular diesel oil, which is confirmed by the experiments mentioned with pure diesel fueling. This chapter deals with the impact of HVO on the operation of DF hydrogen at low loads and speeds. The results confirmed the assumptions. HVO leads to higher energetic H_2 share and allows DF operation at lower load than is possible with regular

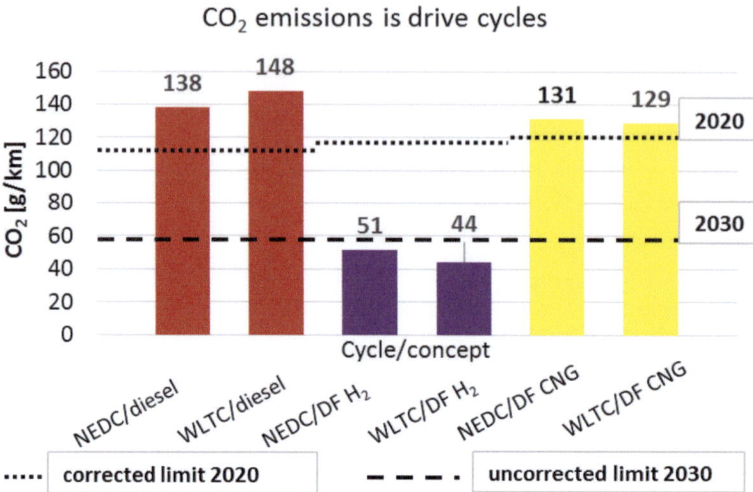

Fig. 8 Simulation of vehicle CO_2 emissions in the NEDC and WLTC driving cycles [15]

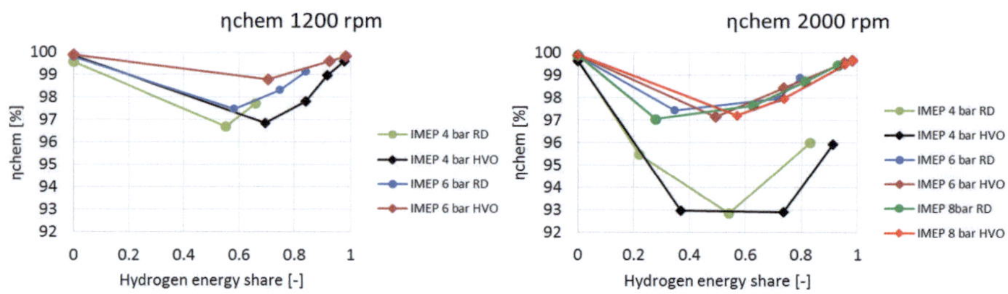

Fig. 9 Chemical efficiency of combustion [22]

diesel oil (Fig. 9). Therefore, HVO highlights the advantageous of dual fuel hydrogen operation, as a further drop in PN and CO_2 emissions. The HVO can extend the opportunities of dual fuel hydrogen operation discussed above [22].

5 Experimental Investigation of Permanent-Lean Hydrogen Combustion Engine (H_2ICE)

H_2ICE represents an attainable path to a hydrogen-fueled transportation sector. Although fuel cells (FC) might be the ultimate goal for hydrogen utilization, internal combustion engines act as a bridging technology which is robust and mature. The following text provides experimental results from a single-cylinder research engine fueled with hydrogen. Primarily focusing on efficiency and emission performance, extending current H_2ICE research with the trends of turbocharging and downsizing. Secondly, the potential of an alternative ignition source is investigated.

5.1 Experimental Facilities and Operating Strategy

The presented data were collected with a single cylinder research engine (Table 3). The ignition source was a spark plug, except where specifically mentioned. The fuel (H_2) was injected into the intake manifold.

The principal pollutant for H_2ICE is NOx and its formation can be efficiently limited with lean operation and is near zero around $\lambda = 2.5$. Maintaining stable combustion at such large dilution is possible due to the excellent ignitability of H_2. Therefore, the control strategy aims to maximize lean operation, but to provide a reasonable power density, the engine is enriched (up to $\lambda = 1.7$) once the peak boost pressure of 2.5 bar is obtained (Fig. 10).

Since the engine was operated lean, turbocharging is required to obtain a reasonable power density. Since the performance of a real turbocharger on a single-cylinder engine is incomparable to a real 4-cyl engine, virtual turbocharging is selected. In this scheme, the intake pressure is maintained by an external compressor, while the

Table 3 Engine parameters

Displacement	375 cm
Bore/stroke	74.5/85.9 mm
Geometric compression ratio	12:1
Ignition source	Spark plug/prechamber
Fueling	PFI H_2

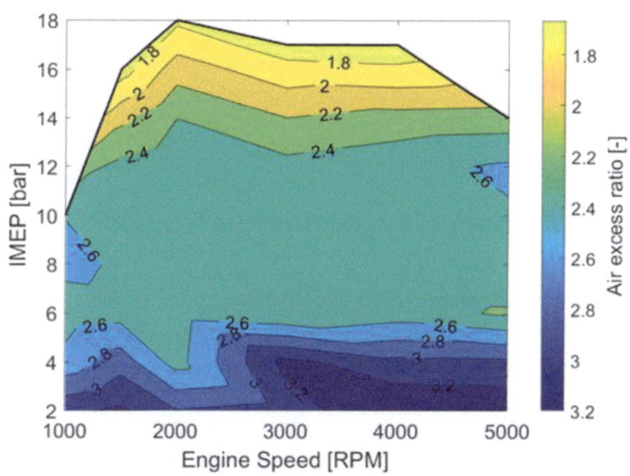

Fig. 10 Air excess ratio (λ) map [23]

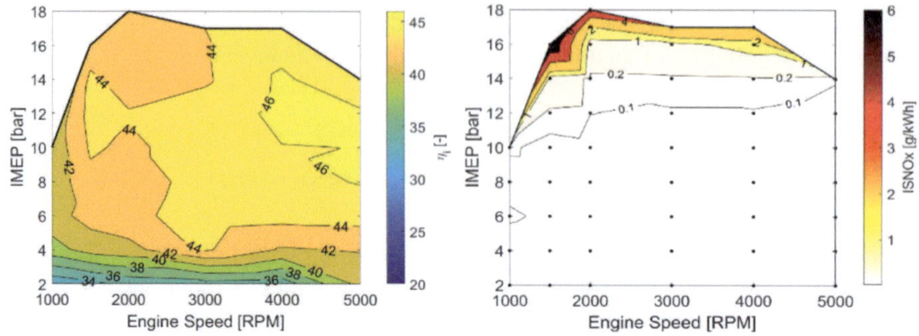

Fig. 11 Indicated efficiency and indicated specific NOx emissions

exhaust backpressure is manipulated with an (exhaust) throttle valve, to achieve an overall turbocharger efficiency of around 45% (details in [23]).

5.2 *Results*

Operating the engine under the specified conditions, complete engine maps were collected for efficiency and NOx emissions (Fig. 11). The left chart shows the obtained indicated thermal efficiency, with peak of 46% at 4000RPM, efficiency above 40% is maintained in most of the map, but at loads below 4 bar IMEP the efficiency starts dropping rapidly due to the necessary throttling at low loads. The right chart depicts engine-out indicated specific NOx emissions. Due to the lean operation ($\lambda = 2.5$), ISNOx of 0.1 g/kWh is attained in a significant part of the engine map. However, since enrichment is required to achieve higher loads, these operating points are accompanied by a rapid increase in NOx emissions—up to 6 g/kWh at 16 bar IMEP and 1500 RPM ($\lambda = 1.7$). The indicated data are reported since the mechanical losses of a 1-cyl engine are incomparable to a real 4-cyl engine. Brake specific data, along with the description of the procedure used, are available in [23].

5.3 *Alternative Ignition Sources for Hydrogen Combustion*

Even with conventional ignition systems using spark plugs, the flammability limits of H_2 mixtures are wide ($\lambda \sim 4$), while the combustion itself is very quick, especially when approaching stochiometric mixtures. It may appear that any more advanced or powerful ignition source is unnecessary; however, looking at the results in Fig. 12 left comparing the spark plug and the pre-chamber ignition at different air excess ratios, we can observe that past $\lambda = 2.8$ the pre-chamber is able to achieve better energy consumption than a spark plug.

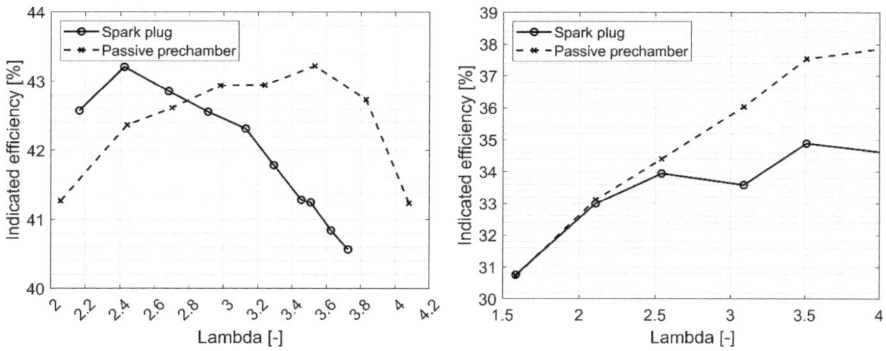

Fig. 12 Experimental indicated specific energy consumption at 6 bar IMEP, 2000RPM for different air-excess ratios [23]—left and efficiency estimation (from thermodynamic model) for pre-chamber ignition at 2 bar IMEP, 2000RPM (spark plug data are evaluated from measurements)—right

Further analysis of the data showed that the differences are mainly due to the variation of combustion duration. Although for both ignition sources the combustion is stable even around $\lambda = 4$, for the spark plug this significant dilution leads to a very slow combustion, which negatively affects the efficiency. This is also the root cause of why H2ICE requires throttling at low loads, slow combustion associated with leaning out the mixture past certain point hurts the efficiency more than the losses caused by throttling.

Meanwhile, for the pre-chamber ignition we see that between $\lambda = 3$–3.8 the energy consumption remains unaffected at its optimal value. This could be utilized to alleviate the throttling losses induced at low loads. Unfortunately, experimental data with pre-chamber at lower loads are not available at the current date. But extrapolating the available data with the use of 1-D CFD simulations, we can estimate the potential of pre-chamber at lower loads. In Fig. 12 right the pre-chamber starts outperforming the spark plug past $\lambda = 2.5$ while at the optimal point ($\lambda = 3.5$) the pre-chamber provides about 2.5 percentage point improvement over the spark plug.

6 High Pressure, Highly Efficient TC Groups and Their Issues

Application of new fuels has its own unique challenges—one of the most important ones is the effect on engine brake efficiency. From the thermodynamics point of view, this is mainly related to combustion speed (i.e., how fast the combustion process is), the amount of pollutants (which are created during the combustion process; e.g., NO_x content in exhaust gases), and possibly some other constraints (e.g., knocking for the case of SI ICE). Optimal conditions for combustion process while achieving prescribed level of pollutants are usually defined by air excess which influences necessary air flow rate, thus leading to needed performance of a boosting system and impacting valve timing as well. All that influences gas exchange process, hence non-linearly affecting engine efficiency—that stresses the complexity of the phenomenon, and it clearly leads to complex optimization process while many design/performance parameters have to be optimized at the same time. Such tasks are usually performed theoretically while adopting suitable simulation tools/models with sufficient predictive capability. This approach was applied to demonstrate certain qualitative trends in the following text while considering large-bore 2-stage turbocharged gas SI ICE, which is typically used in heat & power generation (cogeneration powerplants) and it is operated at constant engine speed. The qualitative trends and conclusions (presented below) are applicable for automotive applications as well.

6.1 Modelling Approach

Engine model was built in 0-D/1-D SW tool [24] while using previous experience with similar engines [25, 26]. The most important model features are the following, based on application of predictive sub-models. Simplified FE (Finite Element) model is used to calculate combustion chamber temperatures while Woschni formula (c.f. [27]) is applied to estimate the heat transfer coefficient between the in-cylinder gas and the walls (boundary condition for FE model). All necessary model constants were estimated by the authors—they are based on authors' experience with similar engines in the class of large-bore turbocharged gas SI ICE. Mechanical efficiency was calculated using default GT-Power formula (based on Chen-Flynn model [28]). Concerning combustion model, no experimental data were available. Simple Vibe function model was applied using the experience from similar engines (c.f. [25, 26]) while considering relatively fast combustion process which his typical for pre-chamber combustion of lean mixtures. Moreover, it is also supported by other literature sources (e.g., [29]) dealing with gas SI ICE equipped with scavenged pre-chamber. Knocking was not considered—however, minimum air excess (depending on engine load) was estimated for methane (CH_4) fuel to avoid this phenomenon. NO_x formation is calculated by means of multi-zone model while using Zeldovich kinetic mechanism [30].

Methane (CH_4) combustion is well known for decades—it is fast and stable up to air excess value of approximately 2.0 when scavenged pre-chamber concept is applied. Based on many literature sources (e.g., [31]), hydrogen (H_2) combustion is supposed to be fast even under very lean mixture operation while knocking is usually not an issue if air excess is larger than 1.7, and scavenged pre-chamber is not needed. Application of ammonia (NH_3) is more problematic as its laminar flame speed is very low (leading to slow turbulent combustion as well) and NO_x formation is significantly higher due to presence of nitrogen in the fuel (e.g., [31]). This knowledge was transferred to the model.

6.2 Effect of Different Fuels

Standard fuel in the class of large-bore gas SI ICE is the methane, which is considered as default fuel in the text and figures. New fuels are represented by hydrogen and ammonia—these were selected mainly due to its zero-carbon content and high potential to produce these fuels from renewable energy sources. Fully optimized results are presented in Fig. 13 (ignore dashed green line for now). The main observation is that new fuels (hydrogen, ammonia) are clearly worse in terms of efficiency when compared with default fuel (methane)—and this trend remains unchanged throughout all simulation work that was carried out within multiple R&D projects. There are thermodynamic reasons behind that. For hydrogen, its low density (mixing device is located upstream of LP compressor) and higher chemical activity (related to increased presence of OH radical) leads to necessity to achieve higher boost pressure levels while applying higher air excess. All that leads to the fact that boosting system has to work harder and it operates on its limit even for the case of very high efficiency 2-stage boost group (its total peak efficiency is more than 72%). To compensate that, combustion phasing is delayed (to increase exhaust gas temperature) while IVC event (strong Miller cycle is usually applied, hence IVC is typically before BDC) needs to be delayed as well, thus using lower millerization effect. All that contributes to decreased efficiency. Regarding ammonia, the main limiting factors are (very) slow combustion and stronger NO_x formation related to ammonia chemistry. Hence, the required boost pressure to compensate these effects is high, leading (again) to the status that boost group operates on its limits, having the same/similar consequences of decreased engine efficiency as for the case of hydrogen fuel. The dominating fact of very slow ammonia combustion is such a negative one that overall efficiency is lower than that of hydrogen—c.f. Fig. 13.

However, if ammonia combustion can be made as fast as the combustion process of methane, engine efficiency can be significantly improved, despite of higher NO_x formation of ammonia chemistry—c.f. green dashed line in Fig. 13. It is slightly better than that of methane—the main reasons behind that are higher density of ammonia (helping gas exchange phase and boost system performance) and additional thermal mass due to presence of nitrogen in the fuel (helping HP indicated efficiency). The authors expect that fast ammonia combustion might be achieved by application of scavenged pre-chamber concept in combination with small amount of hydrogen (injected only into pre-chamber)—this needs to be verified experimentally. Final comment concerns the fact that too slow combustion has (very) strong negative effect on engine efficiency (the fact is well-known for decades)—if that cannot be improved, efficiency will remain low.

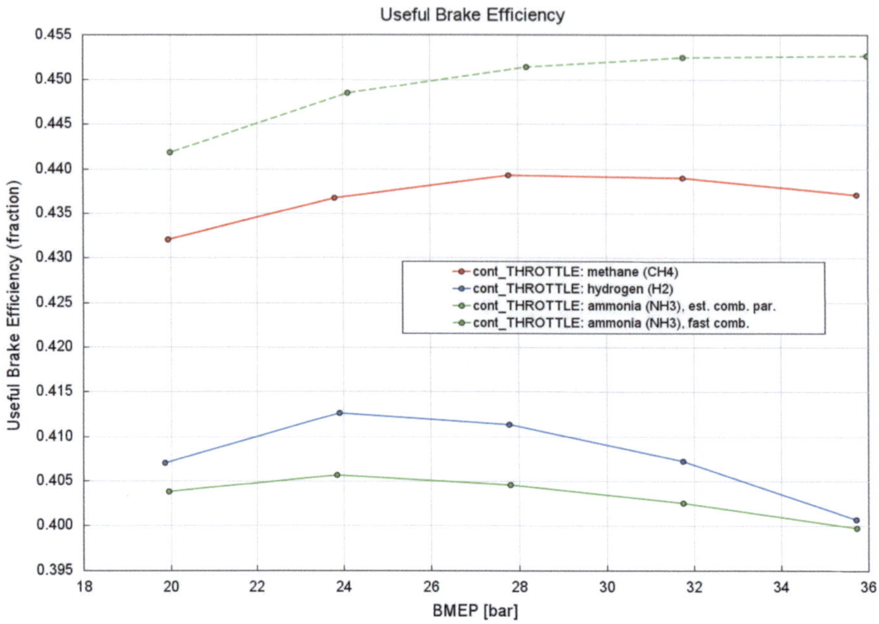

Fig. 13 Influence of applied fuel while considering methane (default), hydrogen and ammonia—fully optimized engine setting to achieve minimum BSFC for each load point (defined by BMEP level) under constant NO_x level of 100% (defined by the TA Luft); dashed green line represents ammonia under assumption of fast combustion (labelled as 'fast comb.')

6.3 Effect of NO_x Level

During last decades there is a clear trend of making the pollutant limits more strict, hence decreasing the NO_x content in exhaust gases. That leads to increase of required air excess, thus making gas exchange process more problematic as boos system has to work harder. Based on that, decrease of engine efficiency is to be expected—this is confirmed in Fig. 14 where the effect of NO_x level (defined by means of scaling factor with respect to the TA Luft norm) is presented. As mentioned above, boost system already operates on its limit even for the standard/default value of NO_x level/limit (scaling factor of 100% with respect TA Luft)—this is especially the case for higher engine loads, thus optimal efficiency is achieved around BMEP of 28 bars. When lower NO_x limit is enforced, engine efficiency drops while optimal engine load decreases—c.f. Fig. 14. The main reason is the limitation of boost system performance (despite very high total efficiency of boost system consisting of 2 high-efficiency turbochargers), which needs to be compensated by delayed combustion and lower millerization level—all that contributes to decreased efficiency. When applying new fuels (hydrogen, ammonia), the effect is even stronger due to facts mentioned in previous sub-section.

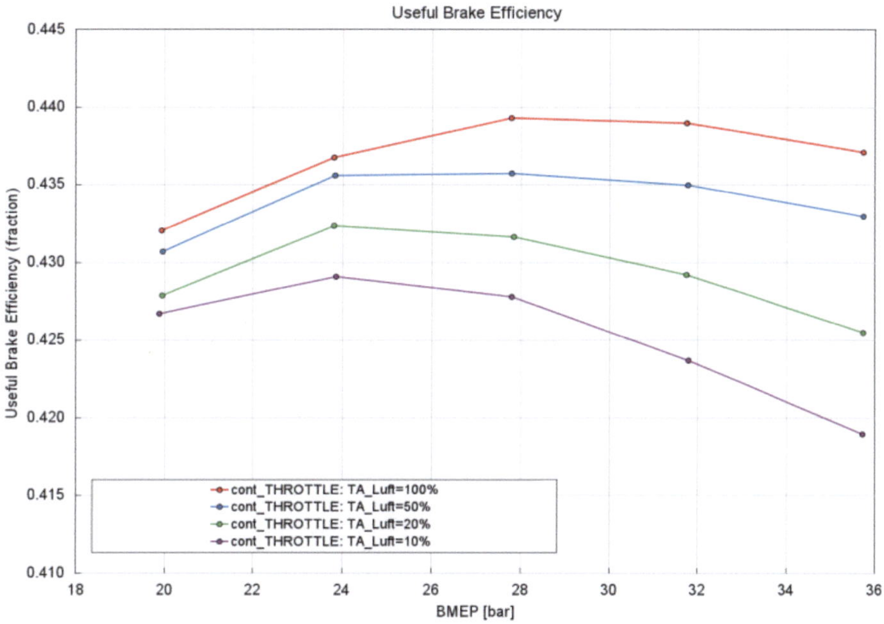

Fig. 14 Influence of NO_x level defined by the scaling factor with respect to TA Luft (considering the values of 100, 50, 20 and 10%) for the case of methane (default fuel)—fully optimized engine setting to achieve minimum BSFC for each load point (defined by BMEP level)

6.4 Application of Electrically Assisted Boosting Systems

Main conclusion of the previous 2 sub-sections is the fact that boost system performance is the most dominant limiting factor of a highly boosted engine, despite very high efficiency of 2-stage turbocharger group. Thus, any improvement in that area is expected to have positive effect on engine efficiency. One possibility is the application of electrically assisted boosting/turbocharging (e-boosting). There are different potential configurations to achieve that—the one considered here is a conservative solution based on electrical energy (for electrical assistance) being provided by electrical machine (e.g., alternator) directly connected to an engine crankshaft. HP turbocharger is equipped with additional e-machine to add/remove power from its shaft.

The results based on full optimization are plotted in Fig. 15, where dashed lines represent cases with electrical assistance (of HP turbocharger) while continuous lines are identical to those from Fig. 14, thus highlighting possible improvement related to e-boosting application. Clear improvement of engine efficiency is visible—and it is stronger when higher engine load is required, and it significantly increases when more strict NO_x levels are targeted. Thus, the improvement is proportional to required boost pressure. Thermodynamic reasons behind that improvement are related to ability to use optimal combustion phasing and high millerization level,

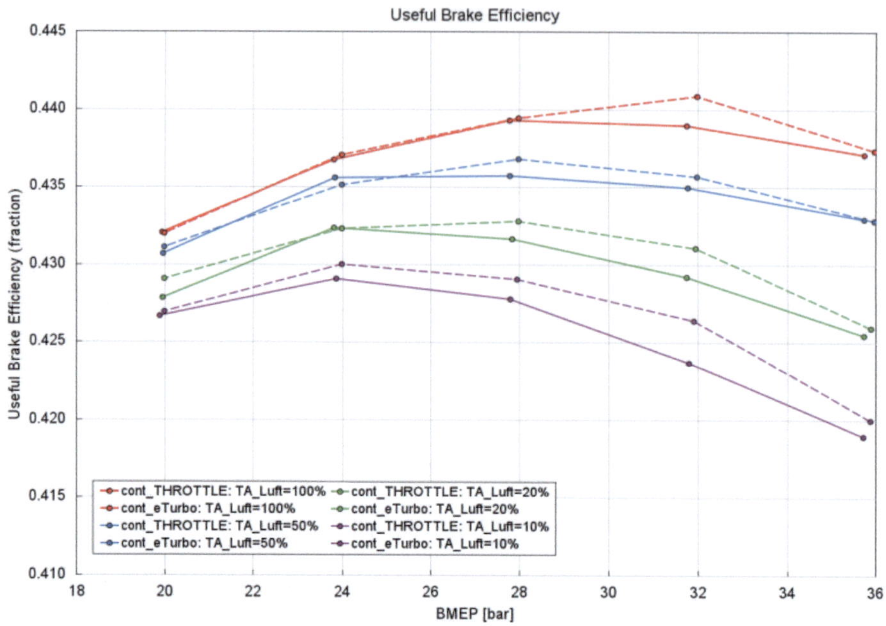

Fig. 15 Effect of electrically assisted boosting (labelled as 'eTurbo', dashed lines) while considering influence of NO_x level defined by the scaling factor with respect to TA Luft for the case of methane (default fuel)—fully optimized engine setting to achieve minimum BSFC for each load point (defined by BMEP level)

despite the fact that there is additional load on crankshaft to power e-machine to help HP turbocharger.

The presence of e-machine will have positive effect on engine transient response as well—that mainly depends on available e-machine power. Empirical experience is that if it is at least 10% of typical turbine power output, then the effect is strong. Moreover, if the e-machine can act as generator (i.e., it is able to send electrical power back to crankshaft), it may have additional positive effect on engine efficiency under low load conditions as such a control is slightly more efficient/effective than classical means, e.g., throttling or waste-gating.

7 Conclusions

Decarbonization of transport calls for new but feasible solutions, offering the best compromise between initial investment, operation costs and market acceptance of new products. Both greenhouse gas emission reduction and induced other harmful side effects should be assessed, well before the new measures are realized. The tool developed and calibrated by the currently available data offers possibilities to do it. The sources of data knowledge combine advanced 1D and 3D simulations and experiments, as well. The examples presented in the text show the yet existing potential of internal combustion engine, fueled by low and zero carbon fuels in a competitive way to the total electrified vehicles.

Application of new fuels usually introduces additional challenges, and these typically lead to even stronger requirements in terms of boost system performance. Based on that, running new fuels is typically associated with decreased engine efficiency—in other words, new fuels may improve situation related to carbon balance, however it is unlikely to improve total ICE brake efficiency. It seems to be a significant challenge to minimize that effect when dealing with new fuels. On the other hand, there are technologies to achieve that. One of them might be the scavenged pre-chamber combustion concept (possibly in dual-fuel mode) while running under (very) lean conditions—this helps both engine cycle thermodynamics and decreased pollutant formation. Boost systems need to be very efficient, most likely using multi-stage systems with potentially electrical assistance.

It is possible due to achieving high engine brake efficiency, especially by reduction of wall-cooling loss using (very) lean mixture, as thermodynamics of ICE cycle is significantly more efficient. Moreover, lean combustion concepts are improved using potentially fully renewable power-to fuel process producing hydrogen. Any lean-burn engine requires high efficiency boost system which becomes one of the most critical components to achieve high engine efficiency.

The symbiosis instead of competition between both electrificating or ICE powering individual transport will contribute to overcoming of huge problems, which would be typical to each of both solutions used as the only one.

Acknowledgements The research for this paper was funded by the Technological Agency of the Czech Republic in the projects TN 0100 0054 BOVENAC (J. Božek's Vehicle National Center of Competence) and TK 040100199 MOSUMO (Model Support of Clean and Sustainable Mobility).

Acronyms, Symbols and Subscripts

batt	Accumulator battery
BEV	Battery electric vehicle
C	Compressor
CA	Crank angle
CA50	50% Mass fraction burnt
CI	Compression ignition engine
CO	Carbon monoxide emissions
CNG	Compressed natural gas
CTU	Czech Technical University
curb	Ready for operation
dens	Per volume
D	Drag resistance

DF	Dual fuel
e	Effective, at brake
ECU	Electronic control unit
e-M	Electric motor
FC	Fuel cell
FCEV	Fuel-cell electric vehicle
flow	Liquid flow
GHG	Greenhouse gas
GTW	Grid-to-wheel
HC	Hydrocarbon emissions
HEV	Hybrid electric vehicle
HVO	Hydrotreated vegetable oil
ICE	Internal combustion engine
IMEP	Indicated mean effective pressure
LCA	Life cycle analysis
load	Cargo load
mix	Mixture
NEDC	New European Driving Cycle
NG	Natural gas
NO_x	Nitrogen oxides emissions
P	Piston
PHEV	Plug-in hybrid electric vehicle
PM	Particulate matter
PN	Particle number
RD	Regular diesel fuel
rel	Relative
RMAX	Maximum rate of pressure rise
road	Specific per distance
roll	Rolling resistance
spec	Mass-specific
T	Turbine
TDC	Top dead center
trans	Transmission
TTW	Tank-to-wheel
veh	Vehicle
WHSC	The World Harmonized Stationary Cycle
wheel	At vehicle wheels
WTW	Well-to-wheel
$\boldsymbol{A}$	Cross-section area [m^2]
$\boldsymbol{a}$	Inertia or slope specific resistance [N/kg]
$\boldsymbol{c}$	Velocity, speed [m/s]
$\boldsymbol{E}$	Specific energy [MJ or kWh/X]
$\boldsymbol{F}$	Force [N]
$\boldsymbol{H_u}$	Ower calorific value
$\boldsymbol{h}$	Specific enthalpy [J/kg]

I	Electric current [A]
i_{cyl}	Number of cylinders [1]
$ISNOx$	Indicated specific NOx emission [g/kWh]
$ISEC$	Indicated specific energy consumption [MJ/kWh]
K	Constant multiplier
L	Lost power [W]
L_t	Stoichiometric air-to-fuel ratio
P	Power [W]
p_e	Brake mean effective pressure
s	Path, distance [m]
t	Time [s]
U	Voltage [V]
η	Efficiency
λ	Air excess
π	Pressure ratio
ρ	Density

References

1. Renewable Energy Directive REDII: EU Regulations, CELEX 32018L2001 CS TXT.pdf, EU (2018)
2. Proposal Renewable Energy Directive REDIII_st08568.en23, EU (2023)
3. GreenNCAP https://www.greenncap.com/
4. Guzzella, L., Sciarretta, A.: Vehicle Propulsion Systems. Springer-Verlag, Berlin Heidelberg (2013)
5. Macek, J.: Simple Comparison of Solutions for GHG Reduction in Transport. GT Suite Conference 2021, Gamma Technologies Inc. (2021)
6. Steinbauer, P., Macek, J., Morkus, J., Denk, P. et al.: Dynamic Optimization of the E-Vehicle Route Profile. SAE Technical Paper 2016–01–0156 (2016). https://doi.org/10.4271/2016-010156
7. Steinbauer, P., Husák, J., Denk, P., Macek, J., Šika, Z., Pasteur, F.: Predictive control of commercial e-vehicle using a priori route information. Int. J. Powertrains **2018**(7), 53–71 (2018). ISSN 1742–4275
8. Macek, J., Steinbauer, P., Morkus, J., Barák, A.: Procedures for determination of the current vehicle parameters. Report to IMPROVE deliverable 3.4, CTU Prague 2014 and Macek, J., Steinbauer, P., et al.: D3.5 Energy management algorithm for vehicle operation. Report to IMPROVE EU FP7 Project, deliverable 3.5, CTU Prague (2015)
9. Macek, J., Morkus, J.: Model of surface vehicle fleet energy consumption suitable for climate-energy policy assessment. In: 50th Conference of Czech and Slovak Combustion Engine Research, pp. 110–123. Mendel University in Brno (2019). ISBN 978-80-7509-668-5
10. Macek, J., Toman, R., Denk, P., Steinbauer, P.: (2020) Hybrid powertrain optimization tools. In: 51th Conference Of Czech And Slovak Combustion Engine Research, pp. 110–129. Czech Technical University in Prague (2020). ISBN 978-80-01-06744-4/

11. Toman, R.: Parallel plug-in HEV topologies evaluation using dynamic programming. In: 51th Conference Of Czech And Slovak Combustion Engine Research, pp. 220–228. Czech Technical University in Prague 2020 (2020). ISBN 978-80-01-06744-4
12. Eward, R., Larivé, J.-F., Beziat, J.-C.: Well-to-wheels analysis of future automotive fuels and powertrains in the European context. In: WELL-to-WHEELS Report Version 3c, pp. 11–31, 33–44, 47–61, 66–70. Publications Office of the European Union, Luxembourg (2011). ISBN 978-9279-21395-3
13. European commission joint research centre, institute for energy and transport. In: Well-To-Wheels Analysis of Future Automotive Fuels and Powertrains in The European Context, Well-To-Wheels Report. Version 4.a JEC WELL-TO-WHEELS ANALYSIS, pp. 13–52, 55–57, 62–64, 73–93. Publications Office of the European Union, Luxembourg (2011). ISBN 978-92-79-33887-8. ISSN 1831-9424 (online). https://doi.org/10.2790/95533
14. Kapica, P.L.: Energy and physics. In: Kapica, P.L. (ed.) Experiment, Teoria, Praktika. Nauka, Moscow (1977) (in Russian)
15. Bortel, I.: Potential of dual fuel compression ignition engine to decrease global warming emissions. Dissertation thesis, CTU in Prague, Prague (2022)
16. Vávra, J., Bortel, I., Takáts, M.: A dual fuel hydrogen—diesel compression ignition engine and its potential application in road transport. In: WCX SAE World Congress Experience, pp. 1–14. SAE International, Warrendale, PA (2019). ISSN 0148-7191
17. Vávra, J., Bortel, I., Takáts, M., Diviš, M.: Emissions and performance of diesel—natural gas dual-fuel engine operated with stoichiometric mixture. Fuel **208**, 722–733 (2017). ISSN 0016–2361
18. Königsson, F., Stalhammar, P., Angstrom, H.-E.: Controlling the injector tip temperature in a diesel dual fuel engine. SAE Technical Paper Series (2012-01-0826). SAE (2012). https://doi.org/10.4271/2012-01-0826
19. Jarolímek, J.: Modification of the combustion chamber of a dual-fuel compression ignition engine in order to increase chemical efficiency. Diploma Thesis, CTU in Prague, Prague (2022)
20. Bortel, I., Vávra, J., Takáts, M.: Effect of HVO fuel mixtures on emissions and performance of a passenger car size diesel engine. Renew. Energy **2019**(140), 680–691 (2019). ISSN 0960-1481
21. Macek, J., Fuente, D., Emrich, M.: A simple physical model of ICE mechanical losses. SAE Technical Paper 2011-01-0610 (2011). https://doi.org/10.4271/2011-01-0610
22. Bortel, I., Vávra, J., Takáts, M.: The extension of opportunities of dual fuel diesel-hydrogen engine by usage of hydrotreated vegetable oil (HVO) instead of regular diesel fuel. Fisita Congress in Prague (2021). https://doi.org/10.46720/F2020-EPV-026
23. Kyjovský, Š, Vávra, J., Bortel, I., Toman, R.: Drive cycle simulation of light duty mild hybrid vehicles powered by hydrogen engine. Int. J. Hydrogen Energy (2023). https://doi.org/10.1016/j.ijhydene.2023.01.137
24. GT-Power User's Manual, GT-Suite version v2022. Gamma Technologies Inc. (2022)
25. Vitek, O., Macek, J.: Engine performance comparison of 2-stage boost group with single-stage one. In: Proceedings of KoKa 2016 (2016). ISBN 978-80-214-5379-1
26. Vitek, O., Macek, J., Klima, J., Vacek, M.: Two-stage turbocharged large-bore gas SI ICE—operation under different NO_x level conditions. In: Proceedings of KoKa 2020 (2020). ISBN 978-80-01-06744-4
27. Woschni, G.: An universally applicable equation for the instantaneous heat transfer coefficient in the internal combustion engine. SAE Trans. **76**, 3065 (1967)
28. Chen, S, Flynn, P.: Development of a single cylinder compression ignition research engine. SAE Technical Paper Series. Paper 650733
29. Wimmer, A., Pirker, G., Zelenka, J., Chmela, F., Zurlo, J., Trapp, C.: The potential of exhaust gas recirculation in large gas engines. In: Proceedings: Conseil International Des Machines A Combustion (CIMAC) Congress, Shanghai. Paper No.: 271 (2013)

30. Zeldovich, Y.B.: The oxidation of nitrogen in combustion and explosions. Acta Physicochim. USSR **21**, 557–628 (1946)
31. Elbaz, A.M., Wang, S., Guiberti, T.F., Roberts, W.L.: Review on the recent advances on ammonia combustion from the fundamentals to the applications. Fuel Commun. **10**, 100053 (2022). ISSN 2666-0520. https://doi.org/10.1016/j.jfueco.2022.100053

Comparison of Properties of Biodiesels Made from Different Feedstocks

Dariusz Kurczyński and Grzegorz Wcisło

Abstract The study investigates new raw materials for biodiesel production, covering oil extraction, ester production, and the evaluation of physicochemical properties and engine performance. Biodiesels from rapeseed, sunflower, camelina, palm oils, and animal fats were produced using the GW-201 reactor developed by Professor Grzegorz Wcisło. The research compares these biodiesels with diesel fuel, following standards PN-EN14214 +A2:2019–05 and PN-EN-590:2018. First-generation biofuels include rapeseed methyl ester (RME), sunflower (SME), camelina (CME), and palm oil (PME), while animal fat methyl ester (AFME) represents second-generation biofuels. The study identifies significant differences in fatty acid composition across biodiesels and assesses key properties such as density, viscosity, calorific value, cetane number, and combustion characteristics. All biodiesels met the EN 14,214 ester content requirement (>96.5% w/w). Results highlight the importance of raw material choice in optimizing biodiesel properties to match diesel fuel for reliable engine performance and reduced environmental impact.

1 Introduction

In the not too distant past, it was believed that the problem for human civilization would be the exhaustion of conventional energy deposits in the form of coal, oil and natural gas. It has currently turned out that this is not the biggest problem of a man. Energy resources are still available in significant quantities. A person may not manage to exhaust them because the world in which they live may drastically

D. Kurczyński (✉)
Department of Automotive Engineering and Transport, Faculty of Mechatronics and Mechanical Engineering, Kielce University of Technology, Kielce, Poland
e-mail: kdarek@tu.kielce.pl

G. Wcisło
Department of Bioprocess Engineering, Energy and Automation, Faculty of Production Engineering and Power Technologies, University of Agriculture in Krakow, Kraków, Poland
e-mail: grzegorz.wcislo@ur.krakow.pl

© The Author(s), under exclusive license to Springer Nature Switzerland AG 2025

A. Marchenko and Z. Koruba (eds.), *Decarbonization of Transport Energy Installations in the Context of Sustainable Development Strategies*, Studies in Systems, Decision and Control 635, https://doi.org/10.1007/978-3-032-05884-3_8

change as a result of climate changes that are a consequence of excessive exploitation of conventional energy sources. These changes are related to the massive emission of carbon dioxide into the atmosphere, which is the most crucial greenhouse gas, responsible for deepening the effect of the greenhouse phenomenon on Earth. This necessitates that people must now take decisive action to limit the emission of carbon dioxide and other greenhouse gases. Limiting carbon dioxide emissions requires the use of energy sources which will not be associated with its emission, or if there is a CO_2 emission, it will not cause an increase in the amount of CO_2 in the atmosphere. One of the possibilities to limit the increase in carbon dioxide concentration in the atmosphere is the use of biofuels, which are fuels obtained from organic matter. Burning biofuels causes the emission of CO_2. The production of biomass by plants is, however, associated with the capture of CO_2 from the atmosphere, necessary for the process of photosynthesis. Carbon dioxide then circulates in closed cycles, and this does not contribute to the increase of its amount in the atmosphere.

An exceptionally large source of carbon dioxide emissions and other harmful compounds into the atmosphere are combustion engines commonly used to power various means of transport, working machines, agricultural machines and various devices. By burning fuels obtained from conventional sources in them, we contribute to the increase in greenhouse gases in the atmosphere. The share of carbon dioxide emissions from transport means is significant. As results from the stoichiometry of combustion, for every litre of burned diesel oil, the engine emits 2.69 kg CO_2 [1]. This causes legal requirements to aim to limit the above-mentioned emission. This necessitates the development of new drives and the search for new fuels, i.e. biofuels. A known and used biofuel for powering internal combustion engines with spontaneous ignition is biodiesel type biofuels, most commonly produced in the transesterification process from vegetable oils and/or animal fats. Currently, biodiesel usually consists of fatty acid esters of vegetable oils. These are definitely less often esters of fatty acids of animal origin. Animal-derived organic matter, which constitutes waste requiring disposal, can be used in the production of biodiesel. There is currently a high interest in finding new raw materials for the production of biodiesel. These novel resources should not compete with food. This is a particularly important condition for the application of the new raw material in biofuel production. The issue of feeding people is prioritised over the production of fuels from organic matter. Therefore, resources are sought which will allow for the future growth of new generation biofuels production, not obtained from resources that can be used for food production. Additionally, resources are being sought which will be dedicated to the production of biodiesel, so the fuel obtained will have beneficial properties in terms of the process of creating a fuel mixture and its effective combustion.

The properties characterising biodiesels obtained from various raw materials affect their potential for use as fuel in compression-ignition engines. They influence the process of creating a fuel–air mixture and the process of its combustion, as well as the emission of harmful exhaust components. The conventional fuel for self-ignition engines is diesel oil. Alternative fuels should have properties as similar as possible to the base fuel to which engines are adapted. It is important to find such a raw material for the production of biodiesel that will allow the fuel to have properties comparable to diesel oil.

Energy consumption is increasing very rapidly. It is related to the increase in the number of people on Earth and results from the economic and civilizational development of an increasing number of societies. Petroleum is the main source of energy. The main consumers of energy, non-renewable, are transport, industry and the electricity generation sector [2]. Transport is a sector that will develop intensively and the demand for energy will increase. It simultaneously has a significant negative impact on the environment. It contributes significantly to climate change. It is necessary to replace fuels obtained from oil with fuels that will not contribute to the increase of carbon dioxide emissions into the atmosphere. Such fuels are biofuels.

The aim of the study was to compare the properties of biodiesel produced from various raw materials. Biodiesel was produced in-house using the GW-201 reactor. The properties of biodiesel from rapeseed oil (RME), sunflower (SME), camelina (CME), palm (PME) and from animal fat waste (AFME) were compared with the properties of conventional diesel oil.

2 Biofuels

Biofuels are fuels that are obtained from organic matter. It can be primary or secondary. The primary organic matter is constituted by plant organic matter. It is formed in the process of photosynthesis, where plants produce organic matter using carbon dioxide in the air, simple mineral compounds dissolved in water, and solar energy for this purpose. Secondary organic matter is animal-derived organic matter created using plant organic matter. Organic matter constitutes a form of stored solar energy reaching the Earth. Organic matter is a renewable energy resource, which humans have used since the beginning of their existence. This material will be renewed in a relatively brief time, meaning it is a source of energy that does not get exhausted. However, this form of energy is currently unable to meet the energy needs of human civilization. This civilization was built on non-renewable energy sources: coal, petroleum and natural gas. They are commonly used and condition the functioning of human economy. However, they pose a significant threat to the biosphere. They contribute to the progressing and increasingly visible climate changes. This causes a person to seek new energy sources. It increasingly uses organic matter as a source of energy.

Organic matter, being stored solar energy, has various applications. Its primary use is to satisfy the nutritional needs of living organisms, especially humans. Additionally, humans use it to obtain various everyday products. For example, in house construction, clothing production, furniture, paper, and many other applications. Additionally, it is used as a source of energy for the implementation of various technological processes, including heating homes. Organic material is increasingly being used to produce fuels that can power internal combustion engines. Various methods of processing organic matter into fuels that can power modern combustion engines used in transport and many other applications are known.

Biofuels can be in solid, gaseous and liquid form. Depending on the state of aggregation and properties, they can have different purposes. Biofuels in gaseous and liquid form, after meeting certain requirements, can be used to power piston combustion engines. The most well-known biofuels are alcohols and biodiesels. Alcohols are most often used, in the case of spark-ignition engines, as an additive to petrol. They can also be used in pure form. Biodiesel is used to power engines with spontaneous ignition. They can be used in pure form or as an addition to diesel fuels. The limitation of using biofuels to power engines is the availability of raw materials for their production. The demand for motor fuels is huge, and the possibilities for biofuel production are limited. A major limitation is the availability of resources. There are currently numerous scientific and development works being conducted, aimed at searching for new raw materials for biofuel production, developing new technologies for their acquisition, improving their properties, and many other detailed objectives.

Biofuels are divided into four generations: first, second, third and fourth. Such classification results from the raw materials used to produce biofuels. The best recognised are first-generation biofuels and relatively well second-generation biofuels. The raw materials for the production of first-generation biofuels are the same materials from which food is produced. They were known, easily accessible and it was from them that fuels began to be produced that could replace conventional fuels for powering internal combustion engines. Its application was promoted in the European Union and around the World. First-generation biofuel resources are primarily used for food production. They are the basis of people's nutrition. Obtaining resources for the production of biofuels can pose serious competition for resources intended for food production. The result of this may be an increase in food prices. Food production is the satisfaction of basic human biological needs. This is causing less and less promotion of raw materials for the production of first-generation biofuels nowadays. There are countries that have given up on promoting first-generation biofuels and control the amount of agricultural production destined for biofuels, so as not to affect food prices. For example, agricultural production surpluses are allocated for fuel production.

Currently, more and more attention is being paid to promoting selected second-generation biofuels. The raw materials for the production of second-generation biofuels are inedible plants and various types of organic waste. These can be waste from agriculture, food preparation facilities, the food industry and other industrial sectors where organic waste is generated, i.e. organic material that is a renewable raw material of plant or animal origin. Additionally, these may be waste from food preparation from homes and restaurants. The resources for the production of second-generation biofuels may not be completely indifferent to food production. Promoting the cultivation of inedible plants for fuel production may lead to these plants occupying land that was previously used for the cultivation of food crops. Therefore, actions are being taken to control whether inedible plants do not occupy agricultural lands previously used for food production. Inedible plants grown for fuel production should occupy agricultural wastelands, less fertile lands, where other plants will not grow or their yields will be uneconomical, that is, production will be unprofitable. Additionally, soils damaged and polluted by human activity should be used for the cultivation of non-edible plants for fuels, on which, for safety reasons, it is not possible to grow plants for human food or animal feed.

Work is being conducted on the development of the implementation possibilities of third and fourth generation raw materials. The raw material for the production of third generation biofuels is algae. They are the subject of numerous studies on their cultivation possibilities and use as a raw material for obtaining biofuels: biodiesel, bioethanol, biomethanol, biohydrogen [3–9]. Algae are built from clusters of individual cells. The size of algae can range from the size of bacteria to several metres in length. Its shape is usually irregular. It may resemble leaves, roots or stems of various shapes in appearance. An important feature of algae is their very rapid biomass increase. They can grow regardless of what time of year it is, and their collection is practically possible every day [10]. They are capable of living in any geographical zone. They live in both salty, fresh, warm and cold waters. According to [11], the biomass of algae that can be cultivated in freshwater is one of the most promising raw materials for the production of biodiesel and other biofuels. Algae are characterized by rapid biomass growth, quick photosynthesis process, and the conversion of significant amounts of carbon dioxide into biomass. They have a high oil content, but above all, they do not require cultivation on agricultural land [12]. Furthermore, they have relatively small nutritional requirements. According to [13] microalgae have been recognised as the most promising raw materials for biodiesel production. However, at the moment their use still requires research, among other things in the area of species selection, large-scale cultivation, harvest technology.

Fourth generation biofuels are supposed to be based on plants that are to be developed using genetic engineering. The raw materials for the production of fourth-generation biofuels are expected to be plants with significantly increased carbon dioxide assimilation from the atmosphere [14, 15]. This will allow for binding more carbon in organic matter. This is to enable the reduction of CO_2 content in the atmosphere by capturing it and then bringing it to a form that can be stored in excavations after oil or coal [16, 17]. This may prove to be a good way to limit the impact of carbon dioxide on the greenhouse effect and climate change. However, there are some environmental safety risks that require particular attention. So such technology may not be introduced quickly.

3 Raw Materials for Biodiesel Production

Biodiesel is a biofuel for powering internal combustion engines with spontaneous ignition. Currently, biodiesel most often consists of methyl or ethyl esters (others can be) of fatty acids of vegetable oils or animal fats. In Europe, the PN-EN 14,214 standard was developed which defines the requirements for fatty acid methyl esters labelled as FAME, which can be used to power self-ignition engines or can be used for heating purposes. They can be used as 100% FAME or as an addition to diesel oil meeting the requirements of the EN 590 standard [18]. Biodiesel should be called clean biofuels intended for powering engines with spontaneous ignition, replacing diesel oils. A mixture of biodiesel and diesel oil should not be called biodiesel.

First generation biodiesels, which are categorised as biofuels, are produced from edible raw materials, which include: rapeseed oil, soybean oil, coconut oil, cottonseed oil, corn oil, palm oil, sunflower oil, mustard oil, olive oil, rice oil, cashew nut oil, cotton seed oil, hazelnut oil, pistachio oil, walnut oil, safflower oil, barley oil, peanut oil, wheat oil, sorghum oil, sesame oil, melon seed oil [19–24]. Inedible vegetable oils, from which biodiesel can be produced that are classified as second-generation biofuels include: neem oil, jatropha oil, nag champa oil, karanja oil, tamanu oil, rubber seed oil, mahua seed oil, tung ford seed oil, babassu oil, chestnut oil, golden chain tree oil, abyssinian oil, Indian neem oil, flue-cured tobacco oil, soapnut oil, peruvian tara oil, Spanish artichoke oil, passionflower oil, saguaro oil, moringa oil, croton oil, horse-chestnut oil, flaxseed oil, Mexican poppy oil, yellow oleander oil, eucalyptus oil [19, 20, 25–31]. Waste oils and fats, which have been used in the production and preparation of food, are also raw materials for the production of second generation esters [32–40]. Biodiesel is also produced from oil obtained from algae, and these are third-generation biodiesels.

Based on the data published in [41], four plant oils produced worldwide account for about 86.4% of the production of all plant oils. The most palm oil is produced. His share in global production is currently around 35.1%. Soybean oil is second in terms of production size, with a share of about 27.4%. Third place is occupied by rapeseed oil with a share of about 14.8%. The next oil is sunflower oil, which accounts for about 9.1% of production. The global production volume in the considered period amounted to 222 million metric tons [41]. There are two main areas of vegetable oils consumption. Fifty seven percent of oils are intended for human consumption [42]. Currently, about 16% of produced vegetable oils are used for biodiesel production. Other uses of vegetable oils include the production of animal feeds, cosmetics and varnishes.

Geographical location is the main factor influencing what raw material is used for the production of esters. The types of plants cultivated and used for raw materials, from which biodiesels will be produced, depend on the climatic conditions. Palm oil is obtained in tropical climate conditions. It is obtained from the pulp of palm fruits. It is produced in Southeast Asia, South America and Africa. Palm oil is used as food in 90% of cases [41]. The rest is used for the production of biofuels and cosmetics. Over 80% of palm oil production comes from Indonesia and Malaysia [42]. Soybean oil is obtained from soy seeds. The largest soybean crops are in Brazil, the USA, Argentina and China [43]. It is used in the food, cosmetic, chemical, and fuel industries. Rapeseed oil is produced in the largest quantities in European Union countries, China, India and Canada [44]. The main use of rapeseed oil is in the food industry. Additionally, it is used in the chemical industry and for bio-diesel production. It is one of the most commonly used cooking oils. Its major advantage is that it contains considerable amounts of linolenic acid (omega-3) and linoleic acid (omega-6) with a low content of saturated fatty acids. Sunflower oil is made from sunflower seeds. It is mainly used for food purposes. The largest producer of sunflower oil in the world is Ukraine. The next producers of this oil are Russia, the European Union and Argentina [43].

All the above-mentioned four vegetable oils, which together account for 86.4% of the production of all vegetable oils, are oils whose main use is to meet food needs. In developing countries, with a large population (India, China, Brazil, Indonesia), the consumption of these oils in food production is increasing [43]. In developed countries, the use of edible vegetable oils for biofuel production has been increasing in recent years. About 70% of biodiesel is based on vegetable oils [42]. These oils, like many others, are a source of food for humans. They cannot therefore be used indefinitely to produce fuels used to power engines. This could lead to an increase in food prices. Food needs are a priority over fuel needs. Fuels obtained from edible plants are classified as first-generation biofuels.

The European Union is the largest producer of biodiesel. Her share in global production amounts to 32.2% [42]. The raw materials used for production are rapeseed oil, palm oil and used cooking oil. In second place in terms of biodiesel production are the United States with a result of 18.3%. The main raw materials in the USA for the production of biodiesel are used cooking oils and soybean oil. Indonesia ranks third in biodiesel production, with a share of 17.6%. In Indonesia, the raw material is palm oil. Brazil comes in fourth place with a result of 12.3%, producing biodiesel from soybean oil. The share of other countries in biodiesel production is below 4%.

The development of biofuel production for use in transport will be fostered by high oil prices. Then their production will be more profitable. The use of vegetable oil as a raw material for biodiesel production largely depends on the policies of countries regarding biofuel production. In some countries, the demand for biodiesel is shaped by the introduction of mandatory addition of biodiesel to diesel fuels. The development of biofuel production is also influenced by the lack of access to oil deposits and the desire to become independent of its supplies. An important factor that will influence the development of biofuels are environmental considerations. By using organic matter as a source of energy, we limit the amount of carbon dioxide released into the atmosphere. Additionally, biofuels are biodegradable and do not pose as big a threat to the environment as in the case of fuels obtained from petroleum. In the study [42], it is predicted that the use of vegetable oil for biodiesel production will increase worldwide in the near future. This will mainly involve markets such as Indonesia, Brazil, and the USA. At the same time, a slight decrease in the use of biodiesel in the European Union is anticipated.

The policy on biofuels in countries that are their largest producers will have a significant impact on the shaping of the biofuel market. It is predicted that the USA will increase its share in global biodiesel production by about 6% by 2032 [42]. Similarly, this will be the case for Brazil, Indonesia or Thailand. A reverse trend is predicted in Europe. This will be the result of the European Union policy, which was expressed in the Directive of the European Parliament and Council (EU) 2018/2001 of 11th December 2018 on the promotion of the use of energy from renewable sources, known as the RED II Directive [45]. In this Directive, among other things, biodiesel produced from palm oil was recognized as a high-risk biofuel. It can be assumed that production from this raw material will decrease. The import of palm oil to Europe will decrease. The Union will not promote first-generation biofuels made from raw materials that are also used to produce food. In the RED II Directive, biofuels produced from waste are highly promoted. The energy of fuel produced from waste is doubled to achieve target indicators compared to the energy of fuel derived from raw materials from which first-generation biofuels are produced. The directive has set a target whereby each member state of the European Union is to ensure at least a 14 percent share of renewable energy in the final energy consumption in the transport sector. Furthermore, the share of new generation biofuels in final energy consumption in the transport sector is to be at least 0.2% in 2022, at least 1% in 2025 and at least 3.5% in 2030. The European Union has also drawn attention to the problem of changing land use for fuel purposes. In the RED II Directive, attention was drawn to the constraints on biofuel production and the high risk of indirect land use change caused by an increase in crops intended for biofuel production in a situation of increased demand globally. The Union does not want to lead to a situation where new areas will be designated for the cultivation of food crops to feed people, in a situation where previously available areas will be allocated for fuel purposes. It is also not possible to designate all wastelands for the cultivation of plants for fuel purposes. This may cause negative effects for the environment, including an increase in the amount of greenhouse gases in the atmosphere. Such restrictions and incentives clearly indicate the direction of development of new generation biofuels in Europe.

4 Biodiesel Production

The production of esters used the GW-201 reactor and technology developed by Professor Grzegorz Wcisło. Both the technology and the reactor design are unique on a global scale and patented by the Patent Office of the Republic of Poland under number P-392963. The mentioned reactor can be used for the production of esters from various raw materials and is mainly dedicated to small refineries and agro-refineries, individuals, farmers, transport companies and others. This reactor allows to produce biodiesel for personal needs. Operating the reactor is simple and safe. The reactor and technology, unique in Europe, allow obtaining FAME in a very short time, specifically 1.5 h/cycle. It allows, among other things, for the production of esters using various alcohols and from various raw materials, such as unrefined and refined oils, used (post-frying) vegetable oils, and animal fats, using various catalysts. The view of the GW-201 reactor is shown in Fig. 1. The basic technical parameters of the GW-201 reactor are presented in Table 1.

In the study, the properties of biodiesel produced from known vegetable oils: rapeseed, sunflower, palm and camelina were compared. The biodiesel obtained from such popular raw materials belongs to the first generation biofuels. A different raw material used to produce biodiesel with the GW-201 reactor was animal waste fats.

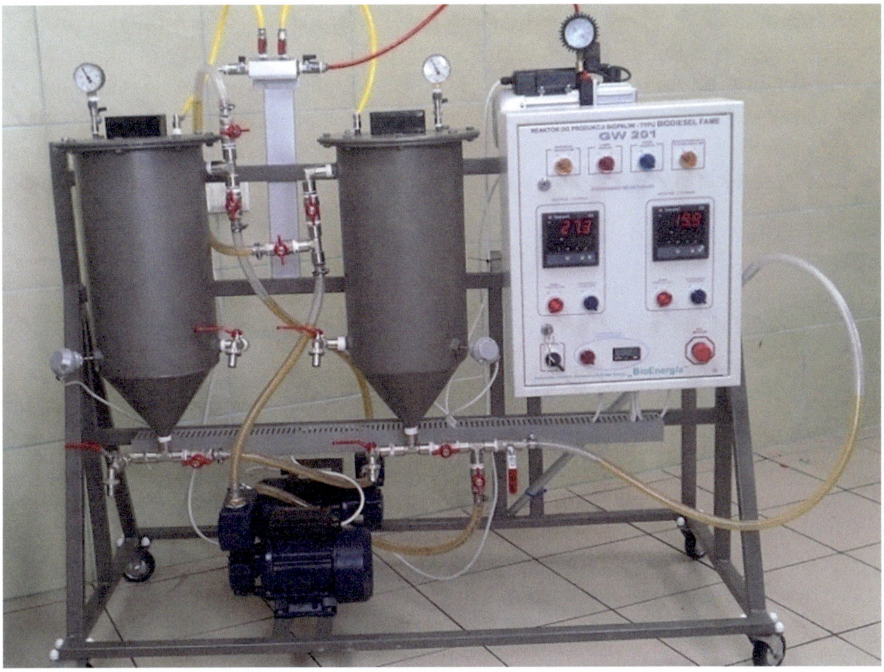

Fig. 1 GW 201 reactor used to produce biodiesel from various raw materials

Table 1 Technical data of GW-201 reactor

Parameter	Unit	Value
Cycle efficiency	dm^3	50
Production time per cycle	h	1.5
Supply voltage	V	230
Rated power	kWh	5.15
Type of catalyst	–	Alkaline/acidic
Process	–	Periodic or semi-continuous
Type of process	–	Single-stage or two-stage

This type of biodiesel obtained from waste belongs to second-generation biofuels. Methyl alcohol and a basic catalyst in the form of potassium hydroxide KOH were used for the production of biodiesel from all five raw materials. The biodiesel production used a proportion in which for every 10 kg of oil, a catalytic mixture was used, created by dissolving 84 g KOH in 1.28 dm^3 of methanol. The catalytic mixtures were prepared using the IKA C-MAG HS 7 magnetic stirrer for this purpose. As a result of the reaction of methanol with potassium hydroxide KOH, a transitional complex is formed, i.e. potassium methanolate CH_3OK, which is a mixture catalysing the transesterification process. The applied solutions and optimised process allowed for the production of very high purity biodiesel within 1.5 h. The biodiesel production process was carried out at a temperature of 60 °C. The quality of all biodiesel produced met the requirements of the EN 14,214 standard for ester content, which is at least 96.5 (wt.%). For each biodiesel, after the appropriate stage of production, the alcohol residues were evaporated by heating the obtained esters to temperatures above the evaporation temperature of methyl alcohol. The evaporation process lasted 20 min.

5 Scope of Testing the Properties of the Biodiesels Produced

The fuel properties of biodiesel obtained depend on the composition and arrangement of fatty acids. In order to define it for the produced biodiesels, chromatographic studies were conducted to determine the fatty acids in their composition. The fatty acid composition analysis was conducted in accordance with the PN-EN 14,103 standard. Applied to gas chromatography research with Thermo Scientific TRACE series flame ionisation detector. Esters from hexanoic acid C6:0 to tetracosanoic acid C:24.1 were identified in biodiesels.

The properties of all biodiesels were determined in accordance with the PN-EN 14,214+A2:2019-05 standard defining the requirements for Fatty Acid Methyl Esters (FAME) intended for use in automotive engines with self-ignition and heating applications. Research for diesel oil was conducted in accordance with the standard PN-EN 590 + A1:2017-06 determining requirements and test methods for diesel oil used for fueling engines with spontaneous ignition driving motor vehicles.

For the studied fuels, properties such as: density, viscosity, calorific value, heat of combustion, ignition temperature, distillation process parameters, cetane number were determined. The density at 15 °C was determined using the oscillating densitometer DA-100M from Mettler Toledo. The dynamic viscosity of the tested fuels was determined at a station equipped with a Reolab QC rotational rheometer from Anton Paar with a thermostatic bath from Grant. Thecombustion heat and fuel value were determined by the calorimetric method using the KL10 Precision BIT calorimeter. The ignition temperature was determined using the semi-automatic Herzog HFP 380 device operating in accordance with the Pensky-Martens method. The distillation process of the examined fuels was conducted using the normal distillation method with the use of the semi-automatic Herzog HDA 620/1 apparatus. The cetane number of the tested fuels was determined in accordance with the standard PN-EN ISO 5165.

6 Research Results and Analysis

Chromatographic tests have been conducted for all esters produced from various raw materials. The composition of fatty acid esters in biodiesels has been determined: rapeseed oil RME, sunflower oil SFME, camelina oil CME, palm oil PME and waste animal fat AFME. The results of these studies are presented in Table 2. In the AFME biodiesel, the highest number of oleic acid esters C18:1 were identified—36.4%(w/w), as well as palmitic acids C16:0 and stearic C18:0, in amounts of 26.8% (w/w) and 27.7% (w/w) respectively. The melting temperature of stearic acid is 67–69 °C [46], of palmitic acid approximately 63 °C [47], and of oleic acid 13 °C [48]. This means that AFME biodiesel under ambient conditions, without the addition of solidifying temperature lowering additives, cannot be used to power the engine. The authors [25] heated the engine powered by AFME biodiesel in the tank to a temperature of about 40 °C. After the engine was finished being powered by AFME biodiesel, the engine was powered by diesel fuel in order to flush the fuel system. Biodiesel produced from PMA palm oil contains about 42% (w/w) of palmitic acid esters. Additionally, it contains large amounts of oleic acid—about 38.2% (w/w). Biodiesel AFME contains esters of linoleic and linolenic acids about 2.2% (w/w). The same acids biodiesel PME contains approximately 12.3% (w/w). Linoleic acid has a melting point of −5 °C, and linolenic acid −11 °C. Linolenic acid (Omega-3) and linoleic acid (Omega-6) are essential in the human diet, as the body cannot produce them itself. This causes these acids to be desirable in food. However, these acids are not desired in biofuel. They have the ability to oxidise quickly. This shortens the storage period of the fuel and worsens its quality. It also increases the price of biodiesel. You need to add more antioxidants to biodiesel. Significant amounts of linoleic and linolenic acids are in sunflower, rapeseed and camelina oils. This is disadvantageous from a fuel perspective. In CME biodiesel, approximately 51.5% (w/w) of linoleic and linolenic acid esters were identified (respectively 15.67% (w/w) and 35.86% (w/w)). The content of linolenic acid esters in CME biodiesel significantly exceeds the permissible value. According to the EN 14,214 standard, the content of linolenic acid methyl ester should not exceed 12% (w/w). Due to its composition, unrefined camelina oil has numerous medicinal properties. Its production is not widespread. According to the standard, it cannot be used as biofuel. In addition, approximately 20.5% (w/w) of eicosenoic acid ester was identified in CME biodiesel. In other biodiesels, its content is clearly below 1% (w/w). In SFME biodiesel, 56.28% (w/w) of linoleic and linolenic acid esters were identified (respectively 54.92% (w/w) and 1.36% (w/w)). From a fuel perspective, this is not beneficial. Only a small content of linolenic acid ester is beneficial here, as with a large reserve it meets the standard requirements. The oleic acid ester content in SFME biodiesel was only about 28.5% (w/w). From a fuel point of view, this is not beneficial, with a high content of linoleic ester. Biodiesel containing a larger amount of oleic ester is characterised by higher heat of combustion and higher calorific value. In RME biodiesel, 27.95% (w/w) of linoleic and linolenic acid esters were identified (respectively 18.67% (w/w) and 9.28% (w/w)). The oleic acid ester in RME was 61.32% (w/w). For fuel purposes, varieties of rapeseed oil with a high content of oleic acid are beneficial, with the smallest possible content of linoleic and linolenic acids. The authors in work [30] studied the operating parameters of a self-ignition engine powered by butyl esters of babassu oil. These esters contain a large amount of short-chain fatty acid esters from C8:0 to C16:0. It has a beneficial effect on the course of the fuel–air mixture creation process, and then on the combustion process and engine performance indicators.

Table 2 Composition of fatty acid esters in biodiesels produced from various raw materials: rapeseed oil RME, sunflower oil SFME, camelina oil CME, palm oil PME and waste animal fat AFME

Fatty acid type	Pattern chemical	Composition of fatty acid esters in biodiesel % (w/w)				
		RME	SFME	CME	PME	AFME
Caproic	C6:0	0	0	0	0	0
Caprylic	C8:0	0	0	0	0	0
Capric	C10:0	0	0	0	0	0
Lauric	C12:0	0	0.11	0	0.32	0.42
Miristic	C14:0	0	0.24	0.46	1.26	3.58
Palmitic	C16:0	4.38	5.89	4.84	42.15	26.81
Palmitoleic	C16:1	0.56	0.31	0.32	1.22	0.65
Stearic	C18:0	1.31	4.81	2.54	0.23	27.73
Oleic	C18:1 (9)	61.32	28.54	13.52	38.21	36.36
Linoleic	C18:2 (6)	18.67	54.92	15.67	10.11	1.81
Linolenic	C18:3 (3)	9.28	1.36	35.86	2.17	0.39
Arachidic	C20:0	0.82	0.38	1.46	0.31	0.37
Eicosenoic	C20:1	0.87	0.32	20.54	0.41	0.35
Behenic	C22:0	0.22	0.22	0.24	0.22	0.22
Erucic	C22:1	0.69	0.26	2.47	0.24	0.32
Lignoceric	C24:0	0.23	0.32	0.29	0.34	0.19
Nervonic	C24:1	0.21	0.21	0.24	0.13	0.11
The degree of identification of fatty acid esters % (w/w)		$\sum = 98.56$	$\sum = 97.89$	$\sum = 98.45$	$\sum = 97.32$	$\sum = 99.2$

Table 3 shows selected properties of biodiesel: RME, SFME, CME, PME and AFME and diesel oil designated in accordance with the requirements of the standards PN-EN 590 + A1:201-06 and EN-PN 14,214 + A2:2019-05. All produced biodiesels contain over 96.5% (w/w) of fatty acid methyl esters. The highest ester content was identified in AFME biodiesel and amounted to 99.2% (w/w). The lowest ester content was obtained by PMA biodiesel, amounting to 97.32. The density of the produced biodiesel is higher than the density of diesel oil. The highest density was achieved by biodiesel AFME (0.889 g/cm^3), and the lowest by biodiesel RME (0.872 g/cm^3). The kinematic viscosity determined for biodiesel is significantly higher than for diesel oil (2.8 mm^2/s). The largest is for PME biodiesel and it is 4.8 mm^2/s. For SMEs, the kinematic viscosity is about 71.4% higher compared to DF. However, for PME the kinematic viscosity is about 17.1% higher compared to RMA. For RME and AFME biodiesel, the kinematic viscosity was 4.1 mm^2/s. This is a value about 46.4% higher compared to the kinematic viscosity of diesel oil. The cetane number of biodiesel is higher than the cetane number of diesel oil. Biodiesels are characterised by better self-ignition abilities compared to DF. This is achieved without introducing additives that increase the cetane number. The flash point temperature is significantly higher for biodiesel compared to the flash point temperature of diesel oil. Considering biodiesels, the highest ignition temperature obtained was for PME at 144 °C, and the lowest for SFME at 112 °C. For DF, the ignition temperature was 58 °C. From a safety application point of view, the values obtained are more favourable for biodiesels. The heat of combustion and calorific value for biodiesel is lower than for diesel oil, which is a result of the oxygen content in the elemental composition of biodiesel. Considering biodiesel, the highest value was obtained for RME, and the lowest for PME. This is due to the high content of oleic acid esters. The calorific value of PME is 6.4% less than the calorific value of RME. The calorific value obtained for SFME and AFME biodiesel was slightly less than that of RME biodiesel.

7 Conclusions

All biofuels of the Biodiesel type produced using technology and a reactor developed by one of the co-authors, Grzegorz Wcisło, are high-quality fuels that meet the basic requirements of the PN-EN 14,214 standard. Particular attention should be paid to the very high content of esters in biodiesels. According to the above-mentioned standard, the ester content in biofuel must be at least 96.5% (w/w). In the produced biodiesels, the ester content ranged from 97.32% (w/w) for PMA to 99.2% (w/w) for AFME.

Four biodiesels were produced from vegetable oils. Three of them, i.e. rapeseed, sunflower and palm, are commonly used in human nutrition. Camelina oil is less known, but it has medicinal properties and should also be used in human diet. It has a high content of linoleic and linolenic acids essential in the human diet. Palm oil has the least beneficial composition when it comes to food use. It contains a small amount of linoleic and linolenic acids. However, the majority of palm oil produced worldwide is used in food production and preparation. Biodiesel AFME is a proposal to meet current requirements for raw materials for biofuel production. It was produced from animal fat waste. Such waste required disposal. The way to get rid of them is to produce eco-friendly biofuel.

Detailed analysis of the obtained results leads to the following conclusions:

1. All produced biodiesels meet the requirements of the PN-EN 14,214 standard in terms of density. It should be 860.0–900.0 kg/m^3 and is slightly higher than diesel oil. This situation causes a greater mass of fuel to be found in the same dose injected into the engine chamber.
2. The same is true for kinematic viscosity. The received values for biodiesel are within the range of 3.50–5.00 mm^2/s and are higher than the viscosity of diesel oil. Higher density and kinematic viscosity can impair fuel fluidity and increase flow resistance in the injection system, which is less advantageous especially at low temperatures.
3. The ignition temperature of biodiesel is about twice as high as that of diesel oil, which is less favourable from the point of view of fuel self-ignition. However,

Table 3 Fractional composition of the examined biodiesels: RME, SFME, CME, PME, AFME and DF diesel oil

% (v/v)] oddestylowania	RME; °C	SFME; °C	CME; °C	PME; °C	AFME; °C [49]	DF; °C
0	272	274	281	292	278	178
10	291	293	298	304	289	202
20	308	309	312	316	296	222
30	315	316	319	322	301	241
40	322	322	326	329	315	255
50	326	328	330	334	320	270
60	331	333	334	337	323	287
70	336	338	339	341	332	305
80	341	343	344	345	337	320
90	344	346	346	348	341	332
95	347	349	348	351	350	338
97 (100)	353	356	355	358	357	343
	Do 250 °C–0% 350 °C–95%	Do 250 °C–0% 350 °C–95%	Do 250 °C–0% 350 °C–95%	Do 250 °C–0% 350 °C–92%	Do 250 °C–0% 350 °C–95.5%	Do 250 °C–36.0% 350 °C–100%

from the perspective of storage and transportation safety, biodiesels are much safer fuels. The ignition temperatures of the studied biodiesels range from 112 to 144 °C. According to the PN-EN 14,214 standard, the ignition temperature should be above 101 °C.

4. The cetane number was higher for biodiesel than for diesel oil. A higher self-ignition ability to some extent reduces the adverse effect resulting from a higher ignition temperature. That's why an engine powered by FAME has similar operating parameters as when powered by diesel fuel.
5. The requirement of a linolenic acid methyl ester content in biodiesel, not exceeding 12% (w/w), is not met in the case of CME biodiesel. This means that this fuel can be stored for a shorter period than other biofuels so as not to lose its properties. This is related to the fact that oleic acid esters are characterized by a greater tendency to oxidize.
6. All biodiesels were characterised by higher initial distillation temperatures, which is disadvantageous compared to diesel oil. The distillation process of diesel oil began at a temperature of 178 °C and ended at a temperature of 343 °C. The distillation process for all biodiesels began at a temperature above 170 °C. The end of the distillation process for biodiesel was in the temperature range from 353 to 358 °C. This allows us to assume that biodiesel fuels, injected into the engine combustion chamber, will burn completely. The most favourable course of distillation was had by RME biodiesel. The process was not significantly different for SFME and AFME biodiesels.
7. All biodiesels were characterised by lower heat of combustion values and calorific values than diesel oil. This is due to the fact that biofuels contain oxygen in their structure, hence less hydrogen and carbon elements, from the oxidation of which energy is obtained. The highest calorific value among biodiesels was characterized by RME. Slightly smaller AFME and SFME.

Considering the analysed fuel properties of biodiesel, biodiesel RME seems to be the most favourable. However, the properties of AFME biodiesel only slightly differ from the properties of RME, and some of them are more favourable. For example, a slightly higher content of fatty acid esters. In addition, AFME contains small amounts of linoleic and linolenic acid esters, which tend to oxidize rapidly, thereby shortening the storage period. Furthermore, the most crucial feature of AFME, in line with the current trends in FAME production, is that they are produced from waste, not from raw materials used for food production, and are a way to dispose of waste. Second-generation biofuels are produced from waste and inedible plants.

Nomenclature

AFME	Animal fats methyl esters
CMO	Camelina oil methyl esters
CO_2	Carbon dioxide

C6:0	Caproic acid
C8:0	Caprylic acid
C10:0	Capric acid
C12:0	Lauric acid
C14:0	Miristic acid
C16:0	Palmitic acid
C16:1	Palmitoleic acid
C18:0	Stearic acid
C18:1	Oleic acid
C18:2	Linoleic acid
C18:3	Linolenic acid
C20:0	Arachidic acid
C20:1	Eicosenoic acid
C22:0	Behenic acid
C22:1	Erucic acid
C24:0	Lignoceric acid
C24:1	Nervonic acid
DF	Diesel fuel
FAME	Fatty acid methyl ester
KOH	Potassium hydroxide
PME	Palm oil methyl esters
RME	Rapeseed oil methyl esters
SFME	Sunflower oil methyl esters
T	Distillation end temperature
TD	Distillation temperature difference for the compared fuels for a specific temperature
TK	Temperature to which the whole volume of fuel should be distilled
T10	Distillation temperature of 10% by volume of fuel
T50	Distillation temperature of 50% by volume of fuel
T90	Distillation temperature of 90% of the fuel volume
T95	Distillation temperature of 95% of the fuel volume
v/v	Volumetric share
w/w	Weight share

References

1. U.S. Energy Information Administration, Carbon Dioxide Emissions Coefficients (2023). Available at: https://www.eia.gov/environment/emissions/co2_vol_mass.php. Access 25 Mar 2024
2. Anwar, M.: Biodiesel feedstocks selection strategies based on economic, technical, and sustainable aspects. Fuel **283**, 119204 (2021). https://doi.org/10.1016/j.fuel.2020.119204

3. Faried, M., Samer, M., Abdelsalam, E., Yousef, R.S., Attia, Y.A., Ali, A.S.: Biodiesel production from microalgae: processes, technologies and recent advancements. Renew. Sustain. Energy Rev. **79**, 893–913 (2017). https://doi.org/10.1016/j.rser.2017.05.199
4. Chen, J., Li, J., Dong, W., Zhang, X., Tyagi, R.D., Drogui, P., Surampalli, R.Y.: The potential of microalgae in biodiesel production. Renew. Sustain. Energy Rev. **90**, 336–346 (2018). https://doi.org/10.1016/j.rser.2018.03.073
5. Posten, C., Shaub, G.: Microalgae and terrestrial biomass as source for fuel—a process review. J. Biotechnol. **142**, 64–69 (2009). https://doi.org/10.1016/j.jbiotec.2009.03.015
6. Saktihivel, R., Ramesh, K., Purnachandran, R., Shameer, P.M.: A review on the properties performance and emission aspects of the third generation biodiesels. Renew. Sustain. Energy Rev. **82**, 2970–2992 (2018). https://doi.org/10.1016/j.rser.2017.10.037
7. Rajak, U., Nashine, P., Verma, T.N.: Assessment of diesel engine performance using spirulina microalgae biodiesel. Energy **166**, 1025–1036 (2019). https://doi.org/10.1016/j.energy.2018.10.098
8. Sun, J., Xiong, X., Wang, M., Du, H., Li, J., Zhou, D., Zuo, J.: Microalgae biodiesel production in China: a preliminary economic analysis. Renew. Sustain. Energy Rev. **104**, 296–306 (2019). https://doi.org/10.1016/j.rser.2019.01.021
9. Akubude, V.C., Nwaigwe, K.N., Dintwa, E.: Production of biodiesel from microalgae via nanocatalyzed transesterification process: a review. Mater. Sci. Energy Technol. **2**, 216–225 (2019). https://doi.org/10.1016/j.mset.2018.12.006
10. Lewandowski, W., Ryms, M.: Biofuels Pro-ecological Renewable Energy Sources. WNT Publishing House, Warsaw (2013)
11. Schroeder, G., Messyasz, B., Łęska, B., Fabrowska, J., Pikosz, M., Rybak, A.: Biomass of freshwater algae as raw material for the industry and agriculture. Przemysł Chemiczny **92**(7), 1380–1384 (2013)
12. Dharmaprabhakaran, T., Karthikeyan, S., Periyasamy, M., Mahendran, G.: Algal biodiesel-promising source to power CI engines. Mater. Today: Proc. **33**(7), 2870–2873 (2020). https://doi.org/10.1016/j.matpr.2020.02.775
13. Chen, H., Ding, M., Li Yuan, Xu.H., Yangyang, L., Wei, Z.: Feedstocks environmental effects and development suggestions for biodiesel in China. J. Traffic Transp. Eng. **7**(6), 791–807 (2020). https://doi.org/10.1016/j.jtte.2020.10.001
14. Bajdor, K., Biernat, K.: Biofuels as alternative energy carriers in internal combustion engines, classification and development perspectives. Arch. Automotive Eng. **1**, 5–19 (2011)
15. Malik, S., Shahid, M., Chen-G, L., Khan, A.Z., Nawaz, M.Z., Zhu, H., Mehmood, M.M.: Developing fourth-generation biofuels secreting microbial cel factories for enhanced productivity and efficient product recovery; a review. Fuel **298**, 20858 (2021). https://doi.org/10.1016/j.fuel.2021.120858
16. Biofuels in the European Vision, a vision for 2030 and beyond. Final report of the Biofuels Research Advisory Council. Available at: https://www.etipbioenergy.eu/images/biofuels_vision_2030.pdf. Access 25 Mar 2024
17. Chisti, Y.: Biodiesel from microalgae. Biotechnol. Adv. **25**(3), 294–306 (2007). https://doi.org/10.1016/j.biotechadv.2007.02.001
18. PN-EN 14214+A2:2019–05 Liquid petroleum products Fatty acid methyl esters (FAME) for use in compression ignition (diesel) car engines and heating applications requirements and test methods
19. Athar, M., Zaidi, S.: A review of the feedstocks, catalysts, and internsification techniques for sustainable biodiesel production. J. Environ. Chem. Eng. **8**(6), 104523 (2020). https://doi.org/10.1016/j.jece.2020.104523
20. Sing, D., Sharma, D., Soni, S.L., Sharma, S., Sharma, P.K.: A review on feedstocks, production processes, and yield for different generations of biodiesel. Fuel **262**, 116553 (2020). https://doi.org/10.1016/j.fuel.2019.116553
21. Kurczyński, D., Łagowski, P.: Performance indices of a common rail-system CI engine powered by diesel oil and biofuel blends. J. Energy Inst. **92**(6), 1897–1913 (2019). https://doi.org/10.1016/j.joei.2018.11.004

22. Singh, D., Sharma, D., Soni, S.L., Sharma, S., Kumari, D.: Chemical compositions, properties, and standards for different generation biodiesels: a review. Fuel **253**, 60–71 (2019). https://doi.org/10.1016/j.fuel.2019.04.174
23. Saladini F., Patrizi N., Pulselli F.M., Marchettini N., Bastianoni S.: Guidelines for energy evaluation of first, second and third generation biofuels. Renew. Sustain. Energy Rev. X **66**, 221–227 (2019). https://doi.org/10.1016/j.rser.2016.07.073
24. Ilkiliç, C.: The effect of sunflower oil methyl ester and diesel fuel blend on the performance of a diesel engine. Energy Sourc. Part A: Recov. Utilization Environ. Effects **30**(19), 1761–1770 (2019). https://doi.org/10.1080/15567030701268443
25. Kurczyński, D., Łagowski, P., Wcisło, G.: Experimental study of fuel consumption and exhaust gas composition of a diesel engine powered by biodiesel from waste of animal origin. Energies **14**, 3472 (2021). https://doi.org/10.3390/en14123472
26. Demirbas, A.: Potential resources of non-edible oils for biodiesel. Energy Source Part B **4**(3), 310–314 (2021). https://doi.org/10.1080/15567240701621166
27. Deepanraj, B., Srinivas, M., Arun, N., Sankaranarayanan, G., Abdul, S.P.: Comparison of Jatropha and Karanja biofuels on their combustion characteristics. Int. J. Green Energy **14**(15), 1231–1237 (2021). https://doi.org/10.1080/15435075.2017.1328420
28. Yadav, A.K., Khan, M.E., Dubey, A.M., Pal, A.: Performance and emission characteristics of a transportation diesel engine operated with non-edible vegetable oils biodiesel. Case Stud. Thermal Eng. **8**, 236–244 (2016). https://doi.org/10.1016/j.csite.2016.08.001
29. Veljkovic, V.B., Lakicevic, S.H., Stamenkovic, O.S., Todorovic, Z.B., Lazic, M.L.: Biodiesel production from tobacco (*Nicotiana tabacum* L.) seed oil with a high content of free fatty acids. Fuel **85**, 2671–2675 (2006). https://doi.org/10.1016/j.fuel.2006.04.015
30. Kurczyński, D., Łagowski, P., Wcisło, G.: Experimental study into the effect of the second-generation BBuE biofuel use on the diesel engine parameters and exhaust composition. Fuel **284**, 118982 (2021). https://doi.org/10.1016/j.fuel.2020.118982
31. Kurczyński, D., Wcisło, G., Łagowski, P., Leśniak, A., Kozak, M., Pracuch, B.: Determination of the effect of the addition of second-generation biodiesel BBuE to diesel fuel on selected parameters of "B" fuels. Energies **16**, 6999 (2021). https://doi.org/10.3390/en16196999
32. Abed, K.A., El Morsi, A.K., Sayed, M.M., El Shaib, A.A., Gad, M.S.: Effect of waste cooking-oil biodiesel on performance and exhaust emissions of a diesel engine. Egypt. J. Pet. **27**, 985–989 (2018). https://doi.org/10.1016/j.ejpe.2018.02.008
33. Borah, M.J., Das, A., Das, V., Bhuyan, N., Deka, D.: Transesterification of waste cooking oil for biodiesel production catalyzed by Zn substituted waste egg shell derived CaO nanocatalyst. Fuel **242**, 345–354 (2018). https://doi.org/10.1016/j.fuel.2019.01.060
34. Fonseca M.J., Teleken J. G., Vitor de Cinque Almeida, da Silva C.: Biodiesel from waste frying oils: Methods of production and purification. Energy Convers. Manage. **184**, 205–218 (2018). https://doi.org/10.1016/j.enconman.2019.01.061
35. Sahar, S.S., Iqbal, J., Ullah, I., Bhatti, H.N., Nouren, S., Habib-ur-Rehman, N.J., Iqbal, M.: Biodiesel production from waste cooking oil: an efficient technique to convert waste into biodiesel. Sustain. Cities Soc. **41**, 220–226 (2018). https://doi.org/10.1016/j.scs.2018.05.037
36. Uddin, M.N., Techato, K., Rasul, M.G., Hassan, N.M.S., Mofijur, M.: Waste coffee oil: a promising source for biodiesel production. Energy Procedia **160**, 677–682 (2019). https://doi.org/10.1016/j.egypro.2019.02.221
37. Gnanasekaran, S.: Prediction of CI engine performance, emission and combustion characteristics using fish oil as a biodiesel at different injection timing using fuzzy logic. Fuel **183**, 214–229 (2016). https://doi.org/10.1016/j.fuel.2016.06.063
38. Adewale, P., Dumont, M.-J., Ngadi, M.: Recent trends of biodiesel production from animal fat wastes and associated production techniques. Renew. Sustain. Energy Rev. **45**, 574–588 (2015). https://doi.org/10.1016/j.rser.2015.02.039
39. Banković-Ilić, I.B., Stojković, I.J., Stamenković, O.S., Veljkovic, V.B., Yung-Tse, H.: Waste animal fats as feedstocks for biodiesel production. Renew. Sustain. Energy Rev. **32**, 238–254 (2014). https://doi.org/10.1016/j.rser.2014.01.038

40. Cunha, A., Feddern, V., De Prá, M.C., Higarashi, M.M., de Abreu P.G., Coldebella, A.: Synthesis and characterization of ethylic biodiesel from animal fat wastes. Fuel **105**, 228–234 (2013). https://doi.org/10.1016/j.fuel.2012.06.020
41. Consumption of vegetable oils worldwide from 2013/14 to 2023/2024, by oil type (in million metric tons). Available at: https://www.statista.com/statistics/263937/vegetable-oils-global-consumption/. Access 12 Mar 2024
42. OECD/FAO: OECD-FAO Agricultural Outlook 2023–2032. OECD Publishing, Paris (2023). https://doi.org/10.1787/08801ab7-en
43. Rosiak, E.: The global market for vegetable oils. Zeszyty Naukowe Szkoły Głównej Gospodarstwa Wiejskiego w Warszawie Problemy Rolnictwa Światowego **17**(1), 173–181 (2017). https://doi.org/10.22630/PRS.2017.17.1.16
44. Czołowi producenci rzepaku. Available at: http://www.rolnikszukaceny.pl/rynek_roslin_oleistych.php?art=284. Access 20 Mar 2024
45. Directive (Eu) 2018/2001 of the European Parliament and of the Council of 11 December 2018 on the promotion of the use of energy from renewable sources (recast)
46. https://pl.vwr.com/store/product/724009/kwas-stearynowy-97. Access 04 April 2024
47. https://www.spirulina.pl/trawa-jeczmienna/kwas-palmitynowy.htm. Access 04 April 2024
48. https://pl.vwr.com/store/product/2360395/kwas-oleinowy-97. Access 04 April 2024
49. Kurczyński, D., Wcisło, G., Leśniak, A., Kozak, M., Łagowski, P.: Production and testing of butyl and methyl esters as new generation biodiesels from fatty wastes of the leather industry. Energies **15**, 8744 (2022). https://doi.org/10.3390/en15228744

Auto-Ignition Delay Period in a Compression-Ignition Engine Running on Dual Fuel Diesel and Gaseous Fuels LPG and CNG

Łagowski Piotr

Abstract This study addresses the challenges of reducing emissions from compression-ignition engines and explores the use of dual-fuel operation with diesel and gaseous fuels (LPG and CNG). The auto-ignition delay period, a critical parameter in combustion processes, was analyzed for a Perkins 1104D-E44TA engine. Comparative tests were conducted for dual-fuel and diesel-only modes, examining maximum combustion pressure and pressure rise rates. Results demonstrate the potential of gaseous fuels to improve engine emissions and efficiency while maintaining reliability. These findings support the use of alternative fuels as viable solutions to meet modern environmental standards.

1 Summary

As a result of today's environmental requirements for reciprocating internal combustion engines in terms of reducing the emission of harmful exhaust components, modern automotive engines must undergo a number of modifications to meet these requirements. In the current era of downsizing, which consists of reducing the capacity of a given engine and increasing its power, this is difficult to achieve, while at the same time ensuring low failure rate of such an engine [1–3]. Taking these aspects into account, a good solution seems to be the use of alternative fuels such as LPG and CNG to power compression-ignition engines. Gaseous fuels are not a novelty in terms of their use to power internal combustion engines; on the contrary, they can be considered as fuels used earlier to power internal combustion engines than petrol or diesel fuels [4–6].

However, the use of gaseous fuels to power engines requires a more in-depth analysis of the impact of these fuels on the combustion process, particularly on the period concerning the auto-ignition delay phenomenon. The auto-ignition delay period is

Ł. Piotr (✉)
Department of Automotive Engineering and Transport, Faculty of Mechatronics and Mechanical Engineering, Kielce University of Technology, Kielce, Poland
e-mail: p.lagowski@tu.kielce.pl

© The Author(s), under exclusive license to Springer Nature Switzerland AG 2025

A. Marchenko and Z. Koruba (eds.), *Decarbonization of Transport Energy Installations in the Context of Sustainable Development Strategies*, Studies in Systems, Decision and Control 635, https://doi.org/10.1007/978-3-032-05884-3_9

a very important parameter determining the combustion process in compression-ignition internal combustion engines [7, 8]. This publication presents the results of tests and analysis of the auto-ignition delay period in a Perkins 1104D-E44TA compression-ignition engine fuelled with dual-fuel diesel and gaseous fuels LPG and CNG. For comparison, the results with dual-fuel operation would be compared to those obtained with diesel-only operation. Parameters such as maximum combustion pressure and maximum pressure rise during the combustion process were also analysed.

2 Introduction

The tradition of using gaseous fuels to power reciprocating internal combustion engines is as old as engine design itself. The development of technologies for obtaining petrol and diesel fuels, abundant oil resources, favourable properties of these fuels, and fewer storage and warehousing problems meant that there was little interest in gaseous fuels for powering engines in cars. Interest in gaseous fuels increased in situations where access to oil resources was limited. This was the case during the First and Second World Wars or the fuel crisis of the 1970s, among others. The current interest in gaseous fuels is mainly driven by the need to find fuels that reduce their negative impact on the environment. Shrinking oil reserves are also an important aspect [9, 10].

The gas fuel commonly used today to power internal combustion engines is the propane-butane gas mixture LPG. Poland is one of the countries where consumption of this fuel is one of the highest in the world and this is due to economic considerations. Another gaseous fuel, also with favourable properties as a fuel for internal combustion piston engines used in transport, is CNG [11, 12]. Among the most important of these properties are the high calorific value of the stoichiometric mixture, high resistance to knocking combustion, lower emission of harmful components with exhaust gases, abundant resources and their availability, and simple chemical structure. The high calorific value of CNG and LPG means that the maximum power loss in gaseous spark-ignition engines does not exceed 10 per cent of that of engines running on standard fuel. In dual-fuel compression-ignition engines, maximum power can be comparable to or even slightly higher than that of diesel engines. The anti-knock resistance of natural gas depends on the admixture content of the higher hydrocarbons in the gas and, depending on the source of the gas, is rated at 120–130 LO. This allows, in smaller spark-ignition engines, to increase the compression ratio by two units—to a value of $\varepsilon = 11$–12, which has a beneficial effect on the overall efficiency of the engine. In medium-sized dual-fuel compression-ignition engines, compression ratios of $\varepsilon = 14$–16 can be used, i.e. the same as in traditionally fuelled direct-injection engines. This allows the engine to be operated interchangeably on diesel alone or in a dual-fuel system, which can be important in the transition period when gas refuelling stations are scarce. [13]. The basic feasibility of a dual-fuel ZS engine must be based on the ability to effectively burn both fuels in the engine combustion chamber. For this to be possible, spontaneous ignition of the mixture must occur, and for this to happen, proper preparation of the fuel–air mixture is required [14].

In addition, for both fuels, there are opportunities to adapt engines already in operation to run on LPG and CNG.

There are two possible ways of supplying compression-ignition engines with CNG or LPG. The first is to adapt the engine to run on gas only. In such an engine, the compression ratio must be reduced and an ignition system installed. A ZS engine adapted to run on mono-fuel natural gas has a higher compression ratio than a ZI engine and can no longer run on diesel. The higher compression ratio of such an engine makes it possible to take full advantage of the properties of natural gas and especially the high octane number, which is higher than for petrol. The second way is the dual-fuel operation of compression-ignition engines with gas fuels and a reduced dose of diesel or diesel alone. In the case of simultaneous fuelling of the engine with LPG/CNG and diesel, the gaseous fuel is fed into the engine intake system or directly into the cylinder. The LPG/CNG air/gas mixture produced is compressed in the engine cylinder and ignited by the injection of a reduced dose of diesel into the cylinder just before the end of the compression process. The supply of natural gas to the engine can be accomplished in various ways. This can be done by means of [10]:

- mixer;
- injector head;
- injector introducing fuel upstream of the intake valve;
- individual injector in the combustion chamber;
- dual-fuel (two-way) injector in the combustion chamber.

In the case of natural gas in vehicle fuelling systems, it can be stored in two forms: compressed CNG or liquid LNG. Natural gas in compressed form is stored at a maximum pressure of around 25 MPa. The disadvantage of this solution is a much lower energy density than diesel. The second way of storing natural gas is possible with special powder-vacuum-insulated cryogenic tanks [15]. Then the energy density is similar to that of diesel. In Poland, interest in natural gas for fuelling engines used in transport is limited by the very small number of stations where vehicles can be refuelled with this fuel. The lack of a network of natural gas filling stations effectively blocks the development of the use of natural gas to power cars. The course of the combustion process itself is very much influenced by the course of the preparation of the fuel mixture, which directly affects the duration of the auto-ignition delay period, which then determines the combustion process and significantly affects, for example, the degree of pressure build-up. Determining the auto-ignition delay period in compression-ignition engines is quite difficult, as it is influenced by various factors. The auto-ignition delay period is the time calculated from the start of fuel injection to the moment of the chain-heat explosion of the pre-ignition reactions manifested in the indicator diagram by the beginning of the rapid increase in pressure and temperature of the working medium caused by the start of fuel combustion α_{SOC} (point 3 in Fig. 1). The increase in these quantities caused by fuel combustion is observed on the indicator diagram as a break in the pressure waveform curve from the curve characterising the compression pressure waveform [7, 16].

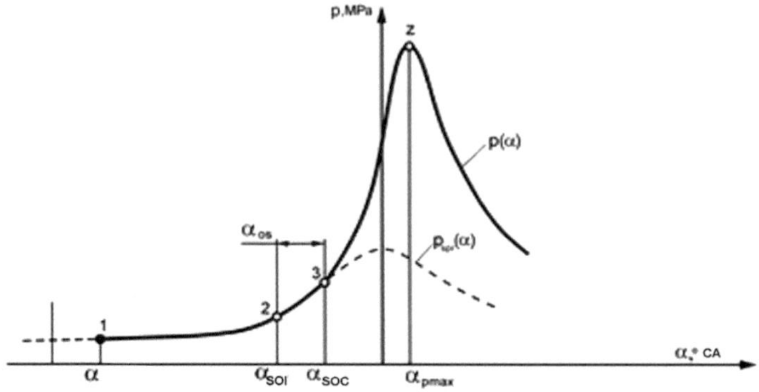

Fig. 1 Auto-ignition delay period on the developed indicator graph, where: α_{SOI}—start of fuel injection, α_{SOC}—start of combustion process, p(α)—combustion pressure run, p_{spr} (α)—compression pressure run, α_{os} (α)—auto-ignition delay period

The total auto-ignition delay time is made up of the physical component τ_f, corresponding to the time lost for the breakdown of the fuel jet into droplets, their partial evaporation and the mixing of fuel vapour with air, and the chemical component τ_{ch}, analogous to the auto-ignition delay of a homogeneous gaseous mixture. Given that the chemical and physical processes occur simultaneously, with a slight shift by phase, assessing the duration of these components is difficult. The two phases are related by the relationship:

$$\tau_s = \tau + \tau_{fch}, \tag{1}$$

where: τ_s—auto-ignition delay time, τ_f—physical component, τ_{ch}—chemical component of auto-ignition delay time.

The value of the auto-ignition delay period α_{os} is calculated from the relationship:

$$\alpha_{os} = \alpha_{soi} - \alpha_{\mathrm{soi}}, ^{\mathrm{o}}\,\mathrm{CA} \tag{2}$$

where α_{soi}—start of combustion; α_{soi}—start of fuel injection.

The auto-ignition delay period relates directly to the element of preparation of the fuel–air mixture that should undergo the auto-ignition phenomenon. During the first phase, the fuel injected through the injector is split into small droplets, which are heated and then vaporised in part or in whole, which also heats the fuel–air mixture to a temperature at which self-ignition through oxidation of the fuel is possible. It should be added that the development of the ignition phenomenon is favourably influenced by the structure of the mixture, or more precisely by its inhomogeneity, which, with a larger amount of fuel supplied to the cylinder, results in a more rapid increase in pressure in the second phase [17–19].

The duration of the auto-ignition delay period is also directly affected:

- the cetane number of the fuel, i.e. the self-ignition capacity of the fuel;

- type of combustion chamber;
- crankshaft speed and engine load;
- fuel injection pressure and temperature;
- fuel atomisation characteristics;

In addition, auto-ignition delay affects the rate of combustion and pressure build-up, temperature, as well as engine starting characteristics, exhaust toxicity and noise.

3 Research Methodology and Materials Used

The tests were carried out on a dynamometer bench at the Thermal Engines Laboratory of the Kielce University of Technology. The scope of the research comprised the determination of the load characteristics of the Perkins 1104D-E44TA engine when it was dual-fuelled with natural gas and diesel, LPG gas and diesel, and for comparison with diesel alone. The tests determined the economic and energy indices and selected fast variables of the tested engine equipped with a common-rail supply system allowing the injected fuel dose to be divided into 2 parts. On the basis of the measurement results, the start of fuel injection and the start of the combustion process were determined. These results allowed the auto-ignition delay period to be determined. During the tests, the engine was operated according to two load curves, at 1400 rpm, at which the engine achieves maximum torque, and at 2200 rpm, at which the engine achieves maximum power.

During the tests, measurements were taken of basic engine operating parameters such as effective power N_e, torque M_o, hourly consumption of diesel and LPG/CNG gaseous fuels, as well as quick-change quantities such as: the course of pressure in the cylinder and injector control current as a function of engine crankshaft rotation angle. In addition, values for the rate of pressure build-up in the cylinder during the combustion process (dp/dα).

The beginning of fuel injection and the beginning of in-cylinder combustion were determined using the injector control current and the curve representing the first derivative of in-cylinder pressure, drawn up as a function of crankshaft angle. By plotting the injector control current waveform on the indicator graph, it was possible to determine the start of fuel injection by reading the moment at which the current value first significantly increases. Determination of the start of combustion, on the other hand, was based on the curve drawn up on the basis of the first derivative of the pressure increase dp/dα. By analysing the course of the determined first derivative of the pressure rise, the point at which the combustion onset occurs is determined as the point at which the value of dp/dα starts to increase sharply. This point represents the angle of onset of combustion in the cylinder. The start of fuel injection is determined from the injector control current and the start of combustion is determined from the pressure rise rate during combustion. This method is illustrated in Fig. 2.

The study used diesel (DF) as the primary fuel and two gaseous fuels, natural gas CNG and LPG fuel. Table 1 shows the basic physico-chemical parameters of the

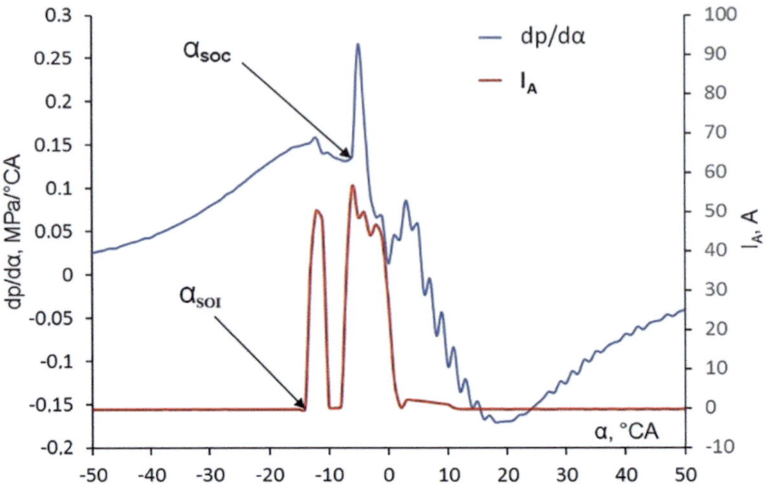

Fig. 2 Methodology of determining the beginning of fuel injection and the beginning of the combustion process based on the indicator diagram with the injector control current and the pressure increase curve dp/dα, where: α_{SOI}—beginning of fuel injection, α_{SOC}—beginning of the combustion process

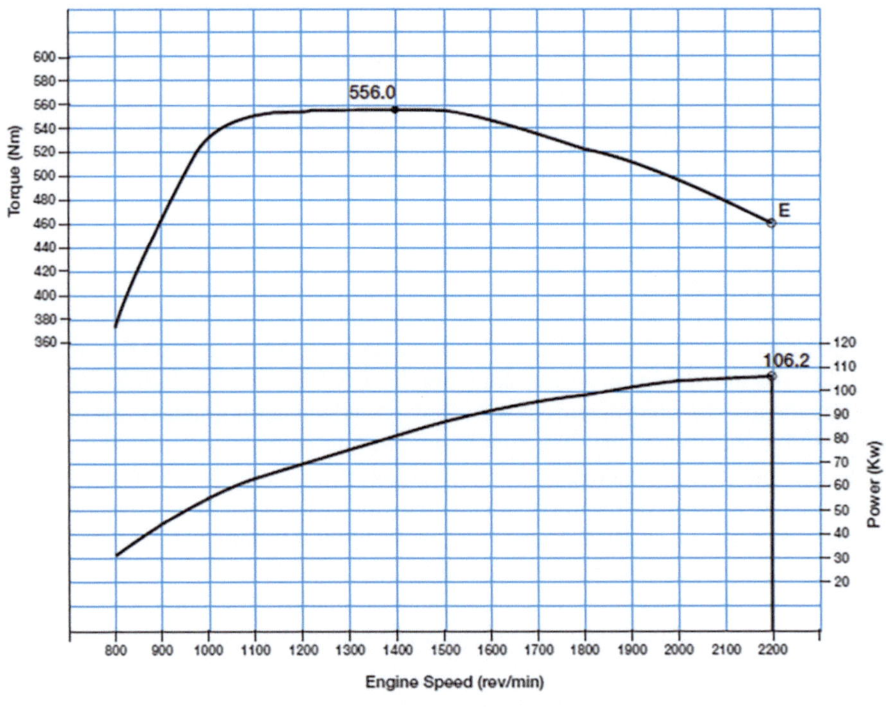

Fig. 3 External speed characteristics of the Perkins 1104D-E44TA engine [20]

Table 1 Selected physical and chemical properties of the fuels used [11]

Property	Fuel		
	DF	LPG	CNG
Density (kg/m)^3	800–860	2,360* *phasealotine	0.7518
Theoretical air demand (kg/kg)	14.5	15.6	16.2
Calorific value (MJ/kg)	42.0–44.0	46.02	45.970
Calorific value of the stoichiometric mixture (MJ/nm)^3	3.745–3.922	3.660	3.473
Heat of vaporisation (kJ/kg)	250–270	406	0
Flammability limits (0.1013 MPa, 0° C)			
Lower, d	0.975	–	1.97
Upper, g	0.18	–	0.59
Auto-ignition temperature (K)	603–623	763	910–923
Rate of flame propagation in a stoichiometric mixture (m/s)		0.305	0.338
Octane number LOM Cetane number LC	50–85	90–94	120–130

aforementioned fuels. The tests were carried out with the engine operating according to the load characteristic at 1400 rpm, at which the engine develops maximum torque.

4 Test Stand and Measuring Apparatus

The test object was a PERKINS 1104D-E44TA four-cylinder compression-ignition engine equipped with a common rail power system with direct fuel injection and electronically controlled solenoid injectors. The engine is equipped with a supercharging system with a turbocharger controlled by a bleeder valve connected to an electro-valve controlled by an electronic engine control unit. This unit, based on information from sensors describing the engine's operating conditions, including its load and crankshaft speed, determines the dose of fuel injected by controlling the pressure value in the fuel tank and the injection time and boost pressure. As a result, it complies with the Tier 3 emission standard for engines with off-highway applications [20]. The basic specifications of the tested engine are shown in Table 2 and its external speed characteristics are shown in Fig. 1. The engine was additionally equipped with a STAG LPG and CNG fuel supply system. The gas supply system enables the engine to be supplied with dual fuel, simultaneously with diesel fuel and ON + CNG, as well as diesel fuel and LPG (ON + LPG). Tests were carried out with the engine running according to the load curve at 1400 rpm, at which the engine achieves maximum torque.

Table 2 Basic technical data of the Perkins 1104D-E44TA compression-ignition engine [20]

Parameter	Unit	Value
Cylinder arrangement	–	Row
Number of cylinders	–	4
Type of injection	–	Direct
Supply system type	–	Common rail
Maximum engine power	kW	96.5
Maximum power speed	rpm	2200
Maximum engine torque	Nm	516.0
Maximum torque speed	rpm	1400
Engine capacity	m^3	$4.4–10^{-3}$
Cylinder diameter	mm	105
Piston stroke	mm	127
Compression ratio	–	16.2
Air supply system	–	Turbocharger, charge air cooler

The gaseous fuel control system is designed for use with compression-ignition engines: naturally aspirated and supercharged. It enables the conversion of a conventional engine supply system to a dual-fuel system with gaseous fuel and diesel, where the task of the diesel dose is to initiate the combustion process. The tests were carried out on an engine dynamometer on which the test engine was installed. In addition to the engine, the test bench consisted of an electro-rotor brake of the AMX—200/6000 type manufactured by ELEKTROMEX CENTRUM with a power of 200 kW [10]. The test bench was equipped with a desktop computer with Automex software for control and monitoring of the test record. Diesel consumption was measured using a mass fuel dosimeter made by Automex. Gas consumption was measured using an Emerson Coriolis mass flow meter. The test bench is also equipped with sensors to measure temperatures in the engine's primary functional systems. The bench has five shell heat exchangers, which are cooled by water in a closed circuit. In addition, the test bench uses five WIKA pressure transducers to measure pressures in the engine's primary functional systems. For the measurement of fast-variable quantities, a measuring system from AVL IndySmart 612 was used to determine the pressure variation waveforms of the working medium in the cylinder. The indexing kit consists of the following components:

- AVL IndiSmart 612 master data acquisition system;
- AVL GH13P/AG04 piezo cylinder pressure sensor;
- Optical crankshaft angle encoder for AVL 365C engine.

The system was operated using the AVL IndiCom Mobile 2012 software, which, among other things, allows indicator graphs to be drawn. The bench was additionally equipped with a system for measuring the injector control current. This system was connected in series to the injector control line of the cylinder from which the cylinder

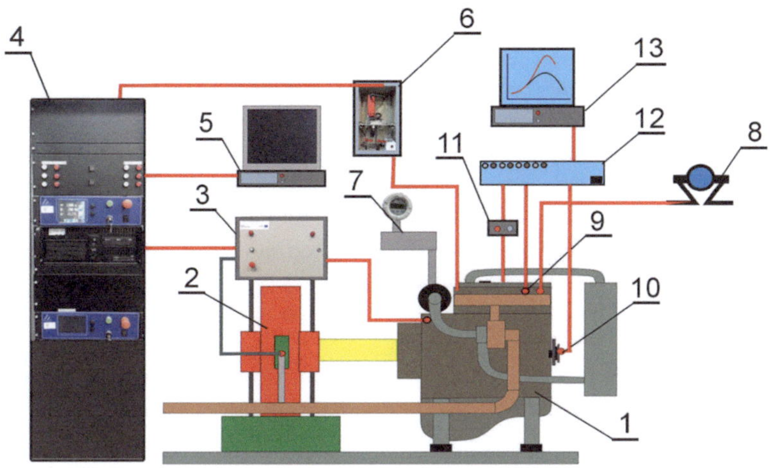

Fig. 4 Diagram of the test bed, where: 1—Perkins 1104D-E44TA engine, 2—Automex AMX 200/6000 brake, 3—measurement module, 4—measurement cabinet with the station control system, 5—computer controlling the test bed parameters and archiving test results, 6—AutomexATMX2040 mass fuel dose meter, 7—ABB mass air flow meter, 8—CNG Coriolis flow meter, 9—AVL engine cylinder pressure sensor, 10—AVL 365C engine crankshaft rotation angle encoder, 11—LA25NP injector operation control current measurement transducer, 12—AVL IndiSmart 612 fast-changing variable indication system, 13—fast-changing variable archiving system

pressure was recorded. In addition to the original engine control system, the test bench is equipped with a NIRA compression-ignition engine controller. This controller provides access to the configuration of engine control parameters selected during ECU tuning to the engine, relating to: power system control, injection times and phases, fuel injection pressure, control of the supercharging system, exhaust gas recirculation and other engine actuators. Parameter changes are possible from a PC located in the test bench control room. A block diagram of the dynamometer bench with the PERKINS 1104D-E44TA engine is shown in Fig. 4.

5 Research Findings and Their Analysis

During testing of the PERKINS 1104D-E44TA engine, it was operated according to load characteristics at a maximum torque speed of 1,400 rpm. Torque was varied from 20 to 550 Nm in 50 Nm increments. During the tests, basic operating indicators such as effective power, torque and hourly diesel and gas fuel consumption were recorded. In addition, cylinder pressure waveforms and injector control current waveforms as a function of engine crankshaft angle were recorded at each operating point. During the tests, 200 duty cycles were recorded and then averaged. Open indicator diagrams were drawn up from the averaged cylinder pressure waveforms and analysed. The maximum values of the cylinder pressure p_{max}, and the angles of their occurrence were determined. Cylinder pressure increment waveforms dp/dα were determined, and the onset of combustion was determined from these.

The gaseous fuel supply system and its control system in dual-fuel operation, reduces the diesel dose by lowering the pressure in the common rail fuel rail. Instead of diesel fuel, gaseous fuel is supplied to the cylinder. The amount of gas supplied to the cylinders is limited due to the possibility of detonation combustion and an excessive rise in the temperature of the working medium in the cylinder.

Figures 5, 6 and 7 show example indicator diagrams of the cylinder pressure of the test engine when it is fuelled with diesel ON (Fig. 5) and dual-fuel DF + LPG (Fig. 6) and DF + CNG (Fig. 7) and its operation according to the load curve at 1400 rpm.

Based on the indicator graphs, the maximum combustion pressures were determined, the values of which are shown in Fig. 8. Figures 9, 10 and 11 show example graphs comparing the degree of pressure build-up during the combustion process dp/dα for the fuels tested. From these graphs, their maximum values were determined and are shown in Fig. 12.

As Fig. 8 shows, the engine load has the greatest influence on the maximum combustion pressure values. For the three loads of 10, 50 and 150Nm, the highest combustion pressures p_{max} were obtained when the engine was fuelled with DF diesel. For the highest loads of $T = 450$ and 500 Nm, the highest combustion pressures occurred when the engine was dual-fuelled with DF + LPG. For medium load values in the range 200 to 400 Nm, the maximum combustion pressures occurred when running the engine on dual-fuel DF + CNG.

Analysing the maximum pressure increments during the combustion process dp/$d\alpha_{max}$, Fig. 12 clearly shows that at loads from 10 to 150 Nm feeding the engine with diesel and dual-fuel DF + LPG significantly higher values occurred compared

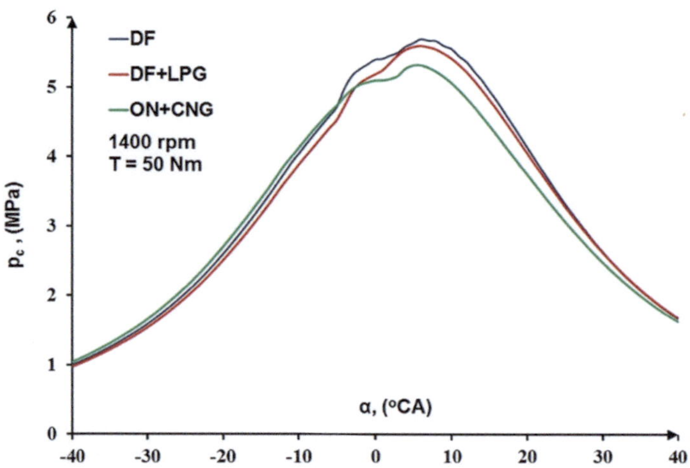

Fig. 5 Pressure waveforms p in the cylinder of a Perkins 1104D-E44TA engine, operating according to a load characteristic of 1400 rpm at $T = 50$ Nm and fed with the fuels tested

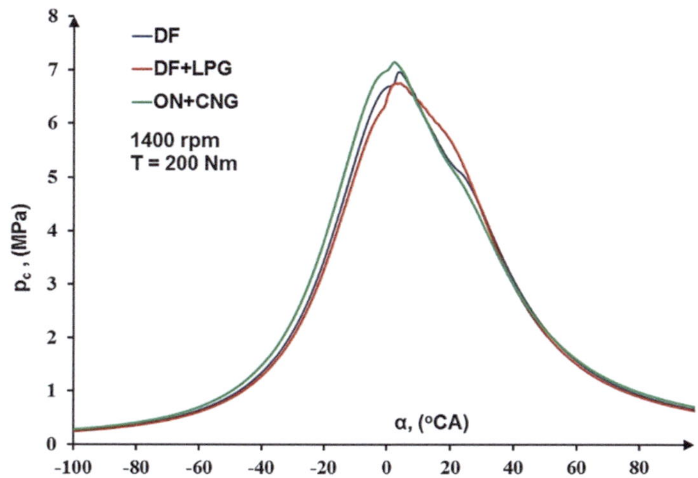

Fig. 6 Pressure waveforms p in the cylinder of a Perkins 1104D-E44TA engine, operating according to a load characteristic of 1400 rpm at $T = 200$ Nm and fed with the fuels tested

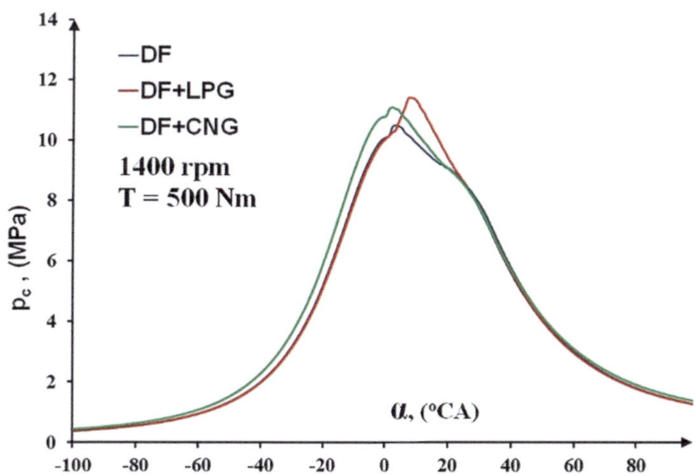

Fig. 7 Pressure waveforms p in the cylinder of a Perkins 1104D-E44TA engine, operating according to a load characteristic of 1400 rpm at $T = 500$ Nm and fed with the fuels tested

to feeding the engine with DF + CNG. For other loads, the trend is reversed where higher values of maximum pressure rise occurred when feeding the engine with DF + CNG compared to those when feeding the engine with the other fuels. Figure 13 shows example diagrams of the pressure rise during the combustion process dp/dα and the injector control current, which allowed the auto-ignition delay period to be determined, the values of which are shown in Fig. 14. Analysing the calculated auto-ignition delay values, it can be clearly seen that feeding the engine with dual-fuel

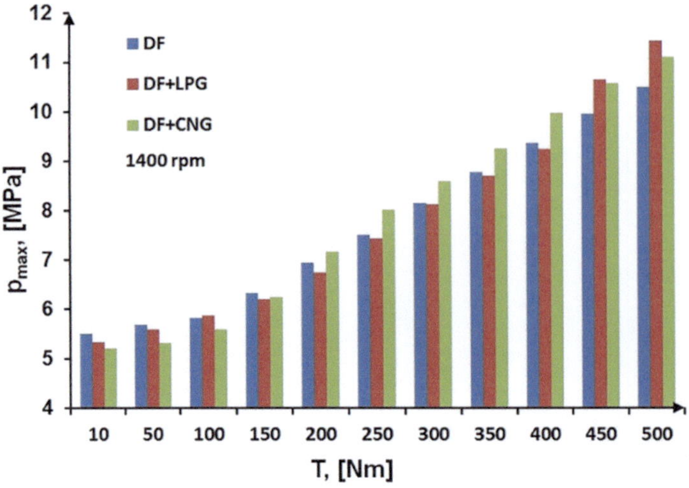

Fig. 8 Values of the maximum pressure p in the cylinder of a Perkins 1104D-E44TA engine, operating at a load characteristic of 1400 rpm and fed with the fuels tested

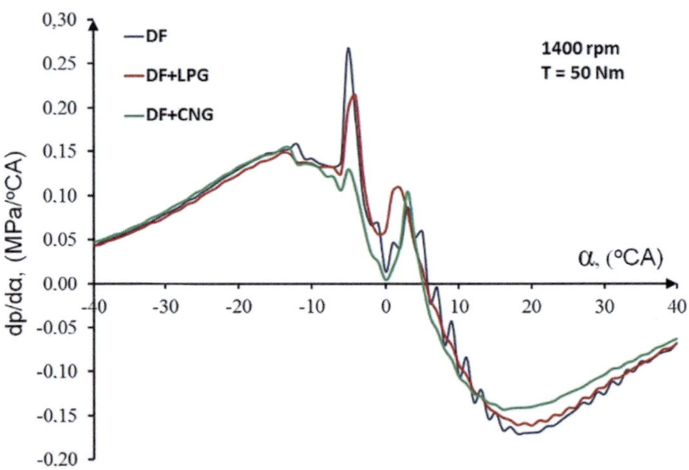

Fig. 9 Pressure rise rate curves during the combustion process dp/dα in the cylinder of a Perkins 1104D-E44TA engine, operating according to a load characteristic of 1400 rpm at $T = 50$ Nm and fed with the fuels tested

DF + CNG at low loads of 10 to 150 Nm resulted in a significant increase in auto-ignition delay compared to DF and DF + LPG, particularly for $T = 10$, 100 and 150 Nm. At most measurement points, the smallest auto-ignition delay values occurred when feeding the engine only with DF diesel, only for loads from $T = 400$ to 500 Nm feeding the engine with DF were the auto-ignition delay values highest.

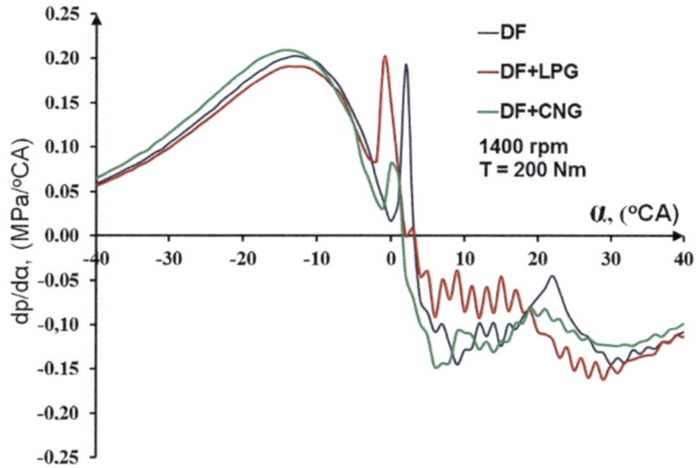

Fig. 10 Pressure rise rate curves during the combustion process dp/dα in the cylinder of a Perkins 1104D-E44TA engine, operating according to a load characteristic of 1400 rpm at $T = 200$ Nm and fed with the fuels tested

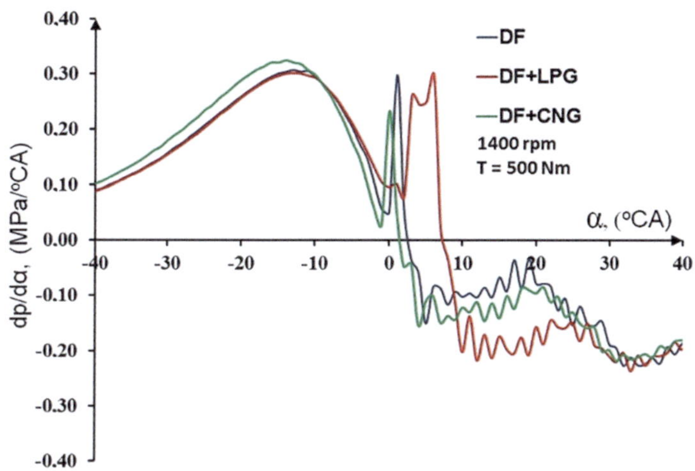

Fig. 11 Pressure rise rate curves during the combustion process dp/dα in the cylinder of a Perkins 1104D-E44TA engine, operating according to a load characteristic of 1400 rpm at $T = 500$ Nm and fed with the fuels tested

When analysing the calculated auto-ignition delay values, it is clear that feeding the engine with dual fuel DF + CNG at low loads of 10 to 150 Nm resulted in a significant increase in auto-ignition delay compared to DF and DF + LPG, particularly for $T = 10$, 100 and 150 Nm. At most measurement points, the smallest auto-ignition delay values occurred when feeding the engine only with DF diesel, only for loads

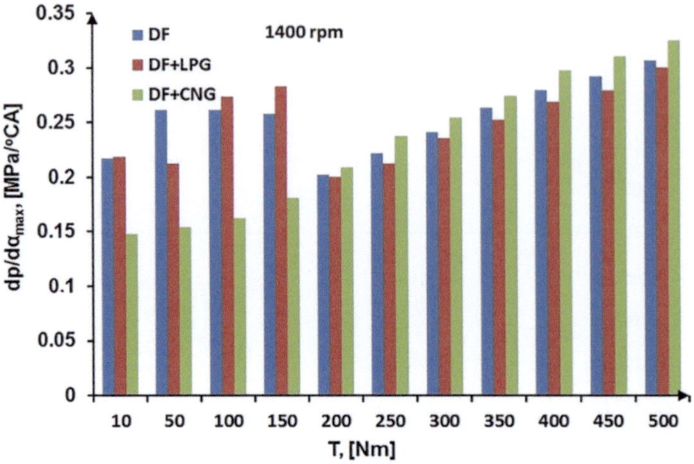

Fig. 12 Values of the maximum pressure rise during the combustion process in the cylinder of a Perkins 1104D-E44TA engine, operating according to a load characteristic of 1400 rpm and fed with the fuels tested

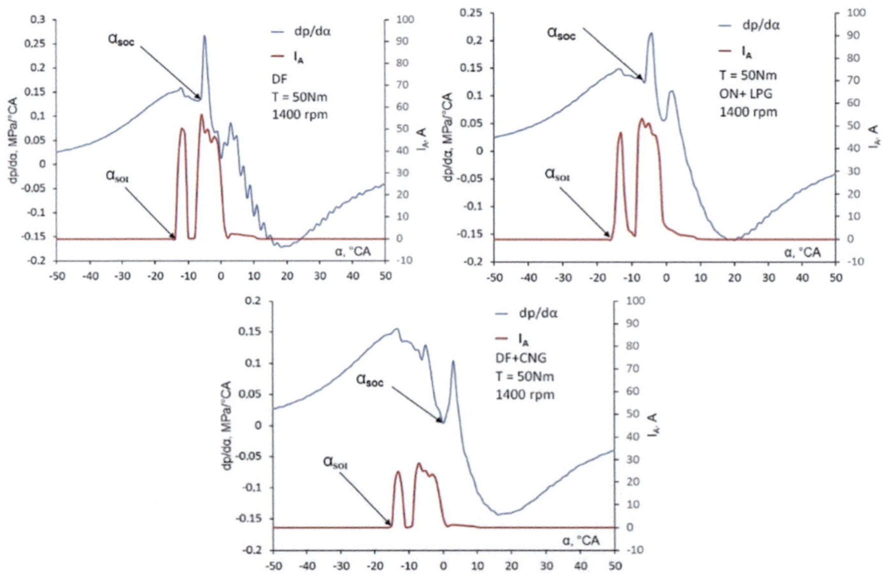

Fig. 13 Plots of the degree of pressure build-up during the combustion process dp/dα and the injector control current of a Perkins 1104D-E44TA engine, operating according to a load characteristic of 1400 rpm at $T = 50$ Nm and fed with the fuels tested

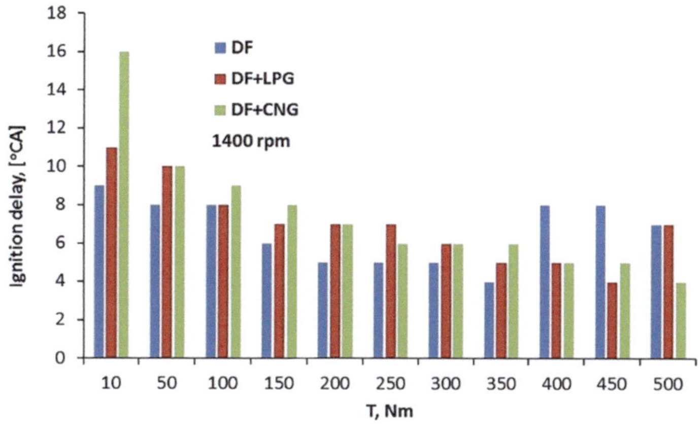

Fig. 14 Values of the auto-ignition delay period in a Perkins 1104D-E44TA engine operating at a load characteristic of 1400 rpm and fed with the fuels tested

from $T = 400$–500 Nm feeding the engine with DF were the auto-ignition delay values highest.

6 Summary

The recently deteriorating state of the environment and the ever-increasing demand for energy in the broadest sense of the term are forcing a growing interest in the combustion of gaseous fuels in engines. This is particularly true for compression-ignition engines, which appear to be irreplaceable for large heavy-duty vehicles used in transport. This paper investigates the effects of dual-fuel feeding of a Perkins 1104D-E44TA compression-ignition engine with diesel and LPG and diesel and natural gas on the auto-ignition delay period and combustion parameters. During the tests, the engine was operated according to the load characteristics for a crankshaft speed of 1400 rpm, which is the maximum torque speed. The test results showed that the maximum combustion pressure values are most influenced by the engine load. At low and medium loads, feeding the engine with diesel DF resulted in the highest combustion pressures p_{max}. For the highest loads, the highest combustion pressures occurred when the engine was dual-fuelled with diesel and LPG. For medium loads in the 200 to 400 Nm range, the maximum combustion pressures occurred when the engine was dual-fuelled with DF + CNG.

For the maximum combustion pressure increments $dp/d\alpha_{max}$ at low and medium loads, feeding the engine with diesel and dual-fuel DF + LPG resulted in more rapid combustion compared to feeding the engine with diesel and natural gas. For the other loads, the trend is the opposite.

Feeding the engine with dual-fuel DF + CNG at low loads resulted in a significant increase in auto-ignition delay compared to the other fuels. At most measurement points, the lowest auto-ignition delay values occurred when the engine was fed with DF diesel alone, except at maximum loads, where the highest auto-ignition delay values were obtained when the engine was fed with diesel alone.

References

1. Reif, K., Dietsche, K.-H.: Kraftfahrtechnisches Taschenbusch. Vieweg+Teubner Verlag, Wiesbaden (2011)
2. Wedler, M.: Alternative Kraftstoffe und Antriebssysteme fuer PKW-Fahrzeuge. GRIN Verlag, Norderstedt (2008)
3. Gruden, D.: Umweltschutz in der Automobilindustrie. Vieweg+Teubner-GWV Fachverlage GmbH, Wiesbaden (2008)
4. Fabiś, P., Flekiewicz, B.: Influence of LPG and DME composition on spark ignition engine performance. Energies **14**, 5583 (2021). https://doi.org/10.3390/en14175583
5. Kurczyński, D., Łagowski, P., Warianek, M.: External velocity characteristics of a Perkins 1104d-E44TA engine running on dual-fuel diesel and natural gas. J. Buses: Tech. Oper. Transp. Syst. **18** (2017)
6. Ambrozik, A., Kurczyński, D., Łagowski, P.: Energy indices of the Perkins 1104D-E44TA engine bifuelled with diesel and natural gas. J. Buses: Tech. Oper. Transp. Syst. **17**(12), 760–767 (2016)
7. Górski, K., Wasilewski, T., Czerwonka, D., Bocho-Janiszewska, A., Longwic, R.: Influence of water-fuel microemulsion on auto-ignition retardation in a compression-ignition engine. Exploit. Test. Buses **6**, 714–717 (2017)
8. Merkisz, J., Pielecha, I.: Alternative Fuels and Vehicle Propulsion Systems. Wydawnictwo Politechniki Poznańskiej, Poznań
9. Monaghan, M.L.: Future Gasoline and Diesel Engines. World Automotive Congress FISITA, F2002PA03, Seoul (2000)
10. Ambrozik, A., Ambrozik, T., Kurczyński, D., Łagowski, P.: Adaptation of the Perkins 1104D-E44TA engine with a common-rail supply system to CNG natural gas supply. Zeszyty Naukowe Instytutu Pojazdów **1**(97), 61–70 (2014)
11. Stelmasiak, Z.: Selected problems of using natural gas to power compression-ignition engines. Archiwum motoryzacji **1**, 13–30 (2006)
12. Stelmasiak, Z.: Experimental study of autoignition and combuestion gas-air mixtures in dual fuel engine. J. KONES Internal Combust. Engines **9**(3–4), 273–283 (2002)
13. Stelmasiak, Z.: Analysis of combustion phenomena in dual fuel engine fed with natural gas (CNG). Fisita 2002 World Automotive Congress, Paper No. F02 V030 (2002)
14. Kowalewicz, A.: Compressed natural gas engines. a review. J. KONES Powertrain Transp. **16**(3) (2009)
15. Cho, H.M., He, B.Q.: Combustion and emission characteristics of a lean burn natural gas engine. Int. J. Automot. Technol. **9**(4), 415–422 (2008)
16. Zhang, Y., Liu, X., Yan, X., Cheng, C.: The studies of an electronically controlled CNG system for dual fuel engines. SAE Paper No. 2001-01-0145 (2001)
17. Kusaka, J., Okamoto, T., Daisho, Y., Kihara, R., Saito, T.: Combustion and exhaust gas emission characteristics of a diesel engine dual-fuelled with natural gas. JSAE Rev. **21**(2000), 489–496 (2000)
18. Pirouzpanach, V., Kashani, B.O.: Prediction of major pollutants emission in direct-injection dual-fuel diesel and natural gas engines. SAE Paper 1999-01-0841 (1999)

19. Mahla, S.K., Dhir, A.: Performance and emission characteristics of CNG-fueled compression ignition engine with *Ricinus communis* methyl ester as a pilot fuel. Environ. Sci. Pollut. Res. **26**, 975–985 (2019)
20. Perkins, Operation and Maintenance Manual. Perkins Engines Company Limited (2012)

Complex Technologies for Reinforcing Stressed Machine Components

Mykola A. Tkachuk, Serhii O. Kravchenko, Grabovskiy Andriy, Mykola M. Tkachuk, Vasiliev Anton, Novikov Maksym, and Subbotina Valeriia

Abstract This paper presents a novel approach to strengthening critical machine components using integrated design and technological solutions. The proposed method combines discrete electric spark alloying and micro-arc oxidation, creating a synergy that enhances durability, load capacity, and energy efficiency. Applications include pistons, crankshafts, and rotors, vital for improving power plants in transportation and other industries. The study details mathematical models for stress–strain analysis of composite structures, demonstrating significant gains in environmental performance, cost-efficiency, and reduced thermal pollution. These innovations align with Industry 4.0 and 5.0 principles, supporting sustainable machine design and operation.

M. A. Tkachuk (✉) · G. Andriy · M. M. Tkachuk · V. Anton
Department of "Theory and Computer-Aided Design of Mechanisms and Machines", National Technical University "Kharkiv Polytechnic Institute", Kharkiv, Ukraine
e-mail: tma@tmm-sapr.org

V. Anton
e-mail: AVasiliev@tmm-sapr.org

S. O. Kravchenko
Department of Engines and Hybrid Powerplants, Educational and Scientific Institute of Power Engineering, Electronics and Electromechanics, National Technical University "Kharkiv Polytechnic Institute", Kharkiv, Ukraine

N. Maksym
Department of "Turbine Engineering", Educational and Scientific Institute of Power Engineering, Electronics and Electromechanics, National Technical University "Kharkiv Polytechnic Institute", Kharkiv, Ukraine
e-mail: novikovmaxim1980@ukr.net

S. Valeriia
Department of "Materials Science", Educational and Scientific Institute of Mechanical Engineering and Transport, National Technical University "Kharkiv Polytechnic Institute", Kharkiv, Ukraine

© The Author(s), under exclusive license to Springer Nature Switzerland AG 2025
A. Marchenko and Z. Koruba (eds.), *Decarbonization of Transport Energy Installations in the Context of Sustainable Development Strategies*, Studies in Systems, Decision and Control 635, https://doi.org/10.1007/978-3-032-05884-3_10

1 Introduction

Civilizational development is characterized by the fact that at different stages, relevant strategies of social activity, and primarily scientific and technological activity, are developed at different times. The current stage, which corresponds to the sustainable development strategies, Industry 4.0 and Industry 5.0, is characterized by a sharp intensification and mutual integration of processes in various fields. In particular, the real sector of the economy is characterized by an upward trend in the requirements for the defining characteristics of machines for various purposes. First of all, this concerns the technical and performance characteristics of energy sector equipment, military and transportation vehicles, etc. Accordingly, the traditional approaches that have been used (Fig. 1) do not currently meet the current requirements.

It involves a gradual linear "conveyor belt". At each stage, a factor Φ_i is considered. By creating mathematical and numerical models of a process or state, it becomes possible to study these processes and states, as well as to develop appropriate recommendations according to the criterion p_i: $T(p) \rightarrow$ max. This set then becomes the starting point for the next (i + 1) stage.

An important aspect of the traditional approach (Fig. 1) is that the sequence of "optimal" solutions for individual criteria, factors, processes, and states does not always lead to an optimal technical solution in general.

Thus, there is a need to develop a new, integrated approach (Fig. 2).

It is assumed here that the search for optimal solutions is carried out on the entire set of parameters p, factors Φ and the entire set of processes and states.

Such an approach (Fig. 2) was implemented in relation to the development of new methods for strengthening machine elements and is presented in this publication.

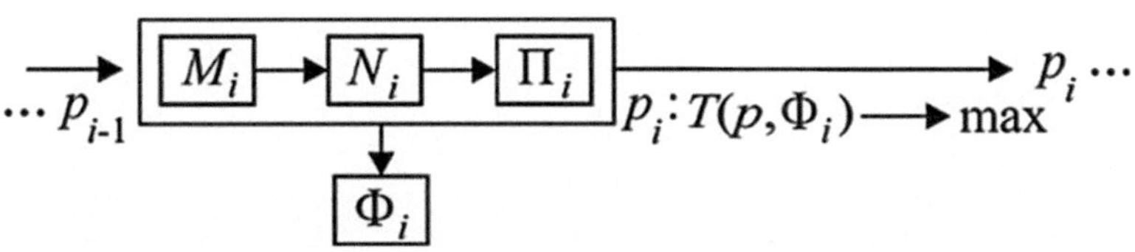

Fig. 10.1 Traditional approach to finding rational design and technological solutions for machine elements: M_i, N_i—mathematical and numerical models of processes ($\Pi_{i)}$ and factors (Φ_i)

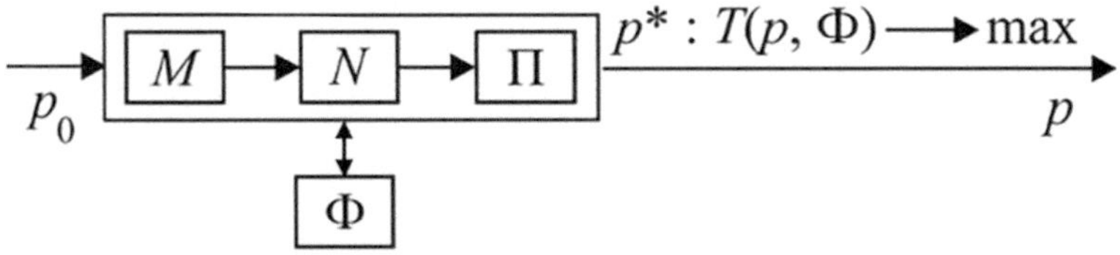

Fig. 2 An improved approach to finding rational design and technological solutions for machine elements: M, N—mathematical and numerical models of processes Π and factors Φ

2 Analyzing Existing Methods of Strengthening and Studying the Strength and Durability of Machine Elements

The problem of creating highly efficient means of economic development, including means of transportation, using cost-effective, energy-efficient and environmentally friendly methods of production, manufacturing and operation is certainly very important at this stage. Sustainable development strategies and technologies, such as Industry 4.0 and Industry 5.0, are becoming increasingly used, enabling the creation of innovative products with higher quality and less time.

It is well known that modern machines are characterized by a variety of factors. These include contact stresses, friction, wear, etc. Thus, the problem of increasing the service life of vehicles is becoming more acute. This resource is mainly regulated by the relevant requirements for increasing the reliability and durability of internal combustion engines (ICEs), as well as other structural elements that determine their strength and durability. At the same time, the main attention should be paid to new advanced technologies for strengthening the surface layers of contacting parts.

Thus, the problem of increasing the strength and wear resistance of machine parts in general is relevant. The main focus is on eliminating or slowing down destructive processes. Such processes naturally occur on the surfaces of parts in contact under load. In turn, the tribomechanical characteristics (strength, friction, wear, etc.) are controlled and optimized. To do this, it is necessary to analyze existing technologies and search for new ones in order to dramatically improve these tribomechanical characteristics.

The following is an analysis of existing technologies for strengthening the surface layers of machine parts in moving contact, their advantages and disadvantages. First of all, the most loaded and critical parts of internal combustion engines (crankshafts, sliding bearing liners, pistons, compression and oil seals, sleeves, etc.) are taken as illustrative objects. These parts mainly limit the service life, performance and efficiency of an internal combustion engine. At the same time, they are vivid examples of highly loaded machine parts that operate under conditions of variable thermomechanical loads on friction surfaces. That is why all the identified problems, circumstances, phenomena, and technologies for their solution can be attributed to the entire class of such parts and technologies for their strengthening in the field of mechanics and mechanical engineering.

Failures of machine parts under normal operating conditions occur as a result of various types of loading, weakening, friction and wear, etc. [1]. Thus, as noted in [1], "the nature of the distribution of failures of new and refurbished internal combustion engines with increasing operating time shows that their greatest intensity is observed" [1] in the initial period of operation. Indeed, an increase in failures was noted at the initial stages of operation (Fig. 3).

In particular, the crankshaft is one of the most complex and stressed parts of an internal combustion engine operating in a wide range of thermomechanical loads

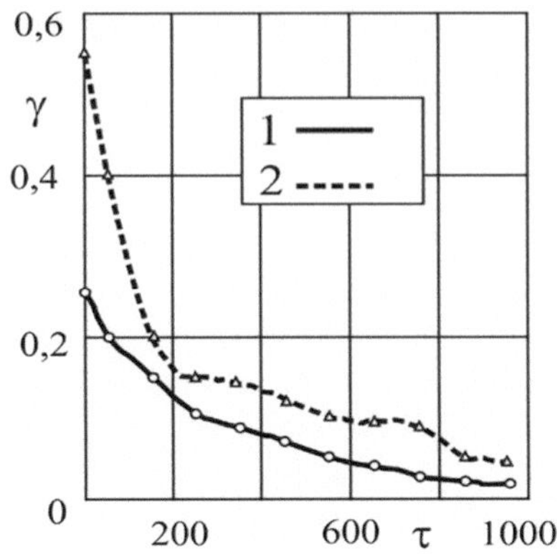

Fig. 3 Distribution of ICE failures γ in the initial period of operation by time (t), hours: 1—new Internal Combustion Engine (ICE); 2—repaired ICE

and rotational speeds. Changes in the technical condition of this ICE element are determined mainly by the characteristics of the "crankshaft neck—sliding bearing liner" [1]. In this case, the working load leads to contact stresses between the specified elements of this tribosystem. Accordingly, friction between them leads to wear of the surface layers. And the gaps thus formed, in turn, lead to an increase in operating loads.

When two contacting surfaces come into contact and move relatively (slip) under load, they wear out (Figs. 4, 5, 6).

Among the surface hardening technologies, there are several main groups, including:—formation of coatings;—change in the chemical composition of the surface layer;—change in the structure or energy state of the surface layer material, etc. [1]. At the same time, about 200 surface engineering technologies are known today.

A significant part of them is the application of various kinds of continuum coatings [4–10].

An analysis of the above-mentioned technologies for the continual strengthening of surface layers makes it possible to identify certain disadvantages they have in common. In particular, the application of coatings in many cases creates a gradient in the physical and mechanical properties of the material in thickness [1]. When loaded, parts strengthened in this way are potentially subject to cracking, peeling, weakening, etc. Thus, there are "birth defects" that are difficult to eliminate.

Due to the stated problems, attention is drawn to those continuum technologies that consist, for example, in growing oxide ceramic structures on the surface of metal parts [11–15]. At the same time, these technologies have to some extent exhausted the possibilities of further increasing the strength and durability of the contacting parts of machines.

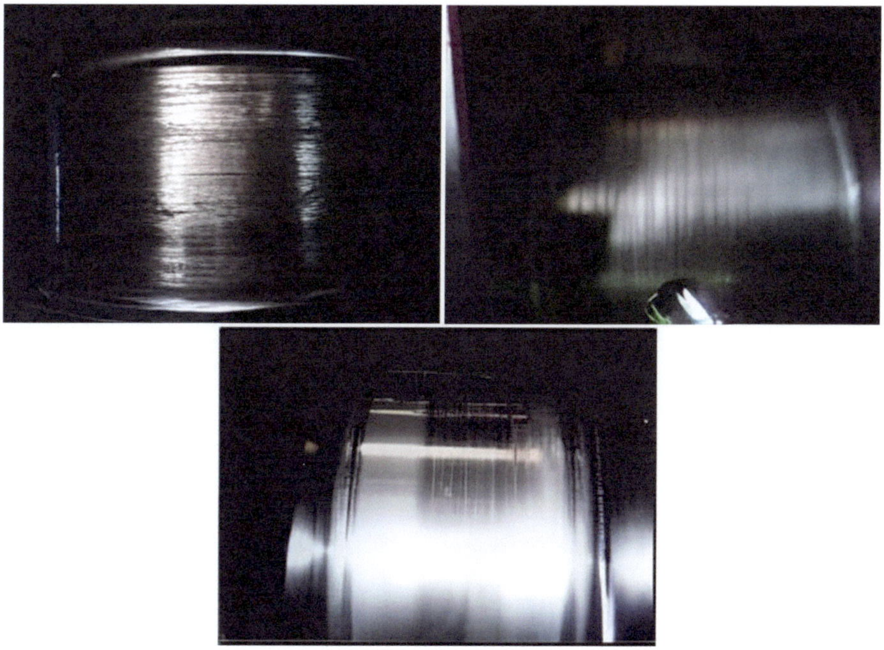

Fig. 4 Abrasive wear of the crankshaft neck of a diesel engine [1]

A certain alternative to the above methods of continuous surface hardening is the processes of discrete surface treatment, which make it possible to create alternating high-strength … microsites in a given sequence (discretely), similar to natural composite materials reinforced with solid and strong phases. It is known that such a structure provides high structural strength of the material [1–3].

At the same time, it is also necessary to reduce the wear of the contacting parts. In this direction, wear studies have laid the foundation for the development of new technologies for applying discrete wear-resistant coatings to parts by electrospark alloying [1].

Based on the above, the development of scientific and technological foundations for increasing the service life of machine elements, in particular, critical parts of engine power units, through the use of discrete hardening and related technologies, is a promising direction for improving their tribotechnical characteristics. However, the capabilities of these technologies are also reaching a certain "ceiling" of their development.

In general, regarding the existing technologies: despite their development, they do not currently provide solutions to the current problems that arise when ensuring the strength and durability of machine elements. This prompts the development of new methods and approaches to their creation.

Fig. 5 Fatigue cracks on the crankshaft neck of a diesel locomotive engine under limited lubrication [1]

3 Ways to Improve the Technical and Tactical Characteristics of Modern Equipment

The problems noted during the review of the state of the art (see above) can be solved on the basis of fundamentally new methods of strengthening, since the resource of constructive changes, as a rule, has long been exhausted. In particular, a number of studies, including those of the authors of this paper, have shown that one of the keys to solving this problem is the creation of so-called discrete, i.e. non-continuous, coatings [1, 2]. It is noted that a so-called archipelago of "load-bearing islands" is being created. These "islands" have the necessary physical and mechanical properties" (strength, wear resistance, contact stiffness, etc.). They can be formed, for example, by electrospark alloying [1–3].

In addition, the authors' attention is also focused on the method of corrosion modification of aluminum alloy parts surfaces—AMDO [11–15], which provides surface modification with the realization of a natural transition from the material core to the ceramic oxide coating created on the surface. This is due to the fact that its electrochemical transformation creates a strong layer with high adhesion and wear resistance [1–3].

To develop these technologies, the authors of [1–3] developed a fundamentally new method that involves connecting an aluminum part hardened by corrosion, on

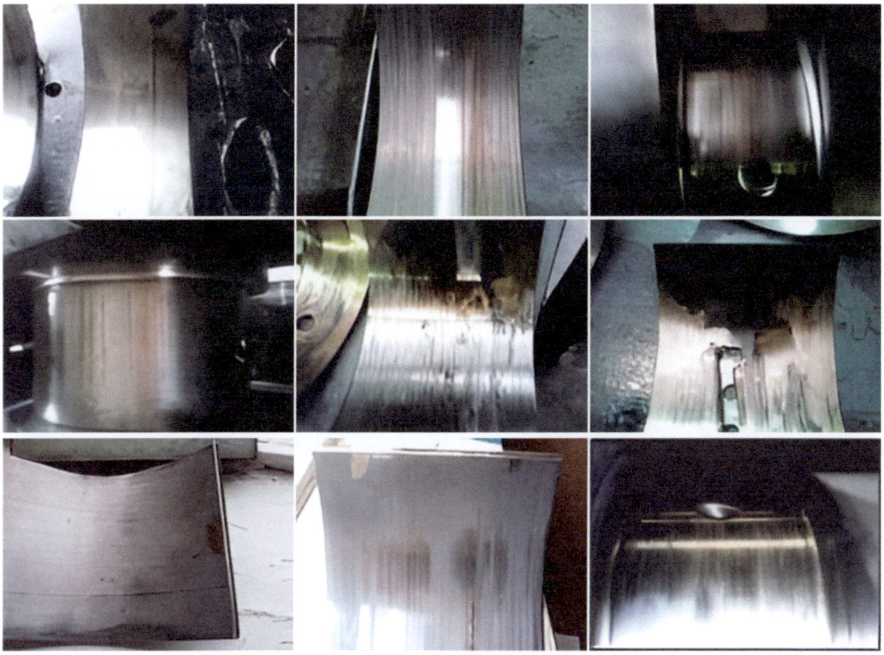

Fig. 6 Fatigue cracks on the crankshaft neck of a diesel locomotive engine under conditions of intensive lubrication and other defects [1]

the one hand, and a steel or cast iron part hardened discretely, on the other, in a friction pair. The result is a unique combination of properties distributed over the contact surface within one part and over the depth within the other, as well as between the contacting parts. "As a result of this combination, a rather productive result is achieved for the entire range of tribotechnical characteristics of the studied pairs of parts" [1–3].

In this paper, as a development of [1–3], the focus is on substantiating the modes of application of the developed technologies, in particular, with respect to internal combustion engines as illustrative objects. This is preceded by a description of the methodological foundations of "synthesis of design and technological solutions based on generalized parametric modeling" [1]. The presentation of the material ends with examples of solving applied problems for specific engineering structures (Fig. 7 [1]).

Fig. 7 Improving the technical characteristics of engines and units of various equipment through design and technological modernization: installations for autonomous power supply of facilities, 2TE10M mainline diesel locomotive with a modernized 1D80B engine, discrete hardening operating moments, corroded pistons (corroded piston after operation and for comparison—serial piston)

4 A Set-Theoretic Approach to Generating Models of Physical and Mechanical Processes and States of Complex and Supercomplex Mechanical Systems

An important circumstance in the creation of new technologies for strengthening machine parts is the fact that this creates the problem of synthesizing technological solutions in an ill-defined parameter space. If we consider the set of characteristics of a particular technical solution T as a function of the set of parameters $P = \cup p_i$, we can write:

$$T = T(P) \tag{1}$$

If the set P is fuzzy, then the inverse of (1) is the synthesis problem, which has the form

$$P = P(T), \tag{2}$$

also has a fuzzy dependency.

In addition, it is important that the parameters P themselves are not just numbers, but also structures, shapes, distributions, etc. That is, they are generalized parameters [1–3]. Moreover, the very composition of the set of parameters is uncertain and subject to search.

In addition, when performing analytical and numerical calculations, the parameters of the mathematical "M" and numerical "N" models begin to play an important role.

On the other hand, the array of variable parameters consists of "design" (geometric) P_G and "technological" P_{Tech} subsets:

$$P = P_G \cup P_{Tech}. \tag{10.3}$$

In this case, in the research cycle, the geometric model "G" is formed depending on all parameters P. The same applies to the "technological" models "Tech", which include various descriptive aspects of the relevant technological operations. Thus, we have:

$$G = G(P), Tech = Tech(P) \tag{4}$$

Relationship (4) reflects the fact that in the presence of a specific set of generalized reflective parameters P, models of certain objects under study are formed in the required format according to certain algorithms. The format of these models is determined in modern conditions by the format of CAD/CAM/CAE systems used in a particular case.

It is clear that relations (4) are simply declarative. First, they cannot be established for any arbitrary research object. However, for a certain defined class of objects, this possibility can be realized. Second, these dependencies are not always reversible.

In other words, in many cases, the description (4) is difficult to formalize. That is, to formulate this description, it is necessary to involve expert efforts in a certain field to which the object under study belongs.

Thus, the developed approach applies not to any objects, but to their individual classes. At the same time, for the implementation of automated research, it is important to acquire functions (4) only for certain classes of structures or technologies under study. This is enough to organize the entire cycle of their design and technological research. Moreover, an important property of this approach is that it can be used to solve both analysis and synthesis problems. Since the researcher has a transparent view of dependencies (4), the organization of synthesis procedures can be built on top of analysis procedures. This is the main advantage of the developed approach. This also provides the possibility of expanding and narrowing the parametric space within which the synthesis of advanced technical solutions of structures and technologies will be carried out.

In particular, highly loaded friction pairs, as demonstrated in previous studies [1–3] by the authors of this paper, can be treated using various hardening technologies. At the same time, it is of interest, both from the point of view of science and practice, to develop the previously proposed hardening methods. They can be improved both individually and by combining them. In this case, we should expect an effect that is more tangible than the total impact of each method. Below we describe one of the methods of their combined hardening, namely, a combination of corrosion hardening of the surface of one part (aluminum) with simultaneous discrete hardening of the surface of the mating part (steel, cast iron). This method can be considered as an illustration of general trends in the application of this type of technology for friction-contacting parts of a wide class of engineering structures.

This paper substantiates the advantages of the new technology of discrete–continuous strengthening of engineering structure parts. This technology, as already mentioned, is a combination of two technologies. In particular, an archipelago of "islands" of stronger and more wear-resistant material is applied to one of the contacting parts (made of steel or cast iron) by the method of electric spark alloying (Figs. 8, 9, 10, 11, 12, 13, 14, 15) [1–3].

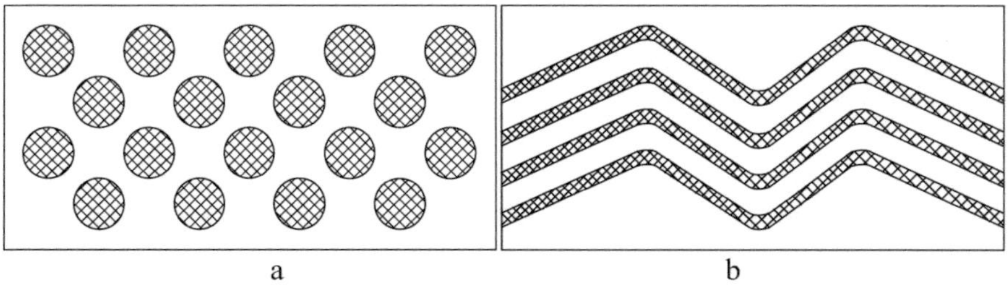

Fig. 8 Varieties of discrete coatings (scheme): **a**—island; **b**—linear

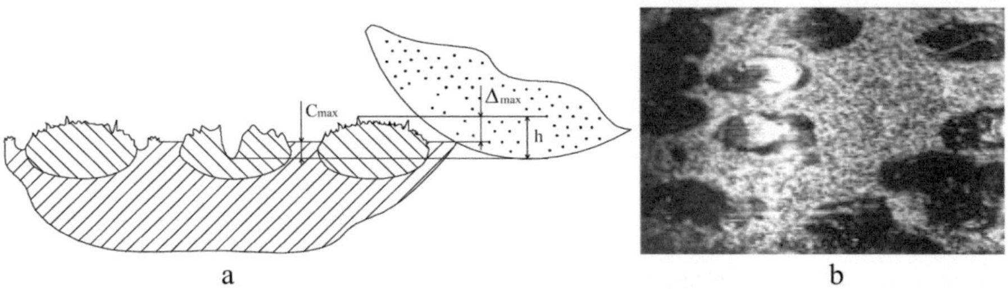

Fig. 9 Surface formation during grinding: **a**—scheme, **b**—location of discrete hardened areas; × 10

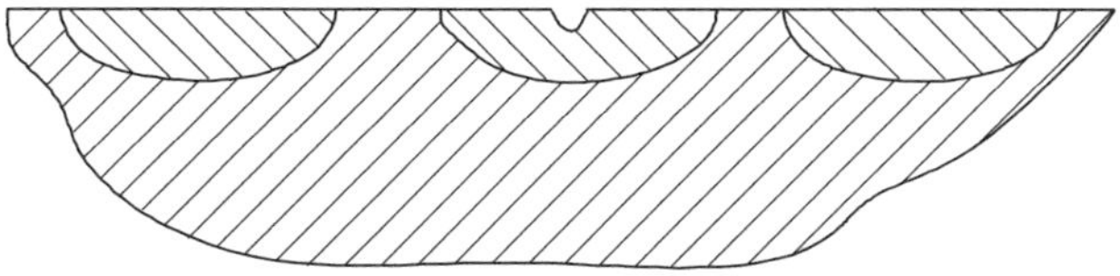

Fig. 10 Surface after grinding with incomplete removal of cavities (diagram)

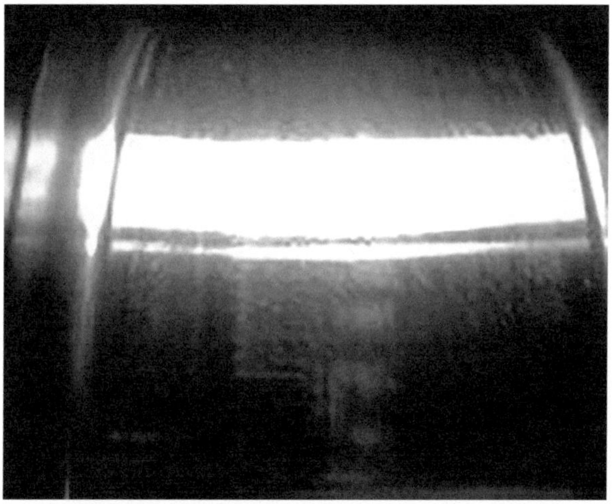

Fig. 11 Appearance of the product after EDM before grinding ×3

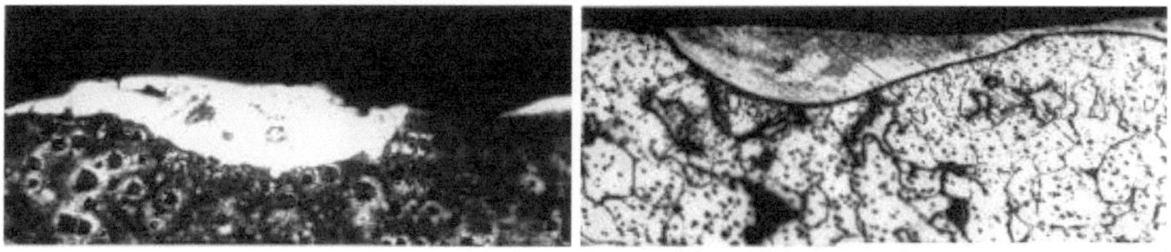

a b

Fig. 12 Microstructure of the cross-sectional grind of high-strength cast iron (**a**) and 40ChMFA steel (**b**) samples in the discharge zone; ×100; electrode—AISI 304 (08X18H10T) steel

The second of the contacting parts (e.g., aluminum alloy) is continuously (continuously) coated with aluminum oxide ceramics. Here, as mentioned above, AMDO technology is used [11–15].

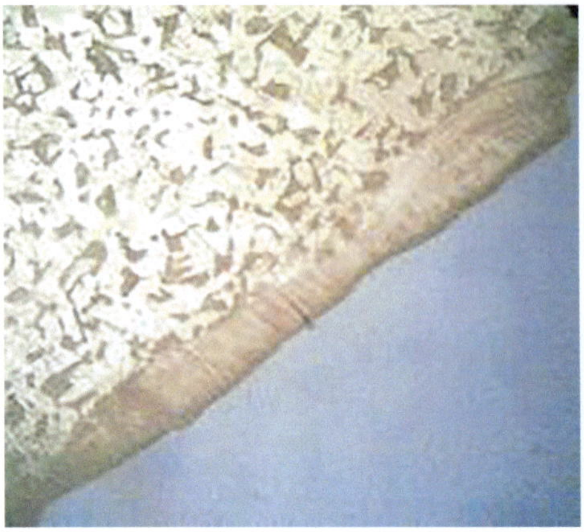

Fig. 13 Microstructure of the cross-sectional grind of the AISI 1045 steel sample (steel 45) in the discharge zone; ×100; electrode—steel (08X18H10T)

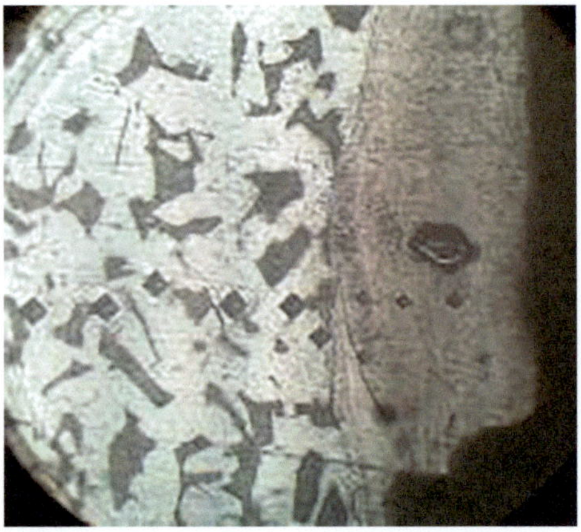

Fig. 14 Structure of AISI 1045 steel (steel 45) in the discharge zone; ×500; electrode—steel (08Kh18N10T)

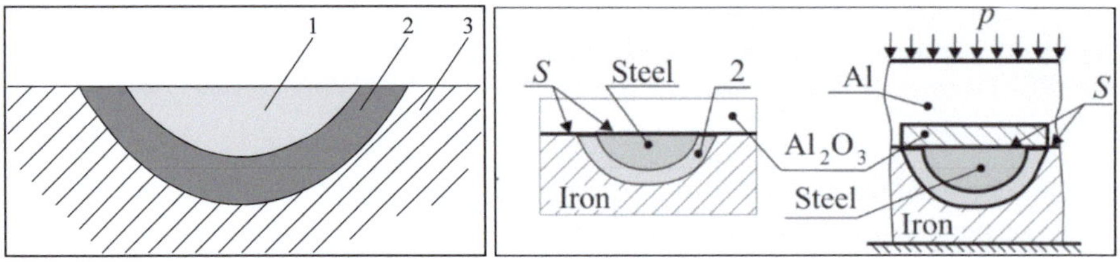

Fig. 15 Schematic representation of the discrete hardening zone after grinding: 1—"white" layer (steel); 2—sublayer (transition layer); 3—base metal (cast iron)

5 Investigation of the Stress–Strain State of Two Parts, One of Which is Discretely Strengthened and the Other is Unstrengthened

Let us consider the contact interaction of two parts, one of which is discretely hardened and the other is unreinforced (Fig. 15). The free surface of the unreinforced body is subjected to a pressure p = 100 MPa. The stress–strain state of the contacting bodies is determined.

The presented patterns of contact pressure distribution (Fig. 16) indicate its significant unevenness. At the same time, a significant part of the effective load of pressing bodies, which falls on a single fragment containing one zone of discrete-strengthened material, is transmitted through this zone. In the area of contact between the unreinforced material and the unreinforced material of another part, the contact pressure is much lower. Accordingly, this area has more favorable conditions for the accumulation of lubricant circulating in the labyrinth between the discrete hardened zones. Taking into account that the surface of these zones has a low coefficient of friction in conjunction with the surface of the base material, we end up with a gain in the integral value of friction forces. In addition, due to the higher mechanical properties of the alloyed material of the hardening zones (compared to the properties of the base material), a certain increase in the stresses acting in it (compared to the case of the absence of discrete hardening zones) is leveled in terms of strength. This can be seen from the distributions of the stress–strain state components (Fig. 17).

In addition, it can also be noted that the top layer of the material of the discretely hardened zones, which contains significant residual stresses after the discrete hardening process, is removed by grinding. This means that as a result, the level of total stresses (from external forces and from the residual stress state) is reduced. All this contributes to the improvement of both tribological and strength characteristics [1–3].

The study of the stress–strain state of two parts, one of which is discretely hardened and the other is not, gave grounds for recording such conclusions:

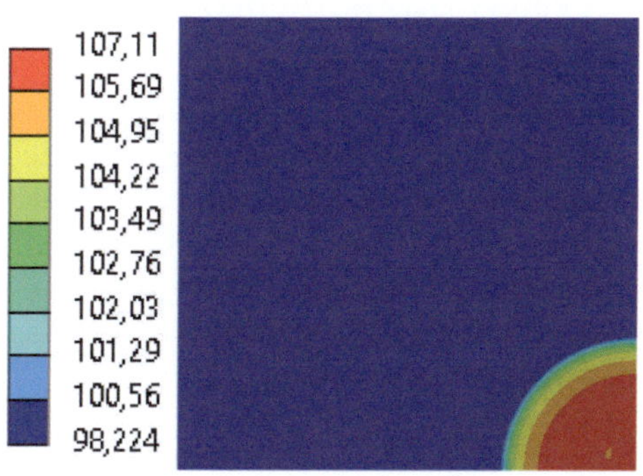

Fig. 16 Distribution of contact pressure q (MPa) when applying discrete hardening at a contact pressure of p = 100 MPa on a free surface (¼ part of the surface is shown due to symmetry)

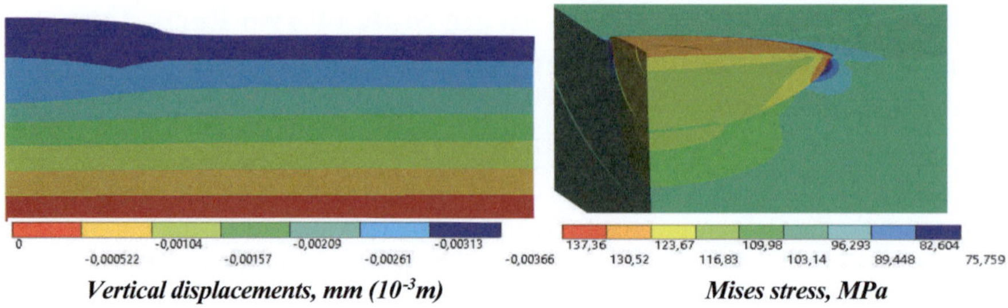

Fig. 17 Components of the stress–strain state in the region of the discretely hardened zone at a pressure of p = 100 MPa (¼ part of the discretely hardened body under study in different projections is shown due to symmetry)

1. The research revealed two effects, conventionally called the "Δ-effect" and the "σ-effect". The "Δ-effect" is that under load, the surface of a discretely hardened part becomes "hilly". These hills are located in the areas of discrete hardening. Consequently, when a contact load is applied, most of it is transferred to these zones. And as already mentioned, the material of these zones is much stronger than the base material of the part. Accordingly, the overall strength of the created composition increases. This property is called the "σ-effect".
2. The uneven transmission of forces between the contacting parts also reduces the total friction force. In addition, overall wear is reduced. This increases the overall efficiency and durability of the pair of contacting parts.

The above factors create positive trends in the entire complex of tribomechanical characteristics in the area of interaction of parts, one of which is subjected to discrete hardening.

6 Investigation of the Stress–Strain State of Two Parts, One of Which is Discretely Strengthened and the Other is Continually Strengthened

As an illustration of the development of the developed approach, we study the influence of technological modes of hardening on the contact interaction of discretely continually hardened structural parts. For this purpose, we consider a representative fragment of this system of bodies (Fig. 18). It shows a continually strengthened part made of an aluminum alloy. A layer of aluminum oxide has been "grown" on it. This part is in contact with a discontinuously hardened part. The main material of this part is cast iron. It has a discretely hardened zone of steel indented into it. The material properties of the elements of the system under study are described by the elastic modulus E and the Poisson's ratio ν.

In particular: for aluminum E = 70 GPa; ν = 0.35; for aluminum oxide E = 370 GPa; ν = 0.35; for cast iron E = 110 GPa; ν = 0.35; for steel E = 210 GPa; ν = 0.3.

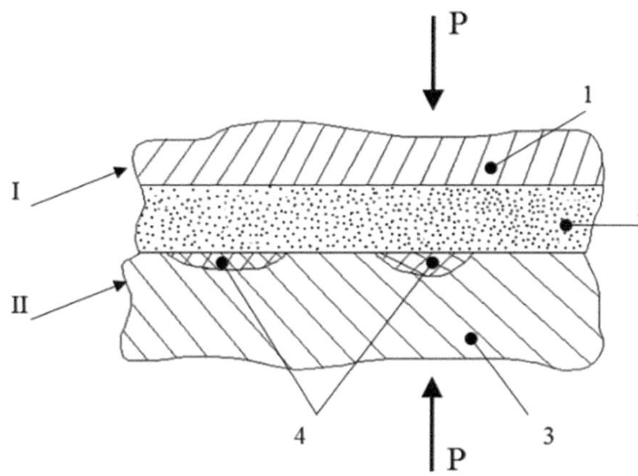

Fig. 18 Diagram of contact interaction of parts:—AK4 alloy part (1), hardened by galvanic-plasma surface transformation with the creation of a corundum layer (2),—part (main material—steel, cast iron (3)), processed by the method of discrete hardening (DHZ—discrete hardened zone (4))

The following variables were varied during the study: p—external pressure on the upper surface of the aluminum part; T—temperature to which the system of contacting bodies is heated. It is important to note that the coefficients of linear thermal expansion α for different materials differ significantly. In particular, the values of α are important for:

$$\text{steel}\ \alpha = 13\ \mu/0\text{C};\ \text{cast iron}\ \alpha = 10\ \mu/0\text{C}$$

As a result of this difference, the profile of a discretely hardened part changes during heating. Thus, an initially flat shape acquires "hills" in the areas of discrete hardening. These "hills" seem to "rise" above the nominal surface when heated, even without a force load. Consequently, the "Δ-effect" and "σ-effect" (see above) are enhanced under load.

Figure 19 shows the characteristic distributions of the contact pressure q when the external pressure and heating temperature T are varied. 20, 21 are partial dependences of the contact pressure level (its maximum value) qmax on p and on T. Figure 22 is an example of the distribution of vertical displacements of a discretely hardened body at different temperatures and external loads. Figure 23 shows an example of the Mises stress distribution in a fragment of discretely continually strengthened bodies.

As can be seen from the data presented in Figs. 19, 20, 21, 22, 23, an increase in p and T leads to an increase in the level of contact pressure q and, accordingly, in the stresses near the discrete hardening zone. It seems natural that an increase in the load p leads to an increase in the level of contact pressure q. However, another effect is more important, namely, the increase in the level of qmax with increasing temperature T. Since in any case the maximum values of q are achieved at the discrete-hardened zone, this effect is favorable in terms of increasing strength, as well as reducing friction and wear of the contacting body system as a whole. By analogy with the "Δ-effect" and "σ-effect", it can be called the "T-effect".

Thus, discrete–continuous hardening retains the "Δ-effect" and "σ-effect" previously identified [1–3], which are characteristic of discrete hardening. In addition, an

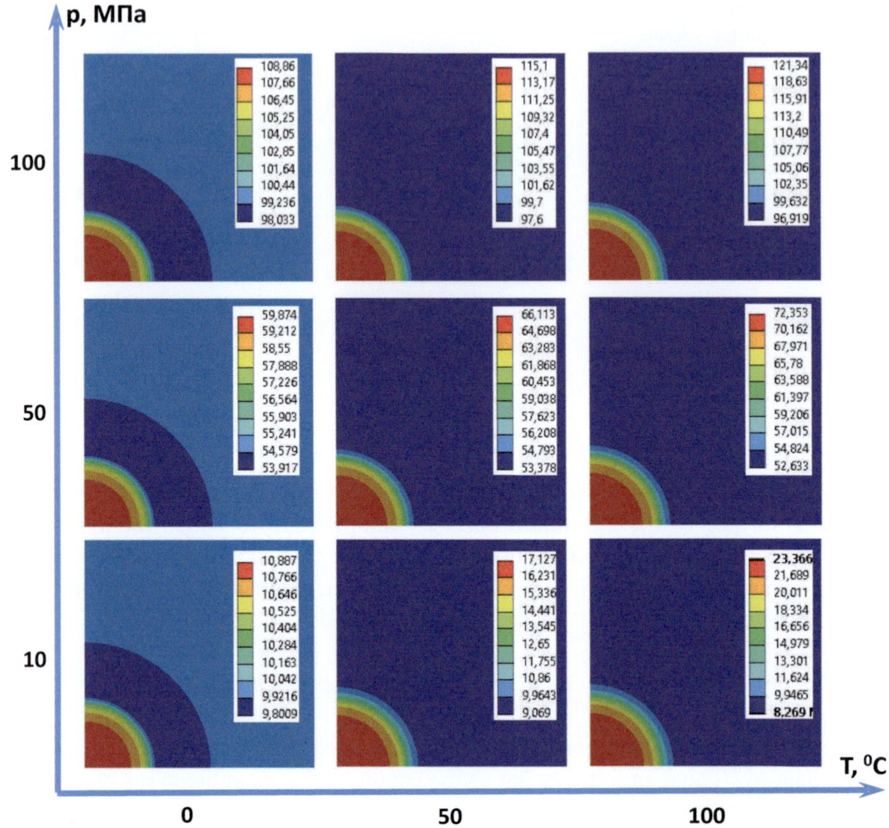

Fig. 19 Dependence of contact pressure distribution q (MPa) on temperature T and external load p

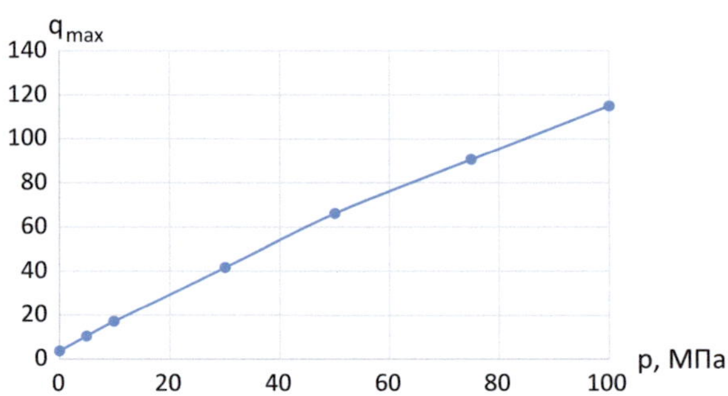

Fig. 20 Dependence of the maximum contact pressure (MPa) on the external pressure at a constant temperature of 50 °C

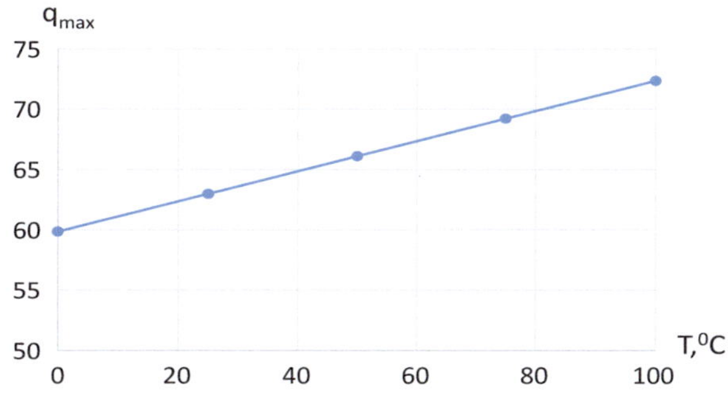

Fig. 21 Dependence of the maximum contact pressure (MPa) on temperature at a constant external pressure of 50 MPa

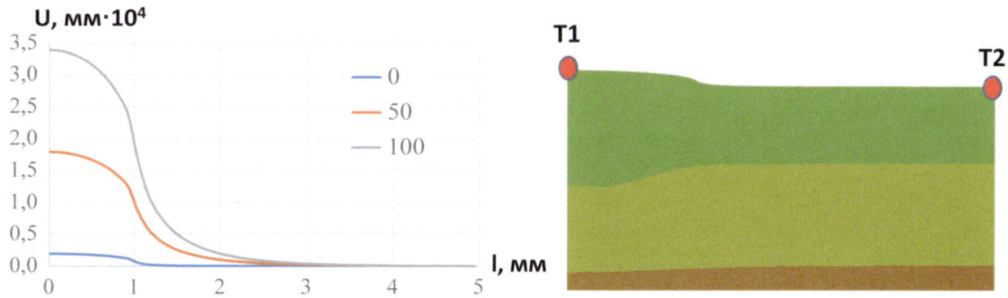

Fig. 22 Diagram of the distribution of vertical displacements of a discretely hardened body at different temperatures T and an external load p = 10 MPa along the coordinate l of the segment between the central point T1 and the angular point T2 on the contact surface with the aluminum oxide layer (displacement U is relative to the point T2, which is taken as the "zero" level)

Fig. 23 Mises stress distribution (MPa) in a fragment of discretely and continually strengthened bodies (1/4 part due to symmetry, on the left—discretely strengthened body, on the right—continually strengthened body) at an external load of 100 MPa and a heating temperature of 50 °C

additional "T-effect" has been identified. It contributes to an increase in the overall strength and durability of contacting discrete-continuously strengthened bodies.

In addition to the change in contact pressure distributions between discrete continuum hardened parts, the friction force distributions also change. Indeed, during mutual sliding, the tangential friction forces will be higher where the contact pressure is higher. That is, we are talking about a zone of discrete hardening. If the coefficient of friction in zone A is lower than in zone B, the total friction force decreases. In this case, zone A is the surface where the material of the discrete hardening zone mates with the aluminum oxide layer (in Fig. 18, pair "2–4"). Zone B is the "3–2" pair.

It should also be noted that zone A has less wear than zone B. This means that the overall wear is reduced. At the same time, the favorable "hilly" shape of the surface of the discretely hardened part is preserved.

Another important circumstance is the mechanism for ensuring a sustainable favorable complex effect from the interaction of the factors "contact pressure distribution"—"friction"—"wear". In traditional technologies, this linear sequence of factors is in a "positive" feedback: the presence of significant contact pressures in the mating of mutually moving parts leads to tangential friction forces that cause wear; during wear, as noted above, loads increase, which in turn leads to an increase in contact pressure, etc. As a result, we get a cyclically progressive process along the "spiral" P (Fig. 24a), which moves away from the nominal (initial cycle) during operation. In contrast to this trend, another, competing trend is realized for the proposed technology (see Fig. 24b).

In the case under consideration (Fig. 18), the action of friction forces causes an increase in wear of the base material. At the same time, the material of the corundum layer wears slightly when it reaches the operating limit. Under these conditions, the contact pressure is redistributed, the total friction force increases more slowly, the friction force in the mating zone with the base material decreases, which leads to a decrease in its wear. Thus, the "negative" feedback scenario is realized, and the development of processes follows the line S (see Fig. 24), which moves away

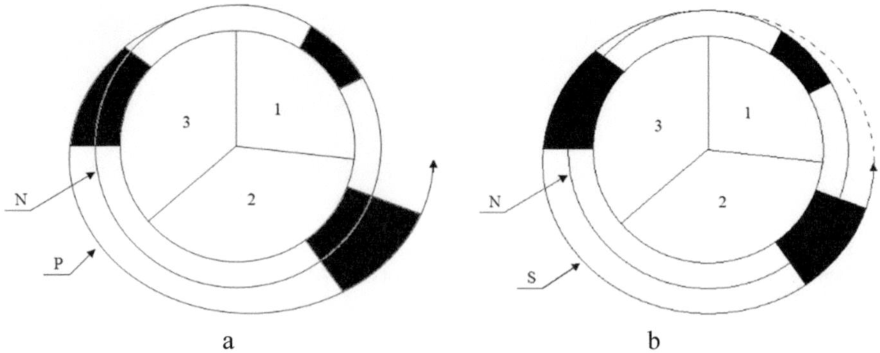

Fig. 24 Mechanisms of development of contact interaction, friction and wear processes: **a**—traditional technologies, **b**—proposed technology; 1—contact interaction, 2—wear, 3—friction

from the line N more slowly than P. Under certain conditions, this can lead to the transformation of S into a practically stable circle, which means the realization of stable processes "contact interaction—friction—wear" (see Fig. 24).

Naturally, under real conditions, actual processes combine trends a and b (see Fig. 24). However, the mechanism of stabilization b itself is important, since it characterizes the tendency to sustainable self-repetition of the studied processes, rather than their intensification (which is typical for traditional processes of processing surface layers of materials of structural elements).

Initial material assessments of the results obtained provide recommendations for the selection of process parameters: the relative area of discrete hardening zones to the total surface area is 65–75%; the thickness of the corundum layer is 0.03–0.3 mm, and its porosity is 10–20%.

7 Conclusions

In general, the developments and studies carried out allow us to draw the following conclusions.

1. The improved method of generalized parametric modeling makes it possible to take into account various variable factors in the course of research. In particular, these are new types of hardening technologies and varied technological regimes of hardening. This creates advantages in substantiating advanced technical solutions for products with improved characteristics.
2. Based on the developed approach, a new hardening method was created. It combines continuous hardening of one contacting part and discrete hardening of the other. This creates tangible advantages. In particular, it avoids the negative qualities of each of the strengthening methods in particular. In addition, the positive properties of each of the methods are dramatically enhanced when they are used together.
3. In the course of studies of the contact interaction of discrete-continuously strengthened bodies, a favorable redistribution of contact pressure between them was confirmed. This is caused by a change in the profile of the contacting surfaces under load (the so-called "Δ-effect"). Accordingly, the stresses are localized in the area of stronger materials (the so-called "σ effect"). This leads to an increase in the overall level of strength and wear resistance of bodies strengthened by this method.
4. Since the set of varying factors can be expanded, this provides additional opportunities to improve the results of discrete–continuous hardening of contacting bodies. In particular, such a factor as the heating temperature of these bodies was introduced for consideration. Due to the uneven temperature expansion, the "Δ-effect" and "σ-effect" described above are enhanced (see Sect. 3). This "T-effect" complements the previously identified positive properties of the developed method of strengthening bodies.

5. The possibility of replenishing the set of variable factors and combinations of variable parameter values expands the space in which progressive technical solutions are sought. That is, the space of generalized parameters is supplemented by "new dimensions". Accordingly, in this expanded space, it is potentially possible to achieve a more significant positive effect than without such an expansion. In particular, we consider the properties of strength and durability. For example, the introduction of the factor of uneven thermal expansion of the elements of the system of hardened bodies is considered (see Sect. 4). The potential positive effect of using this technique (conditionally called the "T-effect") has been established. In addition, it can be considered as a confirmation of the claimed general property. This property has three aspects: (1) the possibility of expanding the set of variable factors within which the desired optimal variant of design and technological solutions is determined, in particular, according to the criteria of strength and wear resistance; (2) the possibility of determining the feasibility of expanding (or narrowing) the generalized parameter space in the synthesis of advanced technical solutions for the system of contacting bodies; (3) the possibility of improving the strength and durability characteristics, for example, by determining the optimal sets and values of generalized parameters in a particular generalized parameter space.
6. An important property of the discrete–continuous strengthening technology is the preservation of the identified positive trends with increasing loads. That is, unlike traditional technologies, the established strengthening effects do not decrease with increasing loads, but are preserved.
7. One of the advantages of discrete–continuous strengthening technology is the improvement of not just one, as in many other cases, but a wide range of characteristics. Thus, the strength is increased, friction and wear are reduced. In general, the functioning of a pair of discrete-continuously strengthened bodies is also stabilized: there is no avalanche-like increase in problems in the chain "load - contact interaction - friction - wear - load increase - …".

Thus, a new approach has been developed that naturally integrates with Industry 4.0 and 5.0 strategies. In particular, we are talking about finding, creating and substantiating the parameters of new breakthrough technical solutions (structures and technologies) using the most powerful modeling tools, as well as with the active participation of an expert researcher.

References

1. Н. А. Ткачук и др. (2015) Континуальная и дискретно-континуальная модификация поверхностей деталей: монография. Нац. техн. ун-т "Харьков. политехн. ин-т", Харьков: Щедра садиба плюс, 259 с. Available at: https://repository.kpi.kharkov.ua/items/cc2db431-b6c8-4138-b345-a98404ee167f

2. Tkachuk, N.A., Kravchenko, S.A., Pylev, V.A., Parsadanov, I.V., Grabovsky, A.V., Veretelnik, O.V.: Discrete and continual strengthening of contacting structural elements: conception, mathematical and numerical modeling. Sci. Tech. [online] **18**(3), 240–347 (2019). https://doi.org/10.21122/2227-1031-2019-18-3-240-347
3. Tkachuk, M.A., Novikov, M., Tkachuk, M.M., Grabovskiy, A., Pakki, G., Tkachuk, H., Podrieza, S., Dereviankin, R.: Areas and stages of design and technological support for enhanced performance of turbo-expander power plants. Bull. Natl. Tech. Univ. 'KhPI' Ser.: Eng. CAD **2**, 106–112 (2023). https://doi.org/10.20998/2079-0775.2023.2.11
4. Sobol, O.V., Postelnyk, H.O., Pinchuk, N.V., Meylekhov, A.A., Zhadko, M.A., Andreev, A.A., Stolbovoy, V.A.: Influence of bias potential magnitude on structural engineering of ZrN-based vacuum-arc coatings. Phys. Chem. Solid State **22**(1), 66–72 (2021). https://doi.org/10.15330/pcss.22.1.66-72
5. Sobol', O., Dur, O.: Structural Engineering of Nanocomposite Coatings Based on Tungsten and Titanium Carbides. Lecture Notes in Mechanical Engineering, pp. 552–561 (2020). https://doi.org/10.1007/978-3-030-50794-7_54
6. Bourebia, M., Laouar, L., Hamadache, H., Dominiak, S.: Improvement of surface finish by ball burnishing: approach by fractal dimension. Surf. Eng. **33**(4), 255–262 (2016). https://doi.org/10.1080/02670844.2016.1232778
7. Pinchuk, N.: Comparative analysis of structure formation and properties of PVD coatings TIN, TI/TIN and tin-mon. The scientific paradigm in the context of technological development and social change (2023). https://doi.org/10.30525/978-9934-26-297-5-19
8. Schalk, N., Tkadletz, M., Mitterer, C.: Hard coatings for cutting applications: physical vs. chemical vapor deposition and future challenges for the coatings community. Surf. Coat. Technol. **429**, 127949–127949 (2021). https://doi.org/10.1016/j.surfcoat.2021.127949
9. Study of advanced nanoscale ZrN/CrN multilayer coatings. East Eur. J. Phys. **2** (2019). https://doi.org/10.26565/2312-4334-2019-2-04
10. Musil, J., Kos, Š, Zenkin, S., Čiperová, Z., Javdošňák, D., Čerstvý, R.: β-(Me1, Me2) and MeNx films deposited by magnetron sputtering: novel heterostructural alloy and compound films. Surf. Coat. Technol. **337**, 75–81 (2018). https://doi.org/10.1016/j.surfcoat.2017.12.057
11. Subbotin, O., Bilozerov, V., Volkov, O., Subbotina, V., Shevtsov, V.: Friction Properties of MAO coatings on aluminum alloys. Bull. Natl. Tech. Univ. 'KhPI' Ser. Eng. CAD **2**, 59–63 (2022). https://doi.org/10.20998/2079-0775.2022.2.07
12. Subbotina, V., Sobol, O., Belozerov, V., Subbotin, A., Smyrnova, Y.: A study of the phase-structural engineering possibilities of coatings on D16 alloy during micro-arc oxidation in electrolytes of different types. Eastern-Eur. J. Enterprise Technol. **4**(12(106)), 14–23 (2020). https://doi.org/10.15587/1729-4061.2020.209722
13. Asquith, D., Yerokhin, A., James, N., Yates, J., Matthews, A.: Evaluation of residual stress development at the interface of plasma electrolytically oxidized and cold-worked aluminum. Metallurgical Mater. Trans. A Phys. Metallurgy Mater. Sci. **44**(10), 4461–4465 (2013). https://doi.org/10.1007/s11661-013-1854-0
14. Dean, J., Gu, T., Clyne, T.W.: Evaluation of residual stress levels in plasma electrolytic oxidation coatings using a curvature method. Surf. Coat. Technol. **269**, 47–53 (2015). https://doi.org/10.1016/j.surfcoat.2014.11.006
15. Martin, J., Leone, P., Nominé, A., Veys-Renaux, D., Henrion, G., Belmonte, T.: Influence of electrolyte ageing on the plasma electrolytic oxidation of aluminium. Surf. Coat. Technol. **269**, 36–46 (2015). https://doi.org/10.1016/j.surfcoat.2014.11.001

Estimating the Efficiency of Electricity Generation Using FAME Bioliquid Produced from Post-frying Vegetable Oils Obtained from the Kiev Area

Grzegorz Wcisło and Dariusz Kurczyński

Abstract This research evaluates the potential of using FAME biofuels, derived from post-frying vegetable oils, as an alternative to diesel in power generators. Utilizing the GW-250 reactor, locally produced biofuels in Kyiv could generate 1.52 GWh of electricity weekly, reducing CO_2 emissions by 42.7 thousand tons annually. The study analyzed physicochemical properties of FAME from frying oils and compared them to diesel. Additionally, the economic benefits of avoided emission certificates within the EU Emissions Trading System were considered. Findings emphasize the feasibility of converting waste oils into sustainable biofuels for electricity generation with significant environmental advantages.

1 Introduction

The modern world must make quick decisions regarding climate change mitigation. They were triggered by a change in the atmosphere composition. It contains greenhouse gases that cause the greenhouse effect. This effect is a natural phenomenon. One of the greenhouse gases is carbon dioxide. The intensive exploitation of natural, non-renewable energy sources has contributed to the increase in CO_2 content in the atmosphere. In addition, the greenhouse gases in the atmosphere are methane, freons, water vapour, nitrous oxide, ozone, hydrofluorocarbons (HFC), perfluorocarbons (PFC), sulphur hexafluoride (SF_6), nitrogen trifluoride (NF_3).

Approximately 75% of greenhouse gas emissions in the European Union are accountable to the production and use of energy [1]. About 80% of greenhouse gases

G. Wcisło
Department of Bioprocess Engineering, Energy and Automation, Faculty of Production Engineering and Power Technologies, University of Agriculture, Krakow, Poland
e-mail: grzegorz.wcislo@ur.krakow.pl

D. Kurczyński (✉)
Department of Automotive Engineering and Transport, Faculty of Mechatronics and Mechanical Engineering, Kielce University of Technology, Kielce, Poland
e-mail: kdarek@tu.kielce.pl

© The Author(s), under exclusive license to Springer Nature Switzerland AG 2025
A. Marchenko and Z. Koruba (eds.), *Decarbonization of Transport Energy Installations in the Context of Sustainable Development Strategies*, Studies in Systems, Decision and Control 635, https://doi.org/10.1007/978-3-032-05884-3_11

emitted worldwide in 2019 were carbon dioxide [2]. This is a result of the widespread use of conventional energy sources: coal, petroleum and natural gas. In the European Union, about one quarter of greenhouse gas emissions in 2019 came from transport [3]. Road transport had the largest share in this emission. It amounted to 71.7%. Of which 60.6% is attributed to passenger cars. In the USA, the transport sector accounted for 28% of carbon dioxide emissions in 2021, and the energy production sector accounted for 25% of its emissions [4]. Humans are responsible for the increase in greenhouse gas emissions in the atmosphere. According to [5], transport has the greatest impact on greenhouse gas emissions.

Energy is needed for the functioning of transport and every other human activity, including the provision of necessary living conditions. Energy is needed to provide heat, generate electricity and perform transport tasks. One way to reduce CO_2 emissions into the atmosphere is to use organic matter to generate electricity and biofuels. Carbon dioxide is absorbed from the atmosphere in the process of producing organic matter by plants. When burning biofuel, CO_2 returns to the atmosphere. According to [6], burning one litre of diesel oil introduces approximately 2.69 kg of carbon dioxide into the atmosphere. A similar result can be achieved based on the data contained in [7]. Minor differences result from the fact that diesel oils come from different raw materials and may have a slightly different elemental composition, fractional composition, density and calorific value.

The European Union has been promoting the acquisition of energy from renewable sources for many years. Defines the requirements for the Community and individual countries in terms of energy consumption from renewable sources. These requirements relate to two areas. The first is the total gross energy consumption in the Community. The second is energy consumption in transport. Directive (EU) 2009/28/EC set a target of achieving 20% share of energy from renewable sources in the final gross energy consumption in European Union countries by 2020 [8]. Furthermore, this Directive established a requirement, according to which 10% of energy used in transport by 2020, in each Member State, should come from renewable sources. These requirements were changed in 2018 [9]. It has been determined that the countries of the European Union are jointly to ensure a 32% share of renewable energy in gross final energy consumption by 2030. Directive (EU) 2018/2001 also increased the share of renewable energy in the transport sector. Every Member State must ensure that fuel suppliers provide a minimum share of renewable energy in final energy consumption in transport of 14% by 2030. The concept of advanced biofuels was introduced in Directive 2018/2001. This concept refers to biofuels produced from raw materials listed in Annex IX of the directive, part A. The raw materials mentioned in this annex include algae and various organic waste materials, such as: straw, manure and sewage sludge, sugarcane pressings, grape pomace and wine yeast sediment, nut shells, seed husks, corn cobs cleaned of grains, and others. The share of advanced biofuels and biogas produced from raw materials listed in Annex IX to Directive 2018/2001, in part A, should be at least 0.2% in 2022, at least 1% in 2025, and at least 3.5% in 2030 in final energy consumption in the transport sector. The contribution of the aforementioned advanced biofuels and biogas to the minimum biofuel inputs in transport, can be considered as equivalent to twice their energy value. Biofuels produced from used cooking oils and from waste animal fats and inedible animal fats are not considered advanced biofuels. The participation of biofuels in the minimum renewable energy inputs in the transport sector can however be considered equivalent to twice their energy value.

Directive (EU) 2009/28/EC defined the terms biofuels and bioliquids. Bioliquids are liquid fuels obtained from biomass and intended for energy purposes, for the production of electricity, heat and cold energy, not used in transport [8, 9]. According to the Renewable Energy Act [10] bioliquids are liquid fuels intended for uses other than transport. Biofuels are defined as liquid or gaseous fuels produced from organic matter and intended for transport. Biofuels and Bioliquids are obtained from organic matter. They differ in application. Biofuels are used to power engines used in motor vehicle propulsion. They must meet rigorous requirements that allow for the reliable operation of internal combustion engines for which they are used. Bioliquids can have quite different properties, depending on the raw material from which they are obtained, and the extraction technology used. They are also used to power engines that drive generators producing electrical energy.

From organic matter in the form of vegetable oils and animal fats, fatty acid esters are obtained in the transesterification process. They offer a biodegradable and non-toxic alternative fuel to diesel. The transesterification process employs alcohol and a catalyst to accelerate the reaction, with methyl alcohol being the predominant choice. This process yields fatty acid methyl esters, although ethyl alcohol is utilized to a lesser extent, and other alcohols may also be employed. In Europe, the EN 14,214 standard delineates the criteria for FAME (fatty acid methyl esters), stipulating that if esters conform to these criteria, they are suitable for powering automotive compression-ignition engines. These esters can be utilized either in their pure form or as an additive to diesel fuels. Furthermore, according to the standard, esters can serve heating purposes. When utilized in transportation, esters are termed biofuels, serving as substitutes for diesel fuels derived from crude oil and utilized to power compression-ignition engines. These esters are then denoted as biodiesel. However, if the same esters are utilized for purposes beyond powering transportation engines, they are referred to as biofuel.

Diesel oils are a non-renewable fuel. Their use is associated with various pollutants being released into the atmosphere and the emission of carbon dioxide. Biofuels and bioliquids are renewable and biodegradable fuels. Burning these fuels also results in carbon dioxide emissions. However, biofuels are obtained from organic matter, the formation of which is associated with capturing carbon dioxide from the atmosphere.

The aim of the work is to present the possibilities of generating electricity using generator sets for power supply, which can use esters obtained from post-frying kitchen oils collected from the city of Kiev's gastronomic establishments. An ester production reactor was used, which can be used locally and does not require complex industrial infrastructure. Attention was also paid to the economic and ecological effect of using methyl esters of post-frying oils instead of diesel oil, for the production of electrical energy.

2 Bioliquids

Interest in fuels obtained from organic matter is increasing worldwide. The main drivers of this interest are the need to reduce carbon dioxide emissions into the atmosphere, to halt the ongoing and visible climate changes, the desire to become independent from oil supplies, the increasing energy demand and the development of a sustainable economy. An increase in fuel production from organic matter is anticipated, including from organic waste material [11]. An example of the use of energy from renewable sources is Brazil. Already in 2017, 43% of the energy consumed in this country came from renewable sources [12]. 80% of the electric energy came from renewable sources. This meant that even then, the Brazilian economy was low emission.

The European Commission in 2019 presented the strategy "The European Green Deal" [13]. This is a package of initiatives aimed at achieving zero net greenhouse gas emissions in Europe by 2050. To achieve this, it will be necessary to move away from conventional energy sources. Directives 2009/28/EC and 2018/2001 promote two types of energy sources generated in organic matter: biofuels and bioliquids [8, 9]. Directive 2018/2001 strongly promotes the production of biofuels from waste.

Bioliquids are fuels obtained from organic matter intended for purposes other than transport. Gil A. shows that bioliquids are clearly less frequently the subject of research than biofuels. Between 2015 and 2020, only 35 research papers were published with the keyword's biomass and bioliquids, and 27,229 research papers were published with the keyword's biomass and biofuels [1]. The use of biofuels for transport applications is definitely more frequently researched. This indicates the need to conduct research in the field of production, properties and use of bioliquids for the generation of electricity. The energy industry will increasingly have to use biomass for electricity production. Therefore, bioliquids can be an interesting proposal.

Bioliquids obtained from various raw materials and based on different technologies may have different properties, which will determine the way they are used. Seljak et al. distinguished two groups of bioliquids [14]. The first group are fuels with properties similar to those of conventional fuels. They are easy to apply to power piston combustion engines and small gas turbines, used in small electric power generation installations. The second group of bioliquids are fuels with high density and viscosity. They require the adaptation of piston engines to be fuelled by this fuel. Its application for powering small turbines is hindered. Esters produced from post-frying vegetable oils can be classified into the first mentioned group of bioliquids. They can be used to power engines with spontaneous ignition.

High viscosity bioliquids can be used as a substitute for heating oil. Park et al. [15] carried out research on bioliquid as a substitute for heavy fuel oil in oil-fired power plants. Bioliquid combustion tests were conducted in oil boilers with a capacity of 75, 90 and 100 MW. The results were compared with the results obtained during the combustion of heavy fuel oil. No problems have been demonstrated in the operation of boilers powered by bioliquid. A clear reduction in nitrogen oxides and sulphur oxides emissions was achieved. It has been demonstrated that with minor changes to the fuel supply system and spraying elements, bioliquid can replace heavy fuel oil in existing boilers. Equivalent results were obtained in the study [16]. However, Pawlak-Kruczek et al. [17] changed the possibility of adding glycerin to coal-fired boiler furnaces. Glycerin is a waste product from ester production. It has been shown that co-firing glycerine with coal dust improves some combustion process parameters. A significant reduction in nitrogen oxide emissions has been observed. The burning of coal dust has improved. A significant reduction in coal consumption has been achieved. Biomass can be co-fired with coal in traditional coal power plants of all types [18].

Tonini et al. [19] presented the possibility of using organic matter found in waste from households. This is possible in so-called waste refineries. Bioliquid can be produced in them from mixed waste. The research conducted by the Authors showed that the waste disposal plant can recover significant quantities bioliquid. The received bioliquid is then converted into electrical energy and heat. Residues of solid organic matter can be incinerated to further recover energy.

A remarkably interesting proposition for generating electricity from organic waste matter was presented by Caetano et al. [20]. The proposal is aimed at farms engaged in animal breeding located far from large power plants and without access to the electricity distribution network. Such farms generate substantial amounts of biological waste from animal breeding. It is planned that these wastes will be used for the production of biogas in a biodigester. Then the biogas will be used to generate electricity using a Stirling engine. It has been determined that for daily energy production to be 50 kWh with continuous engine operation, the population of pigs being bred should be 84 pieces or 26 dairy cows. The daily requirement for organic matter was determined at 180 kg for pig farming and 325 kg for cattle farming. The water demand is 535 and 325 L per day. It was determined that a biodigester with a capacity of 18.7 m^3 would be needed for the processing of pig waste. However, for the processing of cattle waste, a biodigester with a capacity of 11.3 m^3 would be needed. The residues from biogas production will serve as a natural fertiliser.

Biodiesel can be defined as a fuel obtained from organic matter for powering self-ignition engines. If biodiesel are FAME esters, according to the standard, it can be used to power automotive internal combustion engines with spontaneous ignition [21]. Biodiesel is then a biofuel. Biodiesel, or otherwise FAME esters, can be used to fuel the engine that drives the generator unit. Then biodiesel is in accordance with the directives, a biofuel used for the production of electricity. Using biodiesel to generate electricity is a workable solution for meeting local electricity needs. In places where there is no access to electricity supplied from large power plants. These can also be places where resources are available, from which bioliquid in the form of esters can be easily produced. The bioliquid in the form of esters has physicochemical properties similar to the properties of fuel oils. It can be used without any issues to power piston combustion engines with spontaneous ignition and small gas turbines that can power local electricity generators. High viscosity bioliquids cannot be used to power such engines. The technology of obtaining esters is well recognized. Their production, in addition to the industrial scale, is possible on a small and medium scale in places where there is access to local raw materials. Then, there is no need to transport raw materials from great distances to large factories producing esters. The raw materials can be vegetable oils intended for food purposes [22–28], inedible vegetable oils [29–35], oils obtained from algae [35–42], waste animal fats [43–49] and used vegetable oils that are also waste [50–60]. Currently, organic matter waste is increasingly used for the production of esters. About 25% of the biodiesel produced in the world comes from used cooking oils [11]. Significant amounts of biodiesel from waste are produced in the European Union, United States, China, Canada, India, and Singapore.

Chepeliev et al. [61] indicate that biomass can play a significant role in future climate change mitigation. The potential biomass research was conducted for Ukraine, for the scenario of limiting greenhouse gas emissions and mitigating climate change. It was reported that the share of renewable energy sources in the final energy consumption in Ukraine in 2017 was 6.5% and was clearly lower than in neighbouring European Union countries. Furthermore, the share of biofuels and waste in energy consumption in Ukraine is also significantly lower than in other countries. The results of the analyses showed that despite the favourable characteristics of biomass energy production, the potential energy of biomass in Ukraine is not being used. It has been determined that the biomass potential in Ukraine has been used to less than 7.5% since 2015. The research conducted by Chepeliev et al. suggests that biomass could play a significant role in Ukraine's future efforts to mitigate climate change. Biomass can contribute to reducing the use of primary energy to a greater extent than the combined use of solar and wind energy. It has been indicated that tree biomass and biofuels are the most important biofuels in terms of the potential to significantly reduce climate change. The topic presented in this article is a proposal for the use of waste biomass for electricity production.

Interesting theses and conclusions were presented in their work by Moriarty et al. [62]. They stated that biofuels in transport will not be used for long. Withdrawing combustion engine cars from transport will lead to the cessation of biofuels and conventional fuels production. They claim that the production of electricity from biomass to power electric vehicles will result in a greater reduction of greenhouse gases per kilometre travelled than liquid biofuels. A similar proposal concerning the use of biomass was presented by Campbell et al. [63]. They claim that when biomass is used directly in power plants to produce electricity that then charges the battery of an electric vehicle, you will achieve 81% more kilometres travelled and a 108% greater reduction in emissions per unit area of crop land than in the case of cellulose-produced ethanol. The above applications to some extent may also confirm the relevance of using waste oils and fats for electricity production, as presented in this work.

3 Research Equipment and Materials

3.1 Ester Production Reactor

The latest automatic GW-250 reactor, which was designed and operates according to the technology of Professor Grzegorz Wcisło (Fig. 1), was used for the production of FAME. This is a fully automated device. The production process was programmed and controlled using the SIMENS S7-1200 controller and the TIA Portal program. The process has been visualised using the Kinco DTools v.4.2 software and is displayed on the Kinco GL-100E company's touch-operated HMI panel. At any moment of the process, the current states and state changes of reactor elements such as pumps, solenoid valves, temperature sensors and the operation of temperature controllers are signalled on the HMI panel. Three states are signaled on the panel. First, when the reaction substrates are heated during the process, Second, when the desired temperature is reached along with a tolerance window of ±2 °C. Third, when the temperature value for some reason reaches the maximum anticipated in this situation. Then, in order to secure, the power supply to the heaters is switched off. This is a safeguard in case of control elements failure. The reactor allows for the production of esters up to 250 dm^3/cycle, which lasts 2.5 h. The reactor consumes about 8.5 kWh of electrical energy per cycle. Using the GW-250 reactor in question, with 250 dm^3 of vegetable oil and after adding a mixture catalysing the process, approximately 245 dm^3 of FAME (biofuel) and a glycerin fraction are obtained. Table 1 presents the basic technical data of the GW-250 reactor.

3.2 Production of Esters from Fryer Oils (Bioliquid)

The FAME bioliquid was produced using the transesterification method, which involves chemically breaking down large heavy oil or fat particles, i.e. triglycerides, into small molecules similar in size to those in diesel oil. During the reconstruction, alcohols are introduced into the binding of glycerine esters. From one particle of oil plus three particles of alcohol, three particles of higher fatty acid esters are obtained and glycerin as a by-product. If the degree of overreaction is above 96.5% (w/w), high-quality fuel is obtained. If the additionally produced bioliquid meets fuel standards, it can safely power self-ignition engines.

Considering the model and the composition of fatty acids in the oil, for the production of biodiesel, a proportion was used according to which for every 250 dm^3 (230 kg) of oil, a catalytic mixture was used, resulting from the dissolution of 1.75 kg of KOH in about 38 dm^3 (30 kg) of methanol CH_3OH. A mixture was prepared at a station equipped with an IKA C-MAG HS 7 magnetic stirrer. As a result of the reaction of methyl alcohol with KOH, a transitional compound is formed (i.e. potassium methanolate CH_3OK), which is actually a mixture catalysing the process. The implemented solutions and optimised process allowed for the production of FAME

Fig. 1 GW-250 reactor for the production of FAME type biofuels and bioliquids

Table 1 Technical data of the GW-250 reactor

Measured parameter	Unit	Value
Efficiency per cycle	dm^3	250
Production time per cycle	h	2.5
Supply voltage	V	400
Rated power	kWh	8.5
Type of catalyst	–	Basic/acidic
Process	–	Periodic or semi-continuous
Type of process	–	Single-stage or two-stage

of exceedingly high purity, in 2.5 h. The production process was carried out at a temperature of 60 °C. From one cycle with 250 dm^3 of frying oil plus a catalytic mixture, it is possible to obtain about 260 dm^3 of bioliquid, the rest is the waste glycerine fraction, which is used by the chemical and cosmetic industries.

4 Quality Research of the Produced Bioliquid and DF Fuel Oil

After the bioliquid was produced, qualitative analyses were conducted. For comparison, analyses of conventional diesel oil were also performed. The research was conducted in the biofuels laboratory belonging to the Malopolska Centre for Renewable Energy Sources "BioEnergia". The research on bioliquid produced using the GW-250 reactor was conducted in accordance with the PN-EN 14,214 + A2:2019–05 standard, whereas the diesel oil was in accordance with the PN-EN 590 + A1:201–06 [21, 64]. The first test that is carried out in the case of biofuels is the determination of the ester content in biodiesel (bioliquid) using gas chromatography. The PN-EN 14,214 standard stipulates that the content of esters must be at least 96.5% (w/w). Another particularly important parameter is the calorific value, which affects the amount of fuel consumed per hour of operation of the combustion engine. This has special significance from the perspective of replacing diesel fuel with bioliquid produced for powering the generator. To this end, the heat of combustion was first determined by the calorimetric method, and then the calorific value was calculated. The heat of combustion obtained during the burning of a bioliquid sample in a calorimeter was obtained in the unit cal/g. That is why the value of this heat was first converted to the unit J/g according to the following relationship:

$$H_{C_B} = Q_B \cdot 4.1868 = 9863.87\frac{cal}{g} \cdot 4.1868 = 41299.05\frac{J}{g} \tag{1}$$

$$H_{C_B} = 41.3\frac{MJ}{kg}. \tag{2}$$

The lower heating value of the bioliquid was calculated using the following relationship:

$$C_{V_B} = H_{C_B} - r(W_B + 8.94 \cdot H_B)\frac{kJ}{kg}; \tag{3}$$

$$C_{V_B} = 41.3\frac{MJ}{kg} - 2.442\frac{MJ}{kg}(0 + 8.49 \cdot 0.12) = 38.1\frac{MJ}{kg}. \tag{4}$$

The higher heating value for diesel oil was:

$$H_{C_{DF}} = Q_{DF} \cdot 4.1868 = 10835.03\frac{cal}{g} \cdot 4.1868 = 45364.17\frac{J}{g}; \tag{5}$$

$$H_{C_{DF}} = 45.4\frac{MJ}{kg}. \tag{6}$$

The lower heating value of ORLEN Verva B0 diesel oil has been calculated:

$$C_{V_{DF}} = H_{C_{DF}} - r(W_{DF} + 8.94 \cdot H_{DF})\frac{kJ}{kg}; \quad (7)$$

$$C_{V_{DF}} = 45.4\frac{MJ}{kg} - 2.442\frac{MJ}{kg}(0 + 8.94 \cdot 0.13) = 43.2\frac{kJ}{kg}. \quad (8)$$

where Q—calorific value obtained in a calorimeter; cal/g, H_C—heat of combustion; J/g, MJ/kg, C_V—calorific value; MJ/kg, W—moisture content; %(w/w), H—hydrogen content in the fuel; %(w/w), r—heat of vaporisation of water, B—bioliquid, DF—diesel.

The values of selected properties of bioliquid and, for comparison, diesel oil, are presented in Table 2.

Table 2 The properties of the produced WFOME bioliquid and diesel were determined following the requirements of the EN-PN 14,214 + A2:2019–05 and PN-EN 590 + A1:201–06 standards

Property		Standard	WFOME	DF
Content of fatty acid methyl esters, % (w/w)		PN-EN 14,103	98,6	–
Density at 15 °C; g/cm^3		PN-EN ISO 3675	876.41	832.22
Dynamic viscosity 40 °C; mPa·s		EN ISO 3104	4.2	2.6
Cetane number		PN-EN ISO 5165	56.7	51,2
Flash point; °C		PN-EN ISO 3679	143	59
Higher heating value, MJ/kg		PN-C-04375–3	41.3	45.4
Lower heating value, MJ/kg		PN-C-04375–3	38.1	43.2
Fractional composition	up to 250 °C distils, %(V/V) < 65,0	PN-EN ISO 3405	0	33
	up to 350 °C distils, %(V/V) min.85		92	100
	95% (V/V) distils to 360 °C		Yes	Yes
Water content, mg/kg		PN-EN ISO 12937	160	40
Sulfur content, mg/kg		EN ISO 12937	2	8

5 Estimation of the Potential for Electricity Production Using Bioliquid Generated from Used Oil Obtained from Catering Establishments in the Area of Kyiv

5.1 Estimation of the Amount of Bioliquid that Can Be Produced from Frying Oil from Catering Establishments in the City of Kyiv

The amount of oil that can be obtained was determined using data obtained by Mr. Engineer. Artema Riaboshapkoi used in engineering work, which was supervised by one of the co-authors of the work, Grzegorz Wcisło. The thesis was defended at the University of Agriculture in Krakow in February 2024 [65]. The results of this work are particularly important at a time when there is a war in Ukraine and as a result there are particularly severe shortages of electricity supply and/or they can occur at any moment. Therefore, having power generators and for example, biorefineries, you can produce biofuel at any time and use it immediately to power the generator and produce electricity. This is especially important for military strategic objects, hospitals, but it can also be a satisfactory solution for infrastructure cut off from the power grid.

Based on the data [65], it is estimated that there are about 3458 catering points and restaurants in the whole city of Kyiv. These are catering establishments from small to large. Mr Eng. Artema Riaboshapkoi assumed in his thesis that on average, 100 dm^3 of used frying oil can be obtained from a single catering establishment per week. For the purposes of this work, we also assumed 100 dm^3. Additionally, it was assumed for the calculations that the restaurant operates four weeks a month. Based on this, it was estimated that about 345,800 dm^3 can be obtained from all restaurants in one week, which is 69,160 dm^3 daily, assuming that restaurants work five days a week. Based on this amount, it has been estimated how much Biodiesel FAME or bioliquid could be produced weekly. Considering the total amount of oil that can be obtained in a week from the city of Kyiv using a reactor, approximately 359,600 dm^3 of FAME bioliquid can be produced, which is 71,926 dm^3 daily (5 days a week).

5.2 Use of Generators for Electricity Production

As part of the research, two types of power generators have been proposed for electricity production. Knowing the data on the generated power and diesel oil consumption, it is possible to determine how much more bioliquids the power generator would consume. As is known, all biofuels, because they contain oxygen in their structure, are characterised by a lower calorific value compared to diesel oil. Therefore, as part of the work, the equivalent of bioliquid consumption in relation to diesel oil was determined.

Table 3 Volvo FOGO FDT 1000VA generator data

Parameter	value
Nominal power, kVA	1000
Maximum power, kVA	1100
Nominal current, A	1443
Frequency, Hz	50
Voltage, V	400
Fuel consumption—50% load; dm^3/h	106.2
Fuel consumption—75% load; dm^3/h	154.4
Fuel consumption—100% load; dm^3/h	207.4

Table 4 Caterpillar DE50E0 CAT generator data

Parameter	Value
Nominal power, kVA	50
Maximum power, kVA	55
Nominal current, A	72
Frequency, Hz	50
Voltage, V	400
Fuel consumption—50% load; dm^3/h	6.8
Fuel consumption—75% load; dm^3/h	10.1
Fuel consumption—100% load; dm^3/h	13.4

Data for two types of generating units were used to calculate the consumption of fuel and the production of electrical energy. The first one is Caterpillar DE50E0 CAT. The unit at nominal 100% load produces 50 kVAh of current in an hour, consuming 13.4 dm^3 of diesel fuel in the process. The second is a much more powerful Volvo FOGO FDT generator, which is capable of producing 1000 kVAh of electrical energy. The engine of such a unit at 100% load consumes about 207.2 dm^3 of diesel oil per hour. The basic technical parameters of the power generators selected for analysis are presented in Tables 3 and 4. Figure 2 presents the Volvo FOGO FDT 1000VA generator, while Fig. 3 shows the Caterpillar DE50E0 CAT generator.

5.3 *Calculation of the Demand for Bioliquid to Power the Generator and CO_2 Emissions*

Due to the fact that bioliquid has a lower calorific value, the generator engine will consume more bioliquid than it consumes diesel oil. The equivalent of fuel consumption has been calculated, from which it will be possible to determine how much the use of bioliquid will increase. The fuel consumption equivalent e_e was determined from the following dependencies. The Caterpillar DE50E0 CAT generator set at

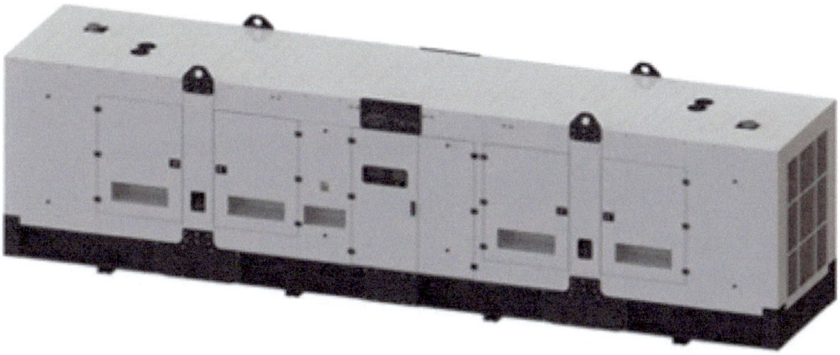

Fig. 2 Volvo FOGO FDT 1000VA generator

Fig. 3 Caterpillar DE50E0 CAT generator

nominal load consumes 13.4 dm^3 of diesel fuel per hour of operation, which had a calorific value of 43.2 MJ/kg. The consumption of bioliquid in relation to diesel fuel supply for the nominal load has been determined from the following relationships:

$$13.4dm^3/h \cdot 43.2\frac{MJ}{kg} = 578.88\frac{MJ}{kg} \cdot dm^3; \tag{9}$$

$$\mathrm{e_e} = \frac{578.88\frac{\mathrm{MJ}}{\mathrm{kg}} \cdot \mathrm{dm}^3}{38.1\frac{\mathrm{MJ}}{\mathrm{kg}}} = 15.19\mathrm{dm}^3 \approx 15.2\ \mathrm{dm}^3/\mathrm{h}. \tag{10}$$

where e_e—equivalent of bioliquid consumption.

The Caterpillar DE50E0 CAT generator, which at nominal load would consume 13.4 dm^3 of diesel fuel per hour, should consume about 15.2 dm^3 of bioliquid after replacement. From the above, it can be seen that if diesel oil is replaced with bioliquid, fuel consumption will increase by 1.8 dm^3 per hour of unit operation.

For Volvo FOGO FDT 1000VA generator:

$$207.4\frac{dm^3}{h}\cdot 43.2\frac{MJ}{kg}=8959.68\frac{MJ}{kg}\cdot dm^3; \tag{11}$$

$$e_e=\frac{8959.68\frac{MJ}{kg}\cdot dm^3}{38.1\frac{MJ}{kg}}=235.16dm^3\approx 235.2dm^3/h; \tag{12}$$

The Volvo FOGO FDT 1000VA generator, which at nominal load would consume 207.4 dm^3 of diesel oil per hour, should consume approximately 235.2 dm^3 of bioliquid after replacement. From the above, it can be seen that in the case of replacing diesel oil with bioliquid, fuel consumption will increase by 27.8 dm^3 per hour of work.

The calculations presented in Table 5 show how much diesel fuel or bioliquid a Caterpillar DE50E0 CAT would consume in a day. The calculations were performed for three variants of daily operation, namely 8, 16, or 24 h.

In Table 6, calculations are presented showing how much diesel fuel or bioliquid a Volvo FOGO FDT 1000VA generator would consume in a day. The calculations were made for three daily operating variants, namely 8, 16 or 24 h.

Tables 7 and 8 show how many generators: The Caterpillar DE50E0 CAT50VA and the Volvo FOGO FDT 1000VA would consume diesel or bioliquid in a month if they ran for 8, 16 or 24 h per day, respectively. The amount of electricity that could be produced was calculated. Additionally, the ecological effect that would occur if bioliquid were used instead of diesel to power the unit was calculated. It has been assumed that an engine powered by diesel, for every 1 dm^3 of diesel burned, emits 2.69 kg of CO_2. If we use a bioliquid produced from used frying oil instead of diesel oil to power the Caterpillar DE50E0 CAT generator, we will also achieve

Table 5 Consumption of bioliquid and DF when powering the Caterpillar DE50E0 CAT generator with a nominal power of 50 kVAh in an 8, 16 or 24 h/day cycle

Operating parameters of the Caterpillar DE50E0 CAT aggregate	Number of working hours per day		
	8	16	24
Amount of bioliquid consumed	121.6 dm^3	243.2 dm^3	364.8 dm^3
Amount of diesel consumed	107.2 dm^3	214.4 dm^3	321.6 dm^3
Amount of produced electrical energy	0.4 MWh	0.8 MWh	1.2 MWh

Table 6 Consumption of bioliquid and DF when powering the Volvo FOGO FDT 1000VA generator with a nominal power of 1000kVAh in an 8, 16 or 24 h/day cycle

Operating parameters of the Volvo FOGO FDT 1000VA aggregate	Number of working hours per day		
	8	16	24
Amount of bioliquid consumed	1881.6 dm^3	3763.2 dm^3	5644.8 dm^3
Amount of diesel consumed	1659.2 dm^3	3318.4 dm^3	4977.6 dm^3
Amount of produced electrical energy	8 MWh	16 MWh	24 MWh

Table 7 Bioliquid and diesel oil consumption and CO_2 emission over 1 month for the Caterpillar DE50E0 CAT50VA power generator working in an 8, 16 or 24 h/day cycle

Fuel consumption and CO_2 emissions when the aggregate is operating over 1 month assuming 21 working days in a month			
Operating parameters of the Caterpillar DE50E0 CAT power generator:	Number of working hours per day		
	8	16	24
Amount of diesel consumed	2251.2 dm^3	4502.4 dm^3	6753.6 dm^3
Amount of bioliquid consumed	2553.6 dm^3	5107.2 dm^3	7660.8 dm^3
Amount of CO_2 emitted at diesel supply	6056 kg	12,111 kg	18,167 kg
Amount of CO_2 emitted at bioliquid supply	0 kg	0 kg	0 kg
Amount of produced electrical energy	8.4 MWh	16.8 MWh	25.2 MWh

Table 8 Bioliquid and diesel oil consumption and CO_2 emission over 1 month for a Volvo FOGO FDT 1000VA power generator working in an 8, 16 or 24 h/day cycle

Fuel consumption and CO_2 emissions when the aggregate is operating over 1 month assuming 21 working days in a month			
Operating parameters of the Volvo FOGO FDT 1000VA power generator	Number of working hours per day		
	8	16	24
Amount of diesel consumed	34,843 dm^3	69,686 dm^3	104,529 dm^3
Amount of bioliquid consumed	39,514 dm^3	79,027 dm^3	118,540 dm^3
Amount of CO_2 emitted at diesel supply	93,728 kg	187,455 kg	281,183 kg
Amount of CO_2 emitted at bioliquid supply	0 kg	0 kg	0 kg
Amount of produced electrical energy	168 MWh	336 MWh	504 MWh

an ecological effect, because at nominal load during 8 h of work for 21 days in a month we will not emit 6056 kg of CO_2. However, in the case of the Volvo FOGO FDT 1000VA unit, when it is powered by bioliquid instead of diesel and operates 8 h a day, for 21 days, we will not emit 93,728 kg of CO_2.

5.4 *Determining the Amount of Additional Income from the Sale of Avoided CO_2 Emission Certificates*

According to EU law, units that use renewable energy sources for electricity production receive certificates of avoided CO_2 emission. Such certificates can be sold as part of the European Emissions Trading System ETS. **Trading of emission rights in the European Union**, is a **tool for reducing greenhouse gas emissions, including CO_2**. ETS is currently the second largest greenhouse gas emissions trading system in the world, after the market in China. It was adopted by the European Parliament and the EU Council in 2003 and came into effect on January 1, 2005. It has been widely applicable since 2017. As of 2019, around 11 thousand entities from 31 European countries, which have high-emission electric power generation facilities and high-emission industry, are participating in emissions trading.

The calculations assumed the value of entitlements to avoided carbon dioxide emissions on April 9, 2024, which according to ETS arrangements was 61.3 € per 1 ton of CO_2. In Table 9, profits from the sale of the above-mentioned CO_2 emission rights for the Caterpillar DE50E0 unit powered by bioliquid are presented, while in Table 10 for the Volvo FOGO FDT 1000VA unit.

By producing your own electricity and having a certificate of avoided CO_2 emissions, which can be sold as part of the European Emissions Trading System ETS, you can earn extra. In Tables 11 and 12, it is presented how much financial resources could be obtained in a year from the sale of emission rights for CO_2 alone.

Table 9 Additional profit from the sale of avoided CO_2 emissions certificates for the Caterpillar DE50E0 power generator

The amount of funds obtained from the sale of CO_2 emission allowances within 1 month and assuming 21 working days in a month			
Parameter	Number of working hours per day		
	8	16	24
Amount of CO_2 emitted at DF supply	6.0 ton CO_2	12.1 ton CO_2	18.1 ton CO_2
Profit	367.8 €	741.7 €	1109.5 €

Table 10 Additional profit from the sale of avoided CO_2 emission certificates for the Volvo FOGO FDT 1000VA power generator

The amount of funds obtained from the sale of CO_2 emission allowances within 1 month and assuming 21 working days in a month			
Parameter	Number of working hours per day		
	8	16	24
Amount of CO_2 emitted at DF supply	93.7 ton CO_2	187.4 tony CO_2	281.1 tony CO_2
Profit	5743.8 €	11,487.6 €	17,231.4 €

Table 11 Annual emissions of dioxide and profit from the sale of avoided emission certificates CO_2 for the Caterpillar DE50E0 generator

The amount of funds obtained from the sale of CO_2 emission allowances within a year—the aggregate works for 12 months			
Parameter	Number of working hours per day		
	8	16	24
Amount of CO_2 emitted at DF supply	72 tony CO_2	145.2 tony CO_2	217.2 tony CO_2
Profit	4413.6 €	8900.7 €	13,314.3 €

Table 12 Annual emissions of dioxide and profit from the sale of avoided emission certificates CO_2 for the Volvo FOGO FDT 1000VA generator

The amount of funds obtained from the sale of CO_2 emission allowances within a year—the aggregate works for 12 months			
Parameter	Number of working hours per day		
	8	16	24
Amount of CO_2 emitted at DF supply	1124.4 ton CO_2	2248.8 ton CO_2	3373.2 ton CO_2
Profit	68,925.7 €	137,851.4 €	206,777.1 €

5.5 *Estimation of How Much Electricity Can Be Produced from Bioliquid, Which Can Be Obtained from Frying Oils from Kyiv*

Based on the calculations made in Chap. 3, it turns out that from the frying oil that can be obtained in one week from the gastronomic establishments in Kyiv, approximately 359,600 dm^3 of FAME bioliquid can be obtained, which is 71926 dm^3 per day (5 days a week). If one were to consider the demand for bioliquid designated for two power generators, one could calculate how much stable and eco-friendly electricity could be produced. Additionally, you can determine how many units such amount of bioliquid would be sufficient to power. This compilation is presented in Table 13. In the case of a lower power unit of 50 kW, assuming 8 h of operation per day, the available bioliquid could power as many as 591 Caterpillar DE50E0 units. If this type of generator were to operate 24 h a day, the available bioliquid would be sufficient to power 197 such generators. In wartime conditions, the power of one unit would be enough to satisfy the needs of about 10 to 20 households, a small manufacturing plant, or a small field hospital. However, a large unit that produces 1 MWh of electric energy would be sufficient to power a neighborhood of houses, one town, a small military unit or a hospital. If it worked 8 h a day, the available bioliquid could power 38 generators. However, if such a unit were to operate 24 h a day, the bioliquid would be enough to power about 13 such units.

Table 13 Estimation of how many units could be powered by the bioliquid that could be produced from waste oil obtained daily in Kyiv (5 days a week)

Quantity of bioliquid produced in 1 day 71,926 dm^3	Number of working hours per day		
Caterpillar DE50E0 generator with a nominal power of 50 kVA			
Parameter	8	16	24
Amount of bioliquid consumed	121.6 dm^3	243.2 dm^3	364.8 dm^3
Number of Caterpillar DE50E0 generator	591	295	197
Total electrical power generated	~236.4 MWh		
Volvo FOGO FDT 1000VA generator with a nominal power of 1000 kVA			
Amount of bioliquid consumed	1881.6 dm^3	3763.2 dm^3	5644.8 dm^3
Number of Volvo FOGO FDT 1000VA generator	38	19	13
Total electrical power generated	~304.00 MWh		

6 Summary

Biofuels and bioliquids are fuels that have many advantages including that they can be produced and used to power vehicle engines or, for example, generators. We can use them to generate electricity, for example. The specific conclusions of the research and analysis carried out are as follows:

1. The bioliquid produced using the company's own technology and equipment belonging to the Małopolskie Centrum Odnawialnych Źródeł Energii "BioEnergia" meets the requirements of the EN-PN 14,214 standard defining the requirements for FAME. From the above, it follows that high-quality bioliquid was used for further analysis, which can safely power a self-ignition engine.
2. The analysis conducted showed that in a large city like Kyiv in Ukraine, there are so many catering establishments that the amount of frying oil would be enough to produce about 71,926 dm^3 of FAME type biofuel (fatty acid methyl esters) every day (five days a week). If such biofuel is used to power engines other than those used in transport, for example for the production of electrical power, such biofuel is referred to as bioliquid. This is particularly important, as excise tax is imposed on all fuels intended for transport in many countries. Meanwhile, the bioliquid is treated differently.
3. The bioliquid is characterised by a lower calorific value, therefore when powering engines with this fuel one has to account for an increase in consumption of about 13%. This is due to the fact that the calorific value of diesel oil as defined is about 43 MJ/kg, while that of bioliquid produced from used cooking oils is about 38.2 MJ/kg.
4. 71,926 dm^3 (five days a week) of bioliquid would be enough to power 591 50kVA generators working for 8 h a day. This implies that in the event of power supply shortages during disasters or war, it would be possible to supply power to about 591 distinct types of facilities, small businesses, or houses, for instance, in a small town.

5. 71 926 dm^3 (five days a week) of bioliquid would be enough to power 38 1000kVA generators working for 8 h a day. This implies that in the event of power shortages, it would be possible to supply a residential area of detached houses, a small hospital or a small military unit.
6. If an ecological bioliquid were used to power a unit with a nominal power of 1000 kVAh working for 8 h a day, it would turn out that over the course of a year a very large amount, as much as about 1124.4 tons of CO_2, would not have been emitted. In such a situation, it is possible to obtain a certificate of avoided emissions, which can be sold on the European ETS exchange and gain approximately 68,925.7 €.
7. If we assume that 71 926dm3 (five days a week) of bioliquid can be obtained from cooking oils per day, and that a Caterpillar DE50E0 generator consumes 121.6 dm3 of bioliquid in 8 h, we find that approximately 0.4 MWh of electricity can be produced in one day. However, since the bioliquid would be enough for 591 units, it turns out that within 1 day, 236.4 MWh of electrical energy could be produced. If 38 units of Volvo FOGO FDT 1000VA were used, which operate with higher efficiency, 304 MWh of electricity could be produced daily.

In summary, it can be concluded that bioliquid is and can be used for so-called green electricity production. In addition, apart from the economic effect, a positive ecological effect can also be achieved.

References

1. Gil, A.: Challenges on waste-to-energy for the valorization of industrial wastes: Electricity, heat and cold, bioliquids and biofuels. Environmental Nanotechnology. Monit. Manag. **17**, 100615 (2022). https://doi.org/10.1016/j.enmm.2021.100615
2. Infografika: Emisje gazów cieplarnianych w Unii Europejskiej wg sektora i krajów [Online]. https://www.europarl.europa.eu/topics/pl/article/20180301STO98928/infografika-emisje-gazow-cieplarnianych-w-unii-europejskiej (dostęp 02.04.2024)
3. European Parliament, CO_2 emissions from cars: facts and figures (infographics), 22–03–2019 [Online]. https://www.europarl.europa.eu/topics/en/article/20190313STO31218/co2-emissions-from-cars-facts-and-figures-infographics (dostęp 02.04.2024)
4. United States Environmental Protection Agency, Sources of Greenhouse Gas Emissions. https://www.epa.gov/ghgemissions/sources-greenhouse-gas-emissions (dostęp 02.04.2024)
5. Kazancoglu, Y., Ozbiltekin-Pala, M., Ozkan-Ozen, Y.D.: Prediction and evaluation of greenhouse gas emissions for sustainable road transport within Europe. Sustain. Cities Soc. **70**, 102924 (2021). https://doi.org/10.1016/j.scs.2021.102924
6. U.S. Energy Information Administration, Carbon Dioxide Emissions Coefficients, September 7, 2023 [Online]. https://www.eia.gov/environment/emissions/co2_vol_mass.php (dostęp 25.03.2023)
7. KOBiZE: Calorific values (WO) and CO_2 emission factors (EC) in 2020 for reporting under the Emissions Trading Scheme for 2023. Warsaw (2022). https://www.kobize.pl/uploads/materialy/materialy_do_pobrania/monitorowanie_raportowanie_weryfikacja_emisji_w_eu_ets/WO_i_WE_do_monitorowania-ETS-2023.pdf (dostęp 15.03.2024)
8. Directive (EU) 2009/28/EC of the European Parliament and of the Council of 23 April 2009 on the promotion of the use of energy from renewable sources (2009)

9. Directive (Eu) 2018/2001 of the European Parliament and of the Council of 11 December 2018 on the promotion of the use of energy from renewable sources (recast)
10. Act of 20 February 2015 on renewable energy sources. Dz.U. 2015 poz. 478
11. OECD/FAO: OECD-FAO Agricultural Outlook 2023–2032, OECD Publishing, Paris (2023). https://doi.org/10.1787/08801ab7-en
12. Country Reports Brazil—2018 update Bioenergy policies and status of implementation. 09 2018. https://www.ieabioenergy.com/wp-content/uploads/2018/10/CountryReport2018_Brazil_final.pdf (dostęp 04.04.2024)
13. European Commission, Communication from The Commission to The European Parliament, The European Council, The Council, The European Economic and Social Committee and The Committee of The Regions, The European Green Deal, Brussels, 11.12.2019 COM(2019) 640 final
14. Seljak, T., Buffi, M., Valera-Medina, A., Chong, C.T., Chiaramonti, D., Katraˇsnik, T.: Bioliquids and their use in power generation—a technology review. Renew. Sustain. Energy Rev. **129**, 109930 (2020). https://doi.org/10.1016/j.rser.2020.109930
15. Park, H.Y., Han, K., Yu, G.S, Jang, J., Park, S., Kim, H. Hee, M.K., Kim, J.-K.: Properties of bioliquids and their impacts on combustion and boiler operation. Energy **193**, 116782 (2020). https://doi.org/10.1016/j.energy.2019.116782
16. Park, H.Y., Han, K., Kim, H.H., Park, S., Jang, J., Yu, G.S., Ko, J.H.: Comparisons of combustion characteristics between bioliquid and heavy fuel oil combustion in a 0.7 MWth pilot furnace and a 75 MWe utility boiler. Energy **192**, 116557 (2020). https://doi.org/10.1016/j.energy.2019.116557
17. Pawlak-Kruczek, H., Ostrycharczyk, M., Zgo´ra, J.: Co-combustion of liquid biofuels in P Cboilers of 200MW utility unit. Proc. Combust. Inst. **34**, 2769–2777 (2013). https://doi.org/10.1016/j.proci.2012.08.010
18. Wolf, J.P., Dong: Biomass combustion for power generation: an introduction. Biomass Combustion Science, Technology and Engineering. A volume in Woodhead Publishing Series in Energy 3–8 (2013). https://doi.org/10.1533/9780857097439.1.3
19. Tonini, D., Dorini, G., Astrup, T.F.: Bioenergy, material, and nutrients recovery from household waste: advanced material, substance, energy, and cost flow analysis of a waste refinery process. Appl. Energy **121**, 64–78 (2014). https://doi.org/10.1016/j.apenergy.2014.01.058
20. Caetano, B.C.N., Hanriot, V.M., Sandoval, O.R., Huebner, R.: Energy conversion of biogas from livestock manure to electricity energy using a Stirling engine. Energy Convers. Manage. **15**, 100224 (2022). https://doi.org/10.1016/j.ecmx.2022.100224
21. PN-EN 14214+A2:2019–05 Liquid petroleum products—Fatty acid methyl esters (FAME) for use in compression ignition (diesel) car engines and heating applications—Requirements and test methods
22. Kurczyński, D.: Eeffect of the RME biodiesel on the diesel engine fuel consumption and emission. Communications **23**(4), 308–316 (2021). https://doi.org/10.26552/com.C.2021.4.B308-B316
23. Kurczyński, D., Łagowski, P.: Performance indices of a common rail-system CI engine powered by diesel oil and biofuel blends. J. Energy Inst. **92**(6), 1897–1913 (2019). https://doi.org/10.1016/j.joei.2018.11.004
24. Issariyakul, T., Dalai, A.K.: Biodiesel from vegetable oils. Renew. Sustain. Energy Rev. **31**, 446–471 (2014). ISSN: 1364–0321. https://doi.org/10.1016/j.rser.2013.11.001
25. Balat, M.: Potential alternatives to edible oils for biodiesel production—a review of current work. Energy Convers. Manage. **52**(2), 1479–1492 (2011). https://doi.org/10.1016/j.enconman.2010.10.011
26. Aldhaidhawi, M., Chiriac, R., Badescu, V.: Ignition delay, combustion and emission characteristics of diesel engine fueled with rapeseed biodiesel—a literature review. Renew. Sustain. Energy Rev. **73**, 178–186 (2017). https://doi.org/10.1016/j.rser.2017.01.129
27. Salimi, Z., Hosseini, S.A.: Study and optimization of conditions of biodiesel production from edible oils using ZnO/BiFeO$_3$ nano magnetic catalyst. Fuel **239**, 1204–1212 (2019). https://doi.org/10.1016/j.fuel.2018.11.125

28. Aleman-Ramirez, J.L., Okoye, P.U., Pal, U., Sebastian, P.J.: Agro-industrial residue of *Pouteria sapota* peels as a green heterogeneous catalyst to produce biodiesel from soybean and sunflower oils. Renewable Energy **224**, 120163 (2024). https://doi.org/10.1016/j.renene.2024.120163
29. Devarajan, Y.: Inedible oil feedstocks for biodiesel production: a review of production technologies and physicochemical properties. Sustain. Chem. Pharm. **30**, 100840 (2022). https://doi.org/10.1016/j.scp.2022.100840
30. Yadav, A.K., Khan, M.E., Dubey, A.M., Pal, A.: Performance and emission characteristics of a transportation diesel engine operated with non-edible vegetable oils biodiesel. Case Stud. Therm. Eng. **8**, 236–244 (2016). https://doi.org/10.1016/j.csite.2016.08.001
31. Kurczyński, D., Wcisło, G., Łagowski, P., Leśniak, A., Kozak, M., Bolesław, P.: Determination of the effect of the addition of second-generation biodiesel BBuE to diesel fuel on selected parameters of "B" fuels. Energies **16**, 6999 (2023). https://doi.org/10.3390/en16196999
32. Kurczyński, D., Łagowski, P., Wcisło, G.: Experimental study into the effect of the second-generation BBuE biofuel use on the diesel engine parameters and exhaust composition. Fuel **284**(2021), 1–16 (2020). https://doi.org/10.1016/j.fuel.2020.118982Received18March2020
33. Demirbas, A.: Potential resources of non-edible oils for biodiesel. Energy Source Part B **4**, 310–314 (2009). https://doi.org/10.1080/15567240701621166
34. Wang, H., Zhou, H., Yan, Q., Wu, X., Zhang, H.: Superparamagnetic nanospheres with efficient bifunctional acidic sites enable sustainable production of biodiesel from budget non-edible oils. Energy Convers. Manage. **297**, 117758 (2023). https://doi.org/10.1016/j.enconman.2023.117758
35. Ulakpaa, W.C., Ulakpab, R.O.E., Egwunyengac, M.C., Egbosiuba, T.C.: Transesterification of non-edible oil and effects of process parameters on biodiesel yield. Clean. Waste Syst. **3**, 100047 (2022). https://doi.org/10.1016/j.clwas.2022.100047
36. Akubude, V.C., Nwaigwe, K.N., Dintwa, E.: Production of biodiesel from microalgae via nanocatalyzed transesterification process: a review. Mater. Sci. Energy Technol. **2**(2), 216–225 (2019). https://doi.org/10.1016/j.mset.2018.12.006
37. Sun, J., Xiong, X., Wang, M., Du, H., Li, J., Zhou, D., Zuo, J.: Microalgae biodiesel production in China: a preliminary economic analysis. Renew. Sustain. Energy Rev. **104**, 296–306 (2019). https://doi.org/10.1016/j.rser.2019.01.021
38. Chen, J., Li, J., Dong, W., Zhang, X., Tyagi, R.D., Drogui, P., Surampalli, R.Y.: The potential of microalgae in biodiesel production. Renew. Sustain. Energy Rev. **90**, 336–346 (2018). https://doi.org/10.1016/j.rser.2018.03.073
39. Rajak, U., Nashine, P., Verma, T.N.: Assessment of diesel engine performance using spirulina microalgae biodiesel. Energy **166**, 1025–1036 (2019). https://doi.org/10.1016/j.energy.2018.10.098
40. Faried, M., Samer, M., Abdelsalam, E., Yousef, R.S., Attia, Y.A., Ali, A.S.: Biodiesel production from microalgae: processes, technologies and recent advancements. Renew. Sustain. Energy Rev. **79**, 893–913 (2017)
41. Brennan, L., Owende, P.: Biofuels from microalgae—a review of technologies for production, processing, and extractions of biofuels and co-products. Renew. Sustain. Energy Rev. **14**, 557–577 (2010). https://doi.org/10.1016/j.rser.2009.10.009 Get rights and content
42. Meraz, R.M., Rahman, Md.M., Hassan, T., Rifat, A. Al, Adib, A.R.: A review on algae biodiesel as an automotive fuel. Bioresour. Technol. Rep. **24**, 101659 (2023). https://doi.org/10.1016/j.biteb.2023.101659
43. Kirubakaran, M., Arul Mozhi Selvan, V.: A comprehensive review of low cost biodiesel production from waste chicken fat. Renew. Sustain. Energy Rev. **82**(1), 390–401 (2018). https://doi.org/10.1016/j.rser.2017.09.039
44. Adewale, P., Dumont, M.-J., Ngadi, M.: Recent trends of biodiesel production from animal fat wastes and associated production techniques. Renew. Sustain. Energy Rev. **45**, 574–588 (2015). https://doi.org/10.1016/j.rser.2015.02.039
45. Kurczyński, D., Łagowski, P., Wcisło, G.: Experimental study of fuel consumption and exhaust gas composition of a diesel engine powered by biodiesel from waste of animal origin. Energies **14**, 3472 (2021). https://doi.org/10.3390/en14123472

46. Kurczyński, D., Wcisło, G., Leśniak, A., Kozak, M., Łagowski, P.: Production and testing of butyl and methyl esters as new generation biodiesels from fatty wastes of the leather industry. Energies **15**(22), 8744–8744 (2022). https://doi.org/10.3390/en15228744
47. Arumugam, A., Ponnusami, V.: Production of biodiesel by enzymatic transestryfication ofwaste sardine oil and evaluation of its engine performance. Heliyon **3**(12), e00486 (2017). https://doi.org/10.1016/j.heliyon.2017.e00486
48. Odetoye, T.E., Agu, J.O., Ajala, E.O.: Biodiesel production from poultry wastes: waste chicken fat and eggshell. J. Environ. Chem. Eng. **9**(4), 105654 (2021). https://doi.org/10.1016/j.jece.2021.105654
49. Chakraborty, R., Sahu, H.: Intensification of biodiesel production from waste goat tallow Rusing infrared radiation: process evaluation through response surfach methodology and artificial neural network. Appl. Energy **114**, 827–836 (2014). https://doi.org/10.1016/j.apenergy.2013.04.025
50. Sahar, S.S., Iqbal, J., Ullah, I., Bhatti, H.N., Nouren, S., Habib-Ur-Rehman, N.J., Iqbal, M.: Biodiesel production from waste cooking oil: an efficient technique to convert waste into biodiesel. Sustain. Cities Soc. **41**, 220–226 (2018). https://doi.org/10.1016/j.scs.2018.05.037
51. Fonseca, M.J., Teleken, J.G., de Cinque Almeida, V., da Silva, C.: Biodiesel from waste frying oils: methods of production and purification. Energy Convers. Manage. **184**, 205–218 (2019). https://doi.org/10.1016/j.enconman.2019.01.061
52. Gad, M.S., Abdel Azizb, M.M., Kayed, H.: Impact of different nano additives on performance, combustion, emissions and exergetic analysis of a diesel engine using waste cooking oil biodiesel. Propuls. Power Res. **11**(2), 209–223 (2022). https://doi.org/10.1016/j.jppr.2022.04.004
53. Kassem, Y., Çamur, H.: Effects of storage under different conditions on the fuel properties of biodiesel admixtures derived from waste frying and canola oils. Biomass Convers. Biorefinery **8**, 825–845 (2018). https://doi.org/10.1007/s13399-018-0339-1
54. Abed, K.A., El Morsi, A.K., Sayed, M.M., El Shaib, A.A., Gad, M.S.: Effect of waste cooking-oil biodiesel on performance and exhaust emissions of a diesel engine. Egypt. J. Pet. **27**, 985–989 (2018). https://doi.org/10.1016/j.ejpe.2018.02.008
55. Borah, M.J., Das, A., Das, V., Bhuyan, N., Deka, D.: Transesterification of waste cooking oil for biodiesel production catalyzed by Zn substituted waste egg shell derived CaO nanocatalyst. Fuel **242**, 345–354 (2019). https://doi.org/10.1016/j.fuel.2019.01.060
56. Lin, Y.C., Hsu, K.H., Chen, C.B.: Experimental investigation of the performance and emissions of a heavy-duty diesel engine fueled with waste cooking oil biodiesel/ultra-low sulfur diesel blends. Energy **36**, 241–248 (2011). https://doi.org/10.1016/j.energy.2010.10.045
57. Adhikesavan, C., Ganesh, D., Charles, A.V.: Effect of quality of waste cooking oil on the properties of biodiesel, engine performance and emissions. Clean. Chem. Eng. **4**, 100070 (2022). https://doi.org/10.1016/j.clce.2022.100070
58. Razzaq, L., Imran, S., Anwar, Z., Farooq, M., Abbas, M.M., Khan, H.M., Asif, T., Amjad, M., Soudagar, M.E.M., Shaukat, N., Fattah, I.M.R., Rahman, S.M.A.: Maximising yield and engine efficiency using optimised waste cooking oil biodiesel. Energies **13**, 5941 (2020). https://doi.org/10.3390/en13225941
59. Niyas, M.M., Shaija, A.: Performance evaluation of diesel engine using biodiesels from waste coconut, sunflower, and palm cooking oils, and their hybrids. Sustain. Energy Technol. Assess.S **53**(Part C), 102681 (2022). https://doi.org/10.1016/j.seta.2022.102681
60. Uddin, M. N., Techato, K., Rasul, M.G., Hassan, N.M.S, Mofijur, M.: Waste coffee oil: a promising source for biodiesel production. Energy Procedia **160**, 677–682 (2019). https://doi.org/10.1016/j.egypro.2019.02.221
61. Chepeliev, M., Diachuk, O., Podolets, R., Trypolska, G.: The role of bioenergy in Ukraine's climate mitigation policy by 2050. Renew. Sustain. Energy Rev. **152**, 111714 (2021). https://doi.org/10.1016/j.rser.2021.111714
62. Moriarty, P., Yan, X., Wang, S.J.: Liquid biofuels: not along–term transport solution. Energy Procedia **158**, 3265–3270 (2019). https://doi.org/10.1016/j.egypro.2019.01.986

63. Campbell, J.E., Lobell, D.B., Field, C.B.: Greater transportation energy and GHG offsets from bioelectricity than ethanol. Science **324**, 1055–1057 (2009). https://doi.org/10.1126/science.1168885
64. PN-EN 590:2022–08 Fuels for motor vehicles—Diesel fuels—Requirements and test methods
65. Riaboshapko, A.: Assessment of the Suitability of Biofuels Produced from Waste Materials for Powering Generators. Engineering thesis, University of Agriculture in Krakow, Krakow (2024)

Strategy for Improving Environmental and Economic Indicators of Road Transport Power Systems

Marchenko Andrii, Parsadanov Igor, Marchenko Natalia, and Rykova Inna

Abstract The future security of humankind is closely tied to reducing environmental pollution caused by harmful emissions and greenhouse gases from road transport. Technical solutions for minimizing diesel engine exhaust emissions have led to notable progress. Over the past 30 years, the efficiency of diesel engines has improved 4.17 times based on fuel and environmental criteria. Further advancements focus on reducing harmful emissions, improving fuel efficiency, energy performance, reliability, using low-carbon hybrid fuels, and implementing decarbonisation. A proposed criterion quantifies decarbonisation measures, allowing comparative calculations for different engine types and modernization stages. Comprehensive assessment of diesel engines is recommended using both fuel and environmental criteria and a new decarbonisation criterion. For Ukraine, enhancing transport diesel engines and adopting strategies to mitigate environmental impacts are crucial for post-war recovery, aiming to rebuild critical economic sectors with advanced technologies.

M. Andrii (✉)
National Technical University "Kharkiv Polytechnic Institute", Kharkiv, Ukraine
e-mail: andrii.marchenko@khpi.edu.ua

P. Igor · R. Inna
Department of Engines and Hybrid Powerplants, Educational and Scientific Institute of Power Engineering, Electronics and Electromechanics, National Technical University "Kharkiv Polytechnic Institute, Kharkiv", Ukraine
e-mail: igor.parsadanov@khpi.edu.ua

M. Natalia
Department of System Analysis and Information and Analytical Technologies, Educational and Scientific Institute of Computer Sciences and Information Technologies, National Technical University "Kharkiv Polytechnic Institute", Kharkiv, Ukraine

© The Author(s), under exclusive license to Springer Nature Switzerland AG 2025
A. Marchenko and Z. Koruba (eds.), *Decarbonization of Transport Energy Installations in the Context of Sustainable Development Strategies*, Studies in Systems, Decision and Control 635, https://doi.org/10.1007/978-3-032-05884-3_12

1 Introduction

Rational use of energy, environmental pollution and climate change caused by the increase in global temperature on our planet are among the main problems of ensuring the security of human existence both now and in the future.

The reasons for the increase in global temperature are associated with a significant increase in greenhouse gas emissions. In turn, the increase in greenhouse gas emissions to the environment, in addition to natural factors, is also a consequence of the implementation of strategies for the economic activity of mankind aimed at increasing production and consumption capabilities, a significant increase in the population of the planet, a complex of factors associated with the negative impact of human activity on the environment.

Despite the fact that each of the known factors of influence can be considered key in terms of its role in increasing global temperature and climate change on our planet, most often both in the press and in scientific circles, problems related to the trends of increasing energy consumption by mankind in production and everyday life are discussed. In addition, significant attention is paid to the above-mentioned problem in transport, where the main factor is internal combustion engines, as the main polluter of the environment. In recent years, the problems of decarbonization when using petroleum-based fuels have become one of the main tasks of the development and use of transport in leading countries of the world.

It is necessary to note the high relevance and complexity of the study of the processes that determine the current warming of the planet's climate and the existing uncertainty in the prognostic assessments of this phenomenon due to the change in the moving balance of greenhouse gases in the atmosphere. Therefore, from the entire complex of these problems, the publication considers the problem of the negative impact on the environment and the planet's climate only of power plants (PPs) with internal combustion engines. The strategies, prospects and directions for improving the effective and environmental performance of ICEs as a significant technological tool that ensures the implementation of sustainable development goals while unconditionally ensuring the ideology and technologies of decarbonization are considered.

It can be argued that the development of civilization in the twentieth century was largely ensured by the use of internal combustion engines as power units in transport, agriculture, construction, etc. Due to a number of advantages related to mobility, economy, reliability, operational life, it is these power plants with internal combustion engines that have found wide application and provide the basic needs of humanity in mechanical energy and, of course, primarily in road transport [1–4]. At the same time, most of the transportation of both freight and passenger vehicles, as well as the performance of agricultural and construction work, is carried out using diesel engines, which, as is known, have the highest efficiency (efficiency) among all thermal engines and the possibility of implementing high capacities in one unit, which ensures the demand for this type of internal combustion engine.

Recently, due to the development and practical implementation of technical solutions aimed at the rational use of natural resources and increasing environmental efficiency, significant progress has been made in reducing emissions of harmful substances with exhaust gases from diesel engines. According to the results of a study conducted at the National Technical University "Kharkiv Polytechnic Institute" [5], it was found that the environmental efficiency of diesel ICE cars has improved more than 4 times over the past three decades. At the same time, the value of emissions of standardized harmful substances—particulate matter, hydrocarbons, nitrogen oxides and carbon, is approaching those minimum levels that can be reliably determined only in special laboratories, especially for the determination of particulate matter.

Taking into account the current problems and trends in motor transport related to the need to reduce greenhouse gas (CO_2) emissions into the environment, it can be predicted that in the future, as now, the strategies for the development, manufacture and operation of internal combustion engines, including diesel engines, will remain relevant. These strategies will involve the widespread use of Industry 4.0 and Industry 5.0 technologies. It can also be predicted that in the near future the trend towards the spread of research and practical use of hybrid fuels with a lower carbon content and the use of "green hydrogen" will become relevant. In this regard, problems arise in the effective use of the methodology for assessing the energy and environmental excellence of internal combustion engines.

It is inevitable that for Ukraine, the implementation of technical solutions aimed at improving diesel engines in transport, along with the introduction of new strategies to reduce the negative impact of power plants on the environment and global warming, will be implemented in conjunction with the implementation of a set of urgent tasks related to post-war recovery and reconstruction, and the tasks of innovative development of a country destroyed by the war.

2 Dynamics and Prospects for Improving the Energy, Economic and Environmental Performance of Diesel Engines for Road Transport

Modern strategies for improving heat engines in transport primarily involve reducing their negative impact on the environment. In addition, trends in the implementation of sustainable development objectives indicate the need to increase the total energy potential in the technologies of human activity, including transport. It should also be noted that the development of transport today is largely determined by the development of internal combustion engines, which in recent years have been "designated" as the main culprits in environmental pollution. And although ICEs are not in the first place in the list of major global environmental pollutants [3], strategies for improving their environmental characteristics are now emerging as the main ones.

Due to a number of advantages in terms of mobility, economy, reliability, and a significant operational resource, these engineering products are today the main source of mechanical energy in transport [6, 7]. The great importance of internal combustion engines is also evidenced by the fact that currently more than 80% of the total volume of world transportation is accounted for by road transport, which includes freight, passenger cars, special vehicles, etc. It is also necessary to take into account the fact that most freight transportation, as well as passenger (bus) transportation, is carried out using reliable diesel engines, which among all thermal engines still hold the lead in terms of fuel efficiency.

The wide use of diesel engines is facilitated by their advantages in comparison with spark-ignition engines running on gasoline.

Unlike gasoline engines, the indicator efficiency of which decreases at low load modes, in diesel engines the indicator efficiency values increase with decreasing load. And as a result, diesel engines provide up to 30% lower fuel consumption in operation compared to gasoline internal combustion engines.

At the same time, despite the above-mentioned advantages of diesel engines, their widespread use is limited. And this limitation is associated with two main factors. The first of these is the above-mentioned legislative bans on the use of diesel internal combustion engines. And the second factor is determined by the amount of diesel fuel that can be obtained from oil (at present, this amount does not exceed 30%), as well as the availability of oil itself. Thus, it can be argued that fuel efficiency, being the main resource-saving indicator, primarily determines the technical level and competitiveness of internal combustion engines, including such an important indicator today as the volume of CO_2 emissions with exhaust gases of internal combustion engines during vehicle operation.

All this determines the relevance of questions about criteria, or markers, which at this extremely important stage of today will allow determining the level of environmental friendliness of transport internal combustion engines, as well as when comparing the values of the criteria and the effectiveness of the selected strategies.

3 Criteria for Assessing the Environmental Friendliness and Economy of Internal Combustion Engines for Land Transport

The authors of the studies propose to divide the use of the main criteria for assessing the environmental friendliness and economy of internal combustion engines for land transport into three periods of the beginning of their main use.

The first stage—from the first stages of using the assessment criteria to the present.

The second stage—the beginning of the implementation of EURO standards (approximately 1990).

The third stage started in 2015 with the adoption of the UN Memorandum on Sustainable Development and the development of state and interstate strategies for decarbonization in transport.

Since the beginning of the era of internal combustion engines, the main criteria have become energy, which characterize fuel consumption per unit of work performed. So, when conducting research, the fuel economy of an internal combustion engine is most often determined by its specific consumption at the rated power ($g_{e\ norm}$) and maximum torque ($g_{e\ mkr}$). Usually, these markers allow you to assess the degree of perfection of the working process and the fuel efficiency of the internal combustion engine in operation.

However, this value differs from the value of specific operating fuel consumption, since it does not take into account fuel consumption at partial loads. Therefore, in research and testing, the fuel efficiency of a car engine, along with the $g_{e\ norm}$ and $g_{e\ mkr}$ indicators, can also be determined by the average operating fuel consumption $g_{e\ ser.ex}$ over the test cycle [5], provided that this cycle sufficiently corresponds to the operating conditions. According to the results of the analysis of numerous tests of diesel engines of trucks [5, 6], it was established that the average operating fuel consumption ($g_{e\ ser.ex}$), determined in accordance with the model of operation of the truck engine over the ESC cycle, is (1.04–1.06) of $g_{e\ norm}$. The above markers also allow for the calculation of the volume of CO_2 emissions with exhaust gases of the internal combustion engine during vehicle operation.

In addition, directly in operation, the fuel efficiency of road transport is determined in kg (liters) per 100 km.

Thus, the above-mentioned criteria for specific fuel consumption of transport internal combustion engines provide the possibility of comparative assessment of both different types and designs of internal combustion engines, and when modernizing specific internal combustion engines. These criteria—markers are still relevant today.

The effectiveness of internal combustion engine development strategies was also evaluated using these criteria, among which the following are still relevant today [1, 6, 7]:

- increasing the efficiency of mixture formation, combustion, and heat generation by increasing pressure and reducing the duration of fuel injection;
- using highly efficient gas turbine supercharging systems and intermediate cooling of charge air;
- ensuring optimal regulation taking into account real operating conditions;
- reducing friction losses in pistons, in bearings, losses in the drive of engine components and assemblies;
- introducing new materials and technologies in the manufacture and organization of engine combustion processes.

The results of the implementation of these measures can be shown as follows: specific fuel consumption at nominal power by years: $g_{e\ norm}$ = 218—222 g/(k•Wh) in 1990; $g_{e\ norm}$ = 215—217 g/(k·Wh) in 2000; $g_{e\ norm}$ = 211—213 g/(k·Wh) in 2020. Accordingly, the operational fuel efficiency indicators of vehicles have also improved.

Thus, the implementation of measures to increase fuel efficiency has allowed over the past thirty years to reduce the specific operational fuel consumption of diesel engines of trucks by more than 5% [5]. However, the authors of the studies increasingly emphasized that the reserves for further improvement of internal combustion engines, and primarily diesel engines, are significantly reduced with the increase in their efficiency [8, 9]. At the same time, experts emphasize that ICEs today have a significant application segment, and in the near future they will still have practical application [10, 11]. These new challenges require the intensification of the entire complex of research on improving the efficiency of internal combustion engines. The problem of decarbonization in transport is becoming a separate and extremely important one. In this case, it is certainly necessary to pay attention first of all to the use of "green" hydrogen [12, 13], the strategies and technologies of which use should be of primary importance both for developers and manufacturers of engines, and for enterprises operating them.

Many factors co ntribute to the solution of the above-mentioned tasks, not only technical progress and competition between manufacturers, but also the requirements for increasing the level of fuel and environmental indicators, which are primarily imposed on modern power plants by international organizations. In recent years, these requirements have increased most significantly for diesel engines of freight vehicles, which, accordingly, has led to significant changes in the level of their fuel and environmental performance.

It should be emphasized that the second stage—the beginning of the implementation of EURO standards raised the relevance of issues of assessing measures to improve internal combustion engines while taking into account their fuel efficiency and environmental friendliness. These issues arose most acutely in relation to diesel internal combustion engines. In the early 2000s, a fuel-ecological criterion was proposed that allowed solving this problem [6, 7].

According to the specified method, a comparative assessment of the level of fuel and environmental indicators of engines is carried out comprehensively, taking into account operating conditions (operating model), climatic features of operation, and also taking into account the cost of fuel.

As the main indicator when using this method, a complex fuel-ecological criterion is taken, which is determined by the formula:

$$\mathcal{K}_{\text{fe}} = \eta_{\text{ av.oper.}} \cdot \beta.$$

In the above formula:

$\eta_{e \text{ av.oper.}}$—average operational effective efficiency;

β—coefficient characterizing the ratio of fuel costs to total fuel costs and compensation for environmental damage from the negative impact of exhaust gases on the environment during the operation of an automobile engine, referred to a unit of power.

The average operational effective efficiency is determined by the relationship

$$\eta_{e \text{ av.oper.}} = 360/H_u \cdot g_{\text{e ser.ex}}$$

where H_u—is the lower calorific value of the fuel; for diesel engines of trucks (in calculations it is taken as 42.7 mJ/kg), and $g_{\text{e ser.ex}}$ kg/(kWh)—is the specific average operational fuel consumption, which can be determined, for example, during bench tests at representative engine operating modes selected in accordance with the model of its operation. Such operation models include engine test cycles (for example, the ESC cycle).

The β coefficient is determined by the expression:

$$\beta = F_c / F_{c.ec.},$$

where F_c and $F_{c.ec.}$ are, respectively, fuel costs and total fuel costs and compensation for environmental damage from the harmful effects of exhaust gases on the environment during vehicle operation, attributed to a unit of power.

Environmental damage is damage in monetary equivalent that is directly related to compensation for the negative impact on the environment during vehicle operation, and which in the calculations are determined per unit of energy produced by the engine.

In accordance with the requirements of the UNECE Regulations, the assessment of the level of environmental performance of diesel engines of freight vehicles is carried out during tests on a motor stand using the ESC cycle—European Steady State Cycle. This cycle includes 13 stationary modes that take into account the real operating conditions of trucks.

According to the ESC cycle, it is envisaged to conduct engine tests with a load at three crankshaft rotation frequencies, shifted towards values lower than the nominal frequency. The results of the tests allow determining the average specific emission value of each of the standardized toxic substances in g/(kW h), and comparing their values with the level established by the standards.

As an example of the practical application of the complex fuel-ecological criterion, data from the results of research on the efficiency of using various types of alternative fuels (Fig. 1) in a modern four-stroke engine with a cylinder displacement of 7.5–10 dm^2 (liters), namely water-fuel emulsion (WFE), rapeseed methyl ether (RME), compressed natural gas (CNG) in comparison with diesel fuel (DF) are presented [14].

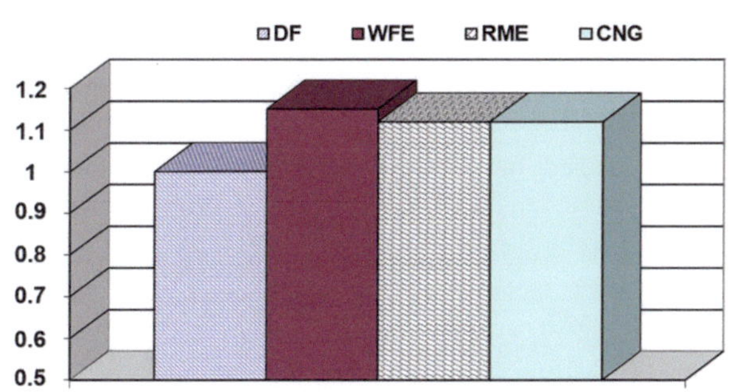

Fig. 1 Relative change in the fuel-environmental criterion of an automobile diesel engine when using different types of alternative fuels compared to diesel fuel [14]

The assessment of the data obtained as a result of the research, in accordance with the determination of the efficiency of using various types of alternative fuels in a diesel engine, allowed us to draw the following conclusions:

- the use of the specified alternative fuels provides a positive effect (mainly due to an increase in environmental efficiency in engine operating modes at maximum load at reduced speeds);
- when using the gas-diesel cycle, the complex criterion of fuel efficiency and toxicity of exhaust gases of a truck diesel engine increases by 11.9%;
- the complex fuel and environmental criterion increase by the same amount (11.9%) when the truck diesel engine operates on rapeseed oil methyl ether.
- when the truck diesel engine operates on a water-fuel emulsion, the complex criterion of fuel efficiency and toxicity of exhaust gases increases most significantly—by 15%.

The main advantage of assessing economic and environmental performance using a complex criterion is that a generalized approach to determining heterogeneous processes during fuel combustion in engines allows us to determine the effectiveness of the implemented technical solutions. In this case, it is additionally possible to provide justification for the validity of the practical implementation of selected solutions aimed at increasing fuel-economic and environmental efficiency, as well as to propose the most effective of the studied options (excluding possible errors), as well as to predict the effect of implementing the proposed solutions.

In this study, using a complex fuel-environmental criterion, the effectiveness of the implemented technical solutions aimed at improving diesel engines of road transport over the past 30 years was assessed.

As the initial data for assessing the level of environmental performance, the change in requirements for the performance of freight vehicles in accordance with the standards of the UNECE Regulations since 1990, starting from Euro-0 to Euro-6 (Table 1). As for the Euro-VII standard, it is currently still at the approval stage.

It can be additionally noted that the specified norms correspond to the level of indicators of the introduced engines (or those that were being prepared for introduction into production by leading companies), that is, on the one hand, the adoption of

Table 1 Requirements for the emission of harmful substances with exhaust gases of diesel engines of mass-produced vehicles according to current and applicable European standards

EURO standards for vehicles weighing more than 3.5 tons	Specific indicators of harmful substances with exhaust gases of ICE, g/(kWh)			
	NOx	CO	CnHm	PM
EURO-0 (1990)	14.4	11.2	2.4	0.7
EURO-6 (2015)	0.4	1.5	0.13	0.01
Emission reduction factor of harmful substances	36	7.5	18	70

Table 2 Change in the coefficient of relative operating environmental costs and fuel-environmental criterion for diesel engines of trucks over the last 30 years

Year	β	K_{fe}
1990	0.227	0.087
2000	0.478	0.186
2010	0.706	0.280
2020	0.912	0.363

strict norms is the implementation of technical solutions to improve the state of the environment, and on the other hand, pressure on competitors engine manufacturers.

From the data provided, it is clear that in recent years the environmental indicators of diesel engines of motor vehicles have changed significantly, emissions of NO_x, CO, C_nH_m with exhaust gases have significantly decreased with a record decrease of 70 times (!) of particulate matter (PM) in exhaust gases.

Of certain interest to specialists are also the results of calculation data, which allow us to assess the change in the coefficient of relative operational environmental costs and the fuel-environmental criterion itself over the past 30 years (Table 2). As noted, the cost of diesel fuel in the calculations (in hryvnias) was assumed unchanged.

The presented data from the comprehensive assessment of operational fuel and environmental indicators of truck engines allow us to state the following:

- the specific average operational fuel consumption of diesel engines is decreasing, however, the dynamics of improving fuel efficiency is decreasing, which is a natural factor associated with increasing engine efficiency;
- emissions of harmful substances with exhaust gases have decreased significantly, primarily this concerns the most aggressive toxic emissions—particulate matter and nitrogen oxides;
- the level of harmful emissions of all standardized toxic substances is approaching the minimum, which can be ensured by improving the engine operating process and exhaust gas cleaning and neutralization systems;
- due to an increase in the coefficient of relative operating environmental costs, as well as to a certain extent the average operating effective efficiency, the fuel-environmental criterion of a diesel engine of a truck has significantly increased—from 0.087 in 1990 to 0.363 in 2020 (i.e. 4.17 times).

Therefore, the reserves for the further evolution of fuel-environmental indicators of diesel engines are limited and are largely related to improving the operating conditions and characteristics of diesel fuels, as well as the use of alternative fuels, provided that their price is not significantly higher than the price of petroleum fuel.

At the same time, it is necessary to note the fact that additional requirements are currently imposed on internal combustion engines related to reducing carbon dioxide (CO_2) emissions into the atmosphere. Carbon dioxide is not a toxic substance [4, 5], but it is a negative factor affecting the environment as one of the components in the formation of the greenhouse effect.

4 Features of the Harmful Effects and Possibilities of Accounting for Carbon Dioxide (Carbon Dioxide) Emissions on the Human Body and the Environment During the Operation of Diesel Engines

Without delving deeply into the problem of the unambiguous understanding of the causes that determine the warming of the planet's climate, the ways and possibilities of humanity to solve this problem, as well as the existing uncertainty in predicting the impact of greenhouse gases and their effluents in the atmosphere on this phenomenon [4], we will consider the possibility of assessing the level of negative environmental impact of CO_2 emissions during the operation of power plants with diesel engines.

It is known that carbon dioxide is a low-toxic gas without odor and color, is one of the gases that make up the atmosphere. Under normal conditions, CO_2 is a relatively insignificant but important component of atmospheric air, which participates in the process of photosynthesis, metabolism, is released by humans and animals, as well as during the processes of fermentation and decay. For the human body, carbon dioxide is no less important than oxygen, and their balance is maintained by the natural processes of photosynthesis and respiration.

And only increased to 0.2–0.4% (2000–4000 ppm) CO_2 concentrations in the air, which is possible in certain types of production (excluding catastrophes and natural disasters), can cause a suffocating effect and weakness in humans. Carbon dioxide is classified as a hazardous substance of hazard class IV.

However, carbon dioxide emissions with exhaust gases of internal combustion engines are associated with degradation processes of the transition in the "matter—energy" system. The consequence of such processes is the greenhouse effect. In the formation of the "greenhouse effect", as a factor of global disruption of the ecological balance and directly related to global warming, a certain role (along with oxygen-containing hydrocarbons, methane and solid particles) is played by carbon dioxide—a product of complete combustion of hydrocarbon fuels.

Global warming, its causes, dynamics, possible methods of reduction, as well as the degree of influence of power plants on the greenhouse effect, are currently causing serious debates and disputes both among the scientific [3, 7, 8] and non-scientific communities, which require a separate analysis and are not considered in this work.

Let us note only one fact—the increase in the level of CO_2 emissions from the combustion products of hydrocarbon fuels over the last five-year cycles is on average less than 6% of the corresponding increase in the CO_2 content in the troposphere [3].

The assessment of the effectiveness of technical solutions for improving diesel engines in order to reduce fuel consumption and emissions of harmful substances with exhaust gases, as shown above, can be effectively carried out using a dimensionless complex criterion, which allows obtaining information about the level of economic and environmental improvement of engines. However, this criterion does not take into account the environmental impact of CO_2 emissions.

In order to determine and further implement effective measures to reduce carbon dioxide emissions with exhaust gases to the environment during the operation of automobile engines, we will consider the possibilities and conditions for using a comprehensive assessment method similar to the fuel-environmental criterion.

For such an assessment, it is necessary to have data on mass emissions (CO_2) for each of the modes of the operating model and determine the relative aggressiveness index for it. The mass emission of CO_2 for each mode of operation of a diesel engine can be determined with sufficient accuracy by calculation if the elemental composition of the fuel, its molecular weight, calorific value and consumption, composition and consumption of air, as well as environmental parameters are known.

The negative impact of CO_2 directly on humans, according to the ideology of a comprehensive assessment of the fuel-ecological criterion, will be characterized by the relative aggressiveness index of carbon dioxide (α_{CO2}), the value of which, as for other harmful substances, can be determined using the relative danger index of the presence of this substance in the air [7, 9], and the corresponding amendments (Table 3). And it is these relative aggressiveness indices that form the basis of the fuel-ecological criterion. The values of the maximum permissible concentrations (MPC) and the relative aggressiveness index of normalized harmful emissions of exhaust gases of diesel engines and carbon dioxide are given in Table 3.

The values of the MPC in the working area, as well as the average daily MPC for carbon dioxide, were taken as maximum in accordance with the level of international standards. If we compare the values of the MPC of harmful substances of exhaust gas emissions of diesel engines, which are regulated, and carbon dioxide, then the impact of carbon dioxide on the human body and the environment is very insignificant.

Therefore, to take into account the direct negative impact of CO_2 on the environment, it is necessary to introduce amendments that characterize the relative aggressiveness indicator and take into account its impact on climate change due to the greenhouse effect, the impact on the biosphere both in the general natural and anthropogenic balance and in the long term.

In addition, it can be noted that, despite the fact that environmental pollution by particulate matter (PM) is a serious problem, currently there are no MPCs for PM in the exhaust gases of automobile diesel engines. This is due to the complex and ambiguous nature of both their fractional structure and the assessment of their impact on the human body. However, taking into account the total high level of toxicity, the value of the relative aggressiveness indicator for PM is taken equal to 200 (mainly

Table 3 Maximum permissible concentrations and indicators of relative aggressiveness of regulated harmful emissions of exhaust gases and carbon dioxide for diesel engines

Diesel engine exhaust gas emissions	MPC, mg/m^3		Relative aggressiveness index (a_i)
	In the working area	Average daily	
Carbon monoxide, CO	20	3	1
Hydrocarbons, C_nH_m	100	1.5	0.63
Nitrogen oxides, NO_x	2	0.04	27.4
Particulate matter, PM	–	–	200
Carbon dioxide, CO_2	1830	730	0.00012

due to the presence of PM (benzo-alpha-pyrene $C_{20}H_{12}$). Other agreed amendments to EURO standards have also been made, the feasibility of which is beyond doubt.

At the same time, the analysis of the data in Table 3 proves the low practical feasibility of using the ideology of the fuel-ecological criterion to assess the perfection of ICE in decarbonization technologies. This is due to the large discrepancy between the permissible concentrations and indicators of relative aggressiveness of normalized harmful emissions of exhaust gases and CO_2.

At the same time, such an undoubtedly useful assessment of environmental pollution associated with exhaust gas emissions of ICE is extremely relevant both for analyzing the effectiveness of decarbonization technologies for power plants in transport, and in the general context—technologies for reducing thermal pollution of the environment. All this creates the need for a new assessment criterion environmental friendliness and economy of internal combustion engines for land transport, which in this publication was referred to as the third stage above. At the same time, together with the criteria for assessing the perfection of internal combustion engines of the first and second stages, this new criterion should only enrich the capabilities of researchers regarding decarbonization technologies.

In the scientific work [12], a new criterion is proposed that allows assessing the perfection of an internal combustion engine according to the main parameters of its working process, namely its energy efficiency, as well as obtaining data on assessing the effectiveness of the implementation of decarbonization processes. It is shown that this criterion can vary from zero to one ($0 < K < 1$).

At the same time

$$K = k_1 * k_2 = (1 - C_{co2}) * \eta_e/\eta_{\max},$$

where,

$$k_1 = (1 - C_{co2}),$$

C_{co2}—the proportion of carbon in fossil fuel ICE (today, for example, for diesel fuel this figure is 0.85);

$$k_2 = \eta_e/\eta_{\max},$$

η_e—effective efficiency of internal combustion engines;

η_{max}—the maximum possible engine efficiency for a given type of fuel, which is defined as $\eta_{max} = E_Q/Q_x$,

here,

E_Q = exergy of heat supplied to the working fluid in the cylinder of an internal combustion engine during fuel combustion;

Q_x—total amount of heat energy supplied to the working fluid in the internal combustion engine cylinder.

In fact, k_1 characterizes the dynamics of reducing the use of carbon in the fuel of internal combustion engines, and thereby the dynamics of changes in CO_2 emissions with exhaust gases. For example, when using a gas-diesel cycle, the share of carbon C_{co2} in hybrid fuel will be smaller. At the same time, the value of k_1 will increase, and of course, the value of the criterion K.

As for k_2, this indicator characterizes the energy efficiency of internal combustion engines. Its growth shows a decrease in fuel consumption for specific work, and thereby shows the dynamics of reducing thermal pollution per unit of work.

Let us consider the value of the proposed criterion at different values of efficiency from 0.3 to 0.5 and different values of the carbon content in hydrocarbon fuel (Table 4).

The data of the table show that the values of the new criterion improve with increasing efficiency and decreasing the share of carbon in fossil fuels. At the same time, the less carbon in the fuel, the greater the role of efficiency in increasing the values of the criterion.

It can be argued that for almost all transport power plants that are currently in operation or will be manufactured in the coming years, according to the data of Table 4 can be identified by this new criterion. Better indicators of specific power plants by this criterion in relation to analogues will also indicate more effective steps by the developers of these internal combustion engines to implement decarbonization strategies. The data of Table 4 can also be recommended for assessing the effectiveness of modernization measures, or when developing new products of transport internal combustion engines. At the same time, the data of Table 4 relate to internal combustion engines, both diesel and gasoline, operating on gas fuel, etc. Finally, it is possible to recommend introducing into the passport database of a specific transport ICE the value of a new criterion, which will allow to determine the position of this ICE in the system of comparisons with analogues and, most importantly, will be a significant factor in preventing the adoption of populist decisions regarding the bans on the use of certain types of transport ICE. It is possible to predict that the use of

Table 4 The value of the new criterion in the range of ICE efficiency from 0.3 to 0.5 and the change in the carbon content in hydrocarbon fuel from 85 to 0%

Percentage of carbon in fuel	Efficiency of a transport power system based on an internal combustion engine					
	0.3	0.33	0.35	0.4	0.43	0.5
0.85	0.0529	0.0582	0.0618	0.0759	0.0756	0.08823
0.75	0.0882	0.970	0.1029	0.1177	0.1265	0.1471
0.65	0.124	0.1359	0.1441	0.1647	0.1771	0.2059
0.55	0.159	0.1747	0.1853	0.2118	0.2277	0.2647
0.45	0.1941	0.2135	0.2264	0.2588	0.2782	0.3235
0.35	0.2294	0.2523	0.2676	0.3060	0.3288	0.3823
0.25	0.2662	0.2912	0.3088	0.3530	0.3794	0.4411
0.15	0.3000	0.3300	0.3499	0.4000	0.4300	0.5000
0.00	0.3529	0.3882	0.4117	0.4706	0.5059	0.5882

the criterion will allow to more effectively implement the strategies of the European Green Deal [13].

It should also be noted that the proposed criterion does not contradict other criteria that have already been used, and there are also new ones. In our opinion, such as mandatory criteria for use should include CO_2 emissions per 1 km. of the vehicle (Table 5), which is already used by a number of companies—manufacturers of transport ICE (Table 5 is based on open data of the manufacturer).

In addition, it will be useful to determine the total energy costs, for example, per 100 km of the route. These data are especially valuable for comparing transport internal combustion engines of the same power. As shown in Table 5, they can provide unexpected results regarding thermal pollution of the environment, which is the main negative factor that decarbonization strategies are aimed at overcoming. Thus, two serial internal combustion engines of the same power of 130 h.p. have different fuel consumption values, different CO_2CO_2 emissions and different overall thermal pollution indicators. Moreover, for a gasoline internal combustion engine, these indicators are significantly worse. This is certainly a separate case; this is a separate unique situation. But this example shows the high relevance of research into internal combustion engines for transport, the high relevance of criteria for objectively assessing the degree of perfection of internal combustion engines. These data are especially important for research centers specializing in the development and modernization of internal combustion engines.

Thus, it can be concluded that the proposed criteria, together with previously known ones, can become an important and significant factor in determining the effectiveness of modernization and the creation of internal combustion engines in transport within the framework of the UN Memorandum on Sustainable Development and decarbonization strategies.

Table 5 New criteria for assessing decarbonization in transport

NEW PEUGEOT 2008, according to the manufacturer of the ICE has 130 h.p. (−97 kW)		
Fuel consumption, l/100 km	Petrol	Diesel fuel
City	8.4	4.4
Road	5.4	3.8
Mixed cyde	6,5 (148 g/km CO_2)	4,0 (110 g /km CO_2)
Calculation of thermal pollution kJ/100 km for AI-95 petrol with a specific mass of 750 kg/m3 and a lower heat of combustion of 44,000 kJ/kg, for diesel fuel, respectively, 840 kg/m3 and 43,000 kJ/kg		
Thermal pollution, kJ/100 km	Petrol	Diesel fuel
City	277 000	159 000
Road	178 000	137 000
Mixed cyde	214 500	144 500

5 Conclusions

Rational and efficient use of energy, taking into account the growing need for it by humanity, is one of the urgent problems that directly determine the further existence of humanity. One of the main sources of mechanical energy is internal combustion engines, which are most widely used in road transport due to the indicators achieved, especially in recent years, fuel economy, environmental friendliness, reliability and resource.

It is shown that at different stages of development of internal combustion engines, various criteria were used to assess their perfection, which are still widely used today. At the first stage, until 1990, criteria for specific fuel consumption of internal combustion engines were used.

At the second stage, with the advent of EURO standards that regulated emissions of harmful substances with exhaust gases, the problem of a comprehensive assessment by fuel and environmental indicators arose. Over time, such a criterion was proposed and showed its high efficiency. The assessment, conducted using the fuel-ecological criterion, allowed us to establish that the efficiency of diesel engines of motor vehicles has increased more than 4 times over the past 30 years. The level of harmful emissions of all regulated toxic substances is approaching the minimum, which can be ensured by improving the engine's working process and systems for cleaning and neutralizing exhaust gases. Together with other design and technological measures, high fuel-ecological indicators have been achieved, which are significantly better than previous models of internal combustion engines. It is also shown that, given the current state of the increase in the average ambient temperature, the state of world economies and, first of all, war-torn Ukraine, the use of "green" hydrogen in energy and transport creates real conditions for the implementation of sustainable development strategies.

At the third stage, an important problem that arose with regard to internal combustion engines was identified as reducing carbon dioxide (CO_2) emissions into the atmosphere. Carbon dioxide affects the strengthening of the greenhouse effect, respectively, that component of global warming of our planet's climate that is associated with the growth of energy consumption. Intergovernmental and state-wide memoranda and declarations adopted in recent years are devoted to the strategies for decarbonization of transport internal combustion engines. In this regard, a problem arose regarding a new criterion for new challenges. For the quantitative assessment of decarbonization measures in internal combustion engines, a new criterion (K) was proposed, which allows us to proceed to specific calculations of the results of innovations carried out in relation to decarbonization. This criterion can be used both for comparing different types and designs of internal combustion engines, and at the stages of modernization of a specific internal combustion engine. At the same time, for a comprehensive assessment of the perfection of internal combustion engines, it is recommended to use all available criteria that are conditionally attributed to the first and second stages of the formation of these criteria.

Data on the values of the new criterion are presented in a wide range of efficiency values and the share of fossil fuel carbon. It is shown that practically all transport power plants that are currently in operation or will be manufactured in the coming years can be identified by this new criterion. Better indicators of specific power plants by this criterion in relation to analogues will also indicate more effective steps by the developers of these ICEs to implement decarbonization strategies. It is recommended to enter the value of the new criterion into the passport database of a specific transport ICE, which will allow determining the position of this ICE in the system of comparisons with analogues.

Thus, the study proposes a set of criteria for assessing the perfection of ICEs and partly of vehicles, which allow at the stage of modernization or creation of new ICE models to obtain a real picture of the effectiveness of design and technological solutions within the framework of decarbonization and sustainable development strategies.

References

1. Reitz, R.D., Ogawa, H., Payri, R., Fansler, T., Kokjohn, S., Moriyoshi, Y., Agarwal, A.K., Arcoumanis, D., Assanis, D., Bae, C., Boulouchos, K., Canakci, M., Curran, S., Denbratt, I., Gavaises, M., Guenthner, M., Hasse, C., Huang, Z., Ishiyama, T., Johansson, B., Johnson, T.V., Kalghatgi, G., Koike, M., Kong, S.C., Leipertz, A., Miles, P., Novella, R., Onorati, A., Richter, M., Shuai, S., Siebers, D., Su, W., Trujillo, M., Uchida, N., Vaglieco, B.M., Wagner, R.M., Zhao, H.: IJER editorial: The future of the internal combustion engine. Int. J. Engine Res. (2019). https://doi.org/10.1177/1468087419877990
2. Liu, Z., Hao, H., Cheng, X., Zhao, F.: Critical issues of energy efficient and new energy vehicles development in China. Energy Policy **115**, 92–97 (2018). https://doi.org/10.1016/j.enpol.2018.01.006
3. Каніло П.М., Туренко А.М., Гриценко А.В., Внукова Н.В. (2020) Глобальні енерго-еколого-кліматичні проблеми та невідкладність їх вирішення: підручник, Харків: ХНАДУ, 388 с
4. Parsadanov, I.V., Rykova, I.V.: Accounting for the harmful effect of carbon dioxide in comprehensive assessment of fuel consumption and exhaust gases toxicity diesel engines. Intern. Combust. Engines (1), 35–40 (2022). https://doi.org/10.20998/0419-8719.2022.1.05
5. Marchenko, A.P., Parsadanov, I.V., Rykova, I.V.: Evolutionary development of motor transport diesels: environmental and economic indicators. Intern. Combust. Engines (1), 3–10 (2023). https://doi.org/10.20998/0419-8719.2023.1.01
6. Марченко А.П., Рязанцев М.К., Шеховцов А.Ф. (2004) Двигуни внутрішнього згоряння: Серія підручників у 6 томах. Т.1 Розробка конструкцій форсованих двигунів наземних транспортних машин. За редакцією проф. А.П. Марченко. та проф. А.Ф. Шеховцова, Х: Прапор, 384 с. Available at: https://repository.kpi.kharkov.ua/handle/KhPI-Press/7800
7. Марченко А.П., Парсаданов І.В., Товажнянський Л.Л., Шеховцов А.Ф. (2014) Двигуни внутрішнього згоряння: Серія підручників у 6 томах, Т. 5 Екологізація ДВЗ—2-е видання. За редакцією проф. А.П. Марченко, Х: Видавничий центр НТУ «ХПІ», 348 с. https://repository.kpi.kharkov.ua/handle/KhPI-Press/9570
8. Zhao, F., Chen, K., Hao, H., Liu, Z.: Challenges, potential and opportunities for internal combustion engines in China. Sustainability **12**(12), 4955 (2020). https://doi.org/10.3390/su12124955
9. Parsadanov, I., Marchenko, A., Tkachuk, M., Kravchenko, S., Polyvianchuk, A., Strokov, A., Gritsuk, I.V., Rykova, I., Savchenko, A., Smirnov, O., Postol, Y., Savchuk, V.: Complex assessment of fuel efficiency and diesel exhaust toxicity. SAE Tech. Pap. CD-ROM/SAE Tech. Pap. Ser. (2020). https://doi.org/10.4271/2020-01-2182

10. WKM (2017) The future of the combustion engine/Assessment of the diesel engine situation. https://www.wkm-ev.de/images/20170708-englisch-Die-Zukunft-des-Verbrennungsmotors.pdf. Accessed 11 Mar 2024
11. Reitz, R.D., Ogawa, H., Payri, R., et al.: IJER editorial: the future of the internal combustion engine. Int. J. Engine Res. **21**(1), 3–10 (2020). https://doi.org/10.1177/1468087419877990
12. Marchenko, A.P., Parsadanov, I.V.: Criteria for assessing the effectiveness of transport power plants decarbonisation in accordance with implementation of the sustainable development concept. Intern. Combust. Engines (1), 3–11 (2024). https://doi.org/10.20998/0419-8719.2024.1.01
13. The European Commission Factsheet. Delivering the European Green Deal (2021). https://commission.europa.eu/publications/delivering-european-green-deal_en. Last updated 9 Oct 2023
14. Marchenko, A., Parsadanov, I., Pylyov, V., Osetrov, O., Linkov, O., Kravchenko, S., Trynov, O., Meshkov, D., Bilyk, S., Savchenko, A., Rykova, I., Aryan, R.: Research and innovation to improve the efficiency of modern diesel engines. IntechOpen eBooks (2022). https://doi.org/10.5772/intechopen.102759
15. Barreiros, R., Siqueira-Filho, E., Silva, A., Oliveira, F.D., Barreiros, P., Dimenstein, M., Sarubbo, L., Martins, R., De, V.: Effect on diesel engine performance parameters using hydrogen and oxygen produced on demand. Energies **15**(19), 7285–7285 (2022). https://doi.org/10.3390/en15197285

Decarbonization of Vehicle Power Plants in the Concept of Sustainable Development

Serhii S. Kravchenko, Anatolii Savchenko, and Viktoras V. Sencila

Abstract Today, power plants with internal combustion engines occupy one of the key positions in industry, energy, agriculture, military affairs, mining, transport and other areas of human activity. Such a wide distribution of internal combustion engines determines the extremely high relevance for humanity of the task of minimizing their negative impact on the environment, which, first of all, consists in the consumption of exhaustible fuel reserves and the emission of harmful substances with exhaust gases. Harmful substances from exhaust gases of internal combustion engines have a detrimental effect on the ecological situation both in Ukraine and in the world and cause such negative consequences as the greenhouse effect, acid rain, and ozone layer depletion. The introduction of alternative fuels for use in internal combustion engines contributes to a partial solution to the above problems. The paper considers the use of the following types of alternative fuels in internal combustion engines: low-calorie gas fuels, water-fuel emulsions, and biofuels. The impact of these types of alternative fuels on its performance and characteristics is shown. The paper analyzes the indicator diagrams of internal combustion engines on alternative fuels, reveals the mechanism of fuel influence on the working process of internal combustion engines. Conclusions are drawn on the feasibility of using these alternative fuels, and recommendations are given for their introduction into use in internal combustion engines.

S. S. Kravchenko · A. Savchenko (✉)
Department of Engines and Hybrid Powerplants, Educational and Scientific Institute of Power Engineering, Electronics and Electromechanics, National Technical University "Kharkiv Polytechnic Institute", Kharkiv, Ukraine
e-mail: anatolii.savchenko@khpi.edu.ua

S. S. Kravchenko
e-mail: serhii.kravchenko@khpi.edu.ua

V. V. Sencila
Lithuanian Maritime Academy, Klaipeda, Lithuania

© The Author(s), under exclusive license to Springer Nature Switzerland AG 2025

A. Marchenko and Z. Koruba (eds.), *Decarbonization of Transport Energy Installations in the Context of Sustainable Development Strategies*, Studies in Systems, Decision and Control 635, https://doi.org/10.1007/978-3-032-05884-3_13

Symbols

ICE	Internal combustion engine
CO_2	Carbon dioxide
LCGF	Low-calorie gas fuels
N/C	Ratio of non-combustible and combustible components
CO	Carbon monoxide
H_2	Hydrogen
CH_4	Methane
N_2	Nitrogen
NO_x	Nitrogen oxides
C/H	Ratio of carbon to hydrogen
N_e	Brake effective power
α	Excess air ratio
pol	Completeness of fuel combustion
WFE	Water-fuel emulsion
EG	Exhaust gases
RPM	Revolution per minutes
p_e	Brake mean effective pressure

1 1. Introduction

In the face of global challenges to humanity regarding environmental and thermal pollution of the environment, which are caused by both negative climate change and destructive natural phenomena, as well as growing anthropogenic pressure, the problem of increasing the economic and environmental efficiency of transport power plants with ICE is currently urgent. This problem is an important component of ensuring the implementation of the global strategy for sustainable development, the solution of which is impossible without efficient transport, almost the entire fleet of which involves the use of ICE. At the same time, leading countries of the world have adopted memoranda and strategies for the implementation of decarbonization technologies for ICE (primarily in transport) as one of the key factors in improving their economic and environmental efficiency [1–3].

The operation of transport power plants directly affects environmental pollution and global warming on the planet. It is well known that the largest amount of mechanical energy (more than 80% of the installed capacity) is produced by ICE, which consume the bulk of the products of oil processing into fuel and some of the natural gas and coal processing. The greatest harm to the environment during the operation of ICE is caused by chemical factors due to atmospheric pollution with toxic substances in the atmosphere by exhaust gases, thermal pollution of the environment and emissions of the greenhouse gas CO_2 [2].

As the experience of engine manufacturing companies around the world shows, today there are every opportunity to solve this problem by using alternative fuels, along with improving the work process and introducing systems for artificially neutralizing toxic emissions both in and outside the engine cylinder.

The expansion of the use of alternative fuels is also an important part of the energy strategy related to the reduction of the use of fossil hydrocarbon fuels. Alternative fuels that have found practical application in ICE include fuels obtained using vegetable oils, WFE and gas fuels. To a large extent, this strategy corresponds to the green policy provided by the use of carbon-free energy [3–7].

It should be noted that the negative anthropogenic pressure is largely the result of the use of fuels in the energy sector that cause emissions of the greenhouse gas CO_2 into the environment. This gas is recognized as the main factor in the increase in the average atmospheric temperature of the planet. As a result, in recent years, at the legislative level, leading countries of the world have developed and approved strategies and programs to expand the use of alternative fuels.

Therefore, along with the restoration of energy resources and ensuring stable transport operations, the most urgent problem is the efficient use of alternative fuels, which solves the problems of energy dependence on the availability of oil resources.

One of the fairly common groups of alternative fuels are low-calorie gases, which are a by-product of industrial and agricultural production. For example, mine gas largely consists of methane CH_4, which in terms of its negative impact on the environment significantly exceeds CO_2. Using them as fuel for ICE allows to simultaneously improve the environmental situation, as well as produce electrical and thermal energy, partially replacing minerals [4–6, 8–14].

Another type of alternative fuel for ICE is WFE, which are a two-phase mixture of traditional diesel fuel and water with specified characteristics of the dispersion of water particles in the fuel. The use of such fuel allows to significantly improve the environmental characteristics of ICE and somewhat improve the fuel efficiency of the engine [15–17].

The use of fuels based on plant-based products, such as sunflower or rapeseed oil, in ICE can also be considered promising. Such fuels are renewable energy sources and therefore allow for a reduction in the consumption of traditional petroleum fuels [7, 18–20].

The aim of the work is to increase the fuel, environmental and energy efficiency of ICE by using various types of alternative fuels.

When solving the above-mentioned problems, taking into account the need to ensure minimal thermal pollution of the environment, as is known, an important role belongs to the effective organization of the working process of a power plant based on ICE. In this case, taking into account the properties of the fuel is of extremely urgent importance. Next, we will consider the main properties of some alternative types of fuel.

2 Low-Calorie Gas Fuels

Secondary processing gases (synthesis gas, generator gas, pyrolysis gas, biogas) and gases obtained as a by-product of industry (coke oven gas, blast furnace gas, mine gas, petroleum gas), i.e. the so-called low-calorie gas fuels, can be effectively used as an alternative fuel for power plants based on ICE. The role of using LCGF is rapidly growing worldwide, since the raw materials for this fuel can be various types of both organic and inorganic raw materials, such as coal, oil, natural gas, peat, wood, household and industrial waste, etc. For example, in the EU countries, programs for the utilization of LCGF are widespread and for more than 10 years, electrical and thermal energy has been produced from these gases [5]. The possibility of using a wide composition of gases in ICE allows obtaining relatively inexpensive thermal and electrical energy from renewable and by-product energy resources.

These gaseous fuels are typically composed of different gases in varying proportions, depending on the type of feedstock and its processing method. Let's take a look at the key characteristics of low-calorie gaseous fuels.

Synthesis gas in industry is produced using steam reforming technology or partial oxidation of methane, gasification or pyrolysis of solid fuels (coal, coke, peat, wood).

The proportions of components in producer gas obtained by gasification can vary significantly depending on the feedstock and the type of conversion (steam or oxygen). The most widely used gasification methods are Lurgi (gas consists of 15–18% CO, 38–40% H_2, 9–11% CH_4, 30–32% CO_2) and Shell (65–66% CO, 30–32% H_2, 0.4% CH_4, 1–2% CO_2). Gasification plants operate in 29 countries, with the largest share of low-calorie gas fuel production—37%—in the Asian-Australian region due to the intensive development of this industry in China.

Regarding the feedstock for synthesis gas production, the most common is coal (51%), followed by oil (25%) and natural gas (22%). Petroleum coke, biomass and waste are also used as feedstock.

In the energy sector, an important role is played by the processing of biomass, which includes organic agricultural and household waste, by methane fermentation to obtain biogas, which makes it a renewable energy source. The macrocomponents of biogas are 40–70% methane (CH_4) and 30–60% carbon dioxide (CO_2). The heat of combustion does not exceed 20–22 MJ/m^3, and the methane number is 110–130, which makes biogas an attractive fuel for engines with spark ignition of the fuel–air mixture.

Wood industry waste can be effectively used to produce **pyrolysis gases**. The technology for producing pyrolysis gas includes the following stages: crushing waste, compacting it, pumping out air, degassing at temperatures above 600 °C and subsequent gasification using pure oxygen. The pyrolysis gas produced by this technology consists of combustible elements: hydrogen (H_2–20–40%), carbon monoxide (CO–34–40%) and methane (CH_4–1–6%), and non-combustible elements—nitrogen (N_2–2–5%) and carbon dioxide (CO_2–20–35%). The calorific value of such gas is 4–10 MJ/m^3.

Coke oven gas is a by-product of coke production. Its composition varies considerably depending on the type of coal used in the process, but the main components include hydrogen ($H_2 \approx 55\%$), methane ($CH_4 \approx 25\%$) and nitrogen ($N_2 \approx 10\%$). Modern steel plants use these fuels to generate electricity in highly efficient internal combustion engine-based plants, and fully cover their own electricity needs, and also sell its surplus. For example, at the Profusa SA coke plant in Bilbao, Spain, 12 GE Jenbacher engines have been operating on coke oven gas since 1995, producing 7,164 kW of electrical power. At the same time, the level of NO_x emissions when operating on coke oven gas is 40% lower than when operating on natural gas.

A promising alternative fuel is currently coal mine gas, which is a mixture of methane and air. The world's coal mine gas reserves are estimated at 260 trillion m^3 [6]. The experience of world engine building indicates the development and use of a methane-air mixture with a methane content of at least 27% in gas engines [8]. Methane, which is the main component of coal mine gas, is one of the gases that cause the greenhouse effect. Its content in the atmosphere increases by 0.6% every year (with an increase in the volume of carbon dioxide by 0.4%) [9]. A large number of methane emissions into the atmosphere occur during coal mining. Therefore, the use of coal mine gas in ICE significantly contributes to decarbonization.

Thus, low-calorie gas fuels, which are waste products of industrial and agricultural production, are released into the environment and pollute the atmosphere. Utilization of these gases will improve the ecological and economic situation. Given the significant difficulties in transporting low-calorie gas fuels, they are poorly suited for use in power plants of vehicles. However, low-calorie gas fuels have found wide application in stationary ICE, which can mostly generate both electrical and thermal energy.

The physicochemical properties of low-calorie gaseous fuels are given in Table 1. The combustible components of such fuels include hydrogen (H_2), carbon monoxide (CO), methane (CH_4) and a small amount of heavy hydrocarbons. The rest (about 10–50% of the total volume) are non-combustible components, such as nitrogen (N_2), carbon dioxide (CO_2) and oxygen (O_2).

The lower calorific value of low-calorie gases varies from 3.5 to 27 MJ/m^3, which is 1.5–10 times lower than the calorific value of natural gas (35.9 MJ/m^3). For example, synthesis gas and generator gas have, respectively, 3.4 and 5.8 times lower volumetric lower calorific values compared to natural gas. The lower calorific value determines the volumetric fuel consumption of the engine. To ensure constant engine power compared to operation on natural gas it is necessary to increase the volumetric cyclic supply of LCGF.

The range of ignition limits determines the possibility of effective leaning or enrichment of the fuel–air mixture: the upper limit corresponds to the maximum enrichment of the mixture, the lower limit to the maximum leaning. An increase in the content of inert components in the fuel leads to a narrowing of the range of ignition limits. It can be seen from Fig. 1b that an increase in the content of inert components in the fuel narrows the ignition limits. The effect of CO_2 on the ignition limits is more significant compared to N_2. H_2 and CO have wider concentration ignition limits (respectively, $\alpha = 0.15$–10 and $\alpha = 0.15$–2.95), and their presence in

Table 1 Properties of gaseous fuels and their components

Gas	Component composition, % vol	Lower calorific value MJ/m^3	Flammability limits		Laminar combustion velocity, cm/s	Required amount of air, m^3/m^3	Methane number	C/H
			Lower	Upper				
Methane	CH_4	35.0	2	0.6	41	9.55	100	3
Hydrogen	H_2	10.8	10	0.15	270	2.38	0	0
Carbon monoxide	CO	12.6	2.94	0.15	45	2.38	73	–
Natural gas	$CH_4 = 92.2$ $C_2H_6 = 4.1$ $C_3H_8 = 1.0$ $C_4H_{10} = 0.36$ $C_5H_{12} = 0.33$ $N_2 = 2$ $CO_2 = 0.1$	34.2	2.0	0.6	40–42	9.96	79.8	3.1
Synthesis gas (Texaco)	$H_2 = 35$ $CO = 52$ $CH_4 = 0.4$ $N_2 = 0.6$ $CO_2 = 12$	10.48	5.63	0.2	70	2.11	56.1	10.48
Pyrolysis gas	$H_2 = 30$ $CO = 35$ $CH_4 = 5$ $N_2 = 5$ $CO_2 = 25$	9.4	5.0	0.37	70	2.03	78	9.7
Mine gas	$CH_4 = 43.8$ $N_2 = 43.2$ $CO_2 = 2.2$ $O_2 = 9.8$	15.47	2.0	0.7	37	3.74	108	3.17
Biogas	$CH_4 = 65$ $CO_2 = 35$	22.75	1.9	0.62	27	6.2	135	4.61
Coke oven gas	$H_2 = 55$ $CO = 8$ $CH_4 = 31$ $N_2 = 3.8$ $CO_2 = 1.2$ $O_2 = 1$	17.8	4.43	0.4	100	4.42	40.5	2.056
Generator gas	$H_2 = 23$ $CH_4 = 2$ $N_2 = 42$ $CO_2 = 9$ $CO = 24$	6.2	2.17	0.6	50	1.31	70.7	7.73

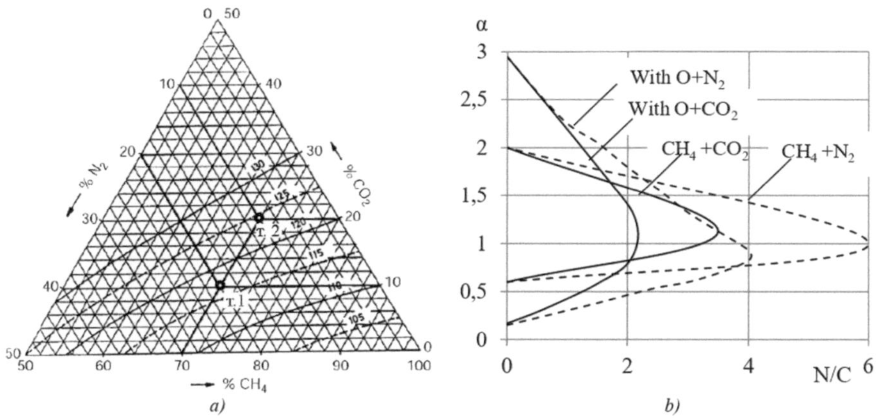

Fig. 1 Effect of the content of inert components on the methane number **a** and ignition limits of the fuel–air mixture **b** [10]: N/C—ratio of non-combustible to combustible components

the LCGF expands the ignition limits. According to Table 1, synthesis gas, pyrolysis, coke oven and generator gases have wider ignition limits compared to other LCGFs. Expanding the ignition limits of the LCGF helps improve the start-up of the internal combustion engine at low temperatures, increases the reliability of ignition and the completeness of combustion of lean fuel–air mixtures.

Inert components do not take an active part in the combustion process, but increase the methane number (MN) of the gas fuel (see Fig. 1a). In this case, the influence of the CO_2 content on MN is three times stronger than the influence of N_2 [10]. For example, for a gas containing 10% CO_2, 20% N_2 and 70% CH_4 (point 1, Fig. 1a), the methane number is 117, and for a gas with 20% CO_2, 10% N_2 and 70% CH_4 (point 2, Fig. 1a) the methane number is 124.

The CO_2 component of the LCGF is a ternary gas with a relatively high heat capacity. For example, the heat capacity of CO_2 at 1500 °C is 44.035 kJ/(kmol°C), which is 1.8 times higher than the heat capacity of N_2 at the same temperature. The increased CO_2 content in the LCGF can lead to a decrease in the maximum pressure and temperature of the cycle, which, accordingly, reduces the thermodynamic efficiency of the cycle.

The laminar combustion rate of the fuel determines the degree of pressure increase in the cylinder and the rigidity of the engine's working process. This characteristic depends on the fuel composition and the degree of leanness of the gas-air mixture. With an increase in the combustion rate, the loads on the crank mechanism parts increase, and the maximum temperature of the working fluid also increases, which causes an increase in nitrogen oxide emissions (NO_x). However, with a decrease in the combustion rate, the incompleteness of fuel combustion increases, which worsens the fuel economy of the engine. The presence of hydrogen (H_2) in the fuel significantly increases the laminar combustion rate, which reduces the incompleteness of LCGF combustion. For mine gas and biogas, the combustion rate is lower compared to natural gas, while for generator, pyrolysis and coke oven gases it is higher. It can be assumed that the completeness of combustion for these fuels will also have a corresponding trend of change.

With an increase in the C/H ratio, i.e. with an increase in the carbon content in the fuel, fuel–air mixtures burn at higher temperatures (under the same combustion conditions). This is explained by the fact that for the complete combustion of a unit mass of carbon, 3 times less air is required than for the combustion of hydrogen. The heat of combustion of fuel is released in a smaller volume of gas, which leads to an increase in temperature. Obviously, the highest temperatures will occur during the combustion of synthesis gas, generator and pyrolysis gases, the lowest—mine gas, biogas and coke oven gases.

Increasing the content of individual components in the composition of the LCGF (in particular, CH_4 and H_2) has an ambiguous effect on the motor properties of the fuel, which leads to both an improvement in some motor properties and a deterioration in others. For example, H_2 has a high combustion rate, which, on the one hand, improves the conditions of fuel combustion in the engine and increases its efficiency, on the other hand, due to an increase in the cycle temperature, leads to an increase in NO_x emissions from the combustion chamber. The presence of CO and CH_4 in the mixture reduces the combustion rate. The presence of CO_2 and N_2 in the fuel creates a solvent effect, reducing the speed and temperature of the flame, but increases the methane number of the gas fuel.

The accumulated experience of operating gas ICE of foreign companies shows that a wide range of low-calorie gases can be effectively used in gas engines.

The results of studies of the influence of various low-calorie gases on the working process and performance of the stationary gas engine 11GD100 (produced in Kharkiv, Ukraine) are shown in Fig. 2. These engines operate on an electric generator with a maximum power of 1100 kW, are used as the main, backup or emergency source of electrical and thermal energy, especially in areas remote from centralized power supply, can be installed on railway locomotives and even on ships. Natural gas is used as the base fuel. This is a vertical engine, ten-cylinder, two-stroke, which has a special unconventional pre-chamber-torch system of ignition of the fuel–air mixture. The fuel and air mixture in the engine cylinder is ignited by the high energy of the torch formed in a special chamber. Due to this, the 11GD100M engines can reliably and efficiently operate on significantly depleted fuel–air mixtures. This, firstly, significantly reduces CO_2 and NO_x emissions with exhaust gases, compared to traditional engine designs, and secondly, it allows these engines to be converted to use gases with low calorific value.

The research was carried out using the developed mathematical model of the working process of the 11GD100 engine, which allows taking into account the component composition of the fuel [11].

Fig. 2 shows the calculated indicator diagrams (p), the temperature of the working gases during combustion (T) and the characteristics of heat release (x) at the nominal operating mode of the 11GD100 engine when using natural gas and LCGF.

From Fig. 2a it is seen that when using synthesis gas, generator gas and pyrolysis gas in the engine cylinder, the ignition delay period is reduced, the combustion rate

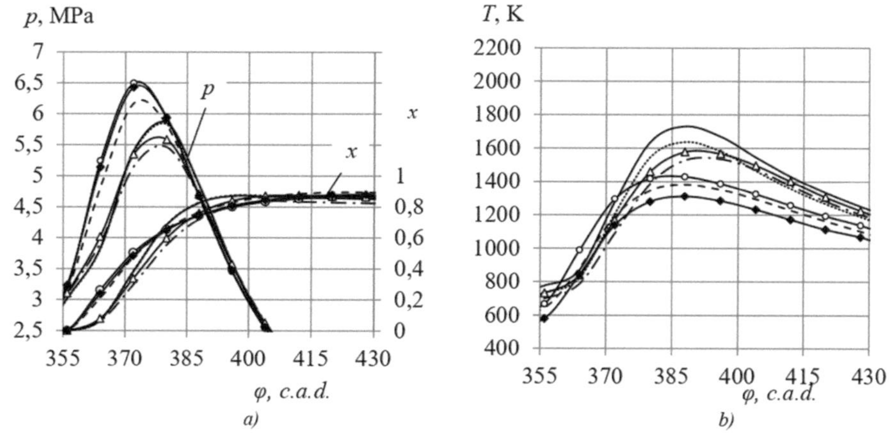

Fig. 2 Indicator diagrams *p*, heat release characteristics *x* **a** and temperature *T* in the cylinder of the 11GD100 engine **b** during combustion using natural gas and LCGF (n = 750 min^{-1}; Ne = 1100 kW):—Natural gas;—Synthesis gas;—Pyrolysis gas;—Mine gas;—Biogas;—Coke oven gas;—Generator gas

increases in the initial combustion period, which in turn increases the maximum cycle pressure by 4.7–9%. This is due to the high hydrogen content in these fuels, which has a high combustion rate and low heat capacity.

In addition, the high hydrogen content leads to an expansion of the limits of ignition and combustion, which increases the completeness of combustion by 4.2–4.6%, as can be seen from the characteristics of active heat release (Fig. 2a).

When using coal mine gas as fuel, the maximum cycle pressure is reduced by 0.5 MPa compared to operation on natural gas. There is also a decrease in the combustion rate and deterioration of the dynamics of heat release.

The process of biogas combustion in the cylinder of the 11GD100M engine proceeds without significant changes in the operating cycle. At the same time, the combustion pressure and temperature of the working fluid almost do not change.

The data presented (Fig. 3) indicate that when using LCGF (except biogas) there is a significant leaning of the fuel–air mixture in the cylinder of the 11GD100 engine in comparison with operation on natural gas (see Fig. 3c). However, the energy efficiency indicators of the working process, the indicator efficiency (Fig. 3a) change to a relatively small extent. This is explained by a significant expansion of the limits of ignition and combustion of the fuel–air mixture (see Table 1).

It is worth noting a significant increase in the volumetric specific fuel consumption (Fig. 3b) when using low-calorie gas fuels in ICE, which can be explained by a decrease in their calorific value. Thus, the largest increase in volumetric fuel consumption (by 5.6 times) is observed when using generator gas, the smallest increase is in the case of using biogas (by 1.5 times), coke oven gas (by 1.95 times), and mine gas (by 2.4 times).

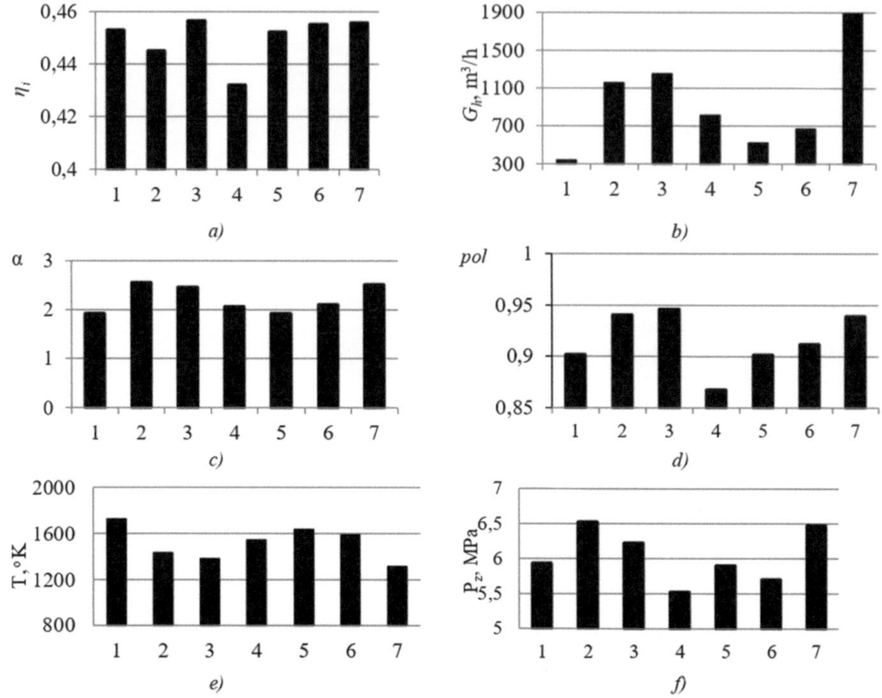

Fig. 3 Engine performance 11GD100 at nominal mode (N_e =1100 kW) when using natural gas and LCGF: **a**—indicator efficiency; **b**—volumetric fuel consumption; **c**—excess air coefficient; **d**—completeness of fuel combustion; **e**—maximum cycle temperature; **f**—maximum cycle pressure; 1—Natural gas; 2—Synthesis gas; 3—Pyrolysis gas; 4—Mine gas; 5—Biogas; 6—Coke oven gas; 7—Producer gas

The only significant difference from the above is the effect of using mine gas, which causes a decrease in the engine's indicator efficiency by 4.6% compared to the engine running on natural gas. Mine gas consists mainly of methane and air, and therefore has properties that are very close to natural gas. Therefore, an increase in the excess air coefficient causes a decrease in the speed and completeness of combustion of mine gas in the ICE cylinder.

Thus, low-calorie gases, which are by-products of industrial and agricultural production, should be used as fuel for stationary power plants with ICE, which will not only improve the environmental situation, but also produce electricity and heat, partially replacing fossil fuels.

3 Water-Fuel Emulsions

Water-fuel emulsion (WFE) is a two-phase mixture containing traditional diesel fuel and water with a specified size of water particles in the fuel. The use of such fuel allows to significantly improve the environmental performance of ICE and to somewhat improve the fuel efficiency of the engine [15].

To determine the influence of WFE on diesel engine performance, a set of diesel engine tests on standard diesel fuel and WFE was conducted. The properties of diesel fuel and WFE used in the study are given in Table 2. The results of the experimental study of the influence of WFE on fuel economy performance and the level of toxic emissions from diesel exhaust are given in Fig. 4. The effective efficiency of the diesel engine is given as a criterion for efficiency [15].

The use of WFE allows to reduce the smoke of exhaust gases by 18…56%. At high RPM, the positive effect of WFE on exhaust gas smoke is higher than at low RPM.

At all studied crankshaft rotation frequencies, the concentration of nitrogen oxides in the exhaust gas when using WFE decreased by 14…66%. The greatest reduction in nitrogen oxide emissions is characteristic of low-load modes.

It is seen that the use of WFE contributes to an increase in the brake thermal efficiency by 0.7…3.4% depending on the load and crankshaft speed. This effect is primarily explained by an increase in the completeness of fuel combustion and intensification of the combustion process during the premixed combustion phase and diffusion combustion phase [15].

The influence of water-fuel emulsion on the working process can be assessed based on the results of diesel engine indication. The indicator diagrams of diesel engine when operating on traditional diesel fuel and water-fuel emulsion are shown in Fig. 5 [16].

It can be seen that the use of water-fuel emulsion causes a noticeable decrease in the maximum pressure level of the cycle. For a more detailed analysis of the influence of water-fuel emulsion on the working process, using the indication data, the differential heat release characteristics were calculated for both types of fuel, which are shown in Fig. 6. The results indicate an increase in the ignition delay period by 2…3 degrees of crank angle when using water-fuel emulsion, which can

Table 2 Physicochemical properties of diesel fuel and WFE used in the study

Parameters	Diesel fuel	Water-fuel emulsion
Density at a temperature of 15 °C, kg/m3	829	857
Heat capacity, kJ/kg·K	2.02	2.37
Specific lower heat of combustion of fuel, kJ/kg	42,500	35,100
The ignition temperature in a closed crucible, °C, is not lower	70.5	Ignition is missing
Water content by mass, %	0	16.3
Kinematic viscosity at a temperature of 20 °C, mm2/s	5.15	6.19

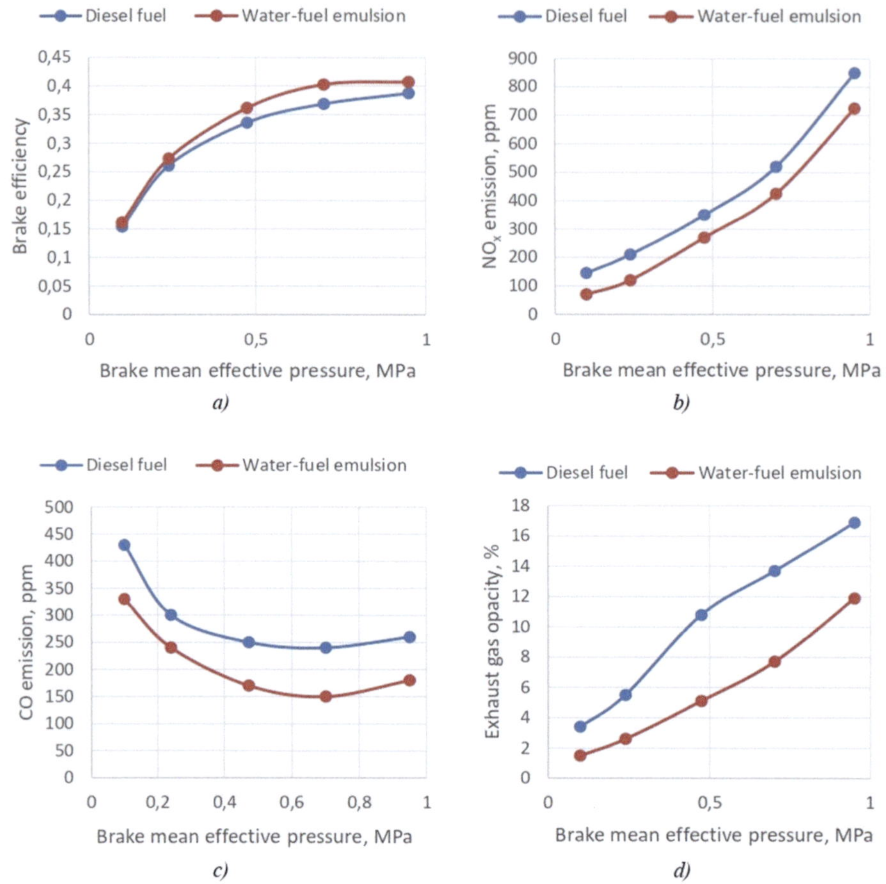

Fig. 4 The effect of load on diesel engine performance when operating on traditional diesel fuel and on water-fuel emulsion: **a**—brake efficiency; **b**—NO_X emissions; **c**—CO emission; **d**—exhaust gases opacity

primarily be explained by an increase in the activation energy of the water-fuel emulsion [17] and a decrease in the temperature and pressure inside the cylinder at the beginning of the fuel injection process compared to traditional diesel fuel [16].

The heat release rate during the flash phase is higher when using a water-fuel emulsion. This can be explained by the fact that a larger amount of fuel enters the cylinder and partially evaporates during the longer ignition delay period when using a water-fuel emulsion. During the diffusion combustion phase, a noticeable increase in the heat release rate when using a WFE compared to the characteristics for diesel fuel was noted [15, 17].

There are several hypotheses that reveal the mechanism of the influence of WFE on combustion processes in diesel engines. Within the framework of one of the hypotheses, the influence of WFE is explained primarily by the catalytic effect of

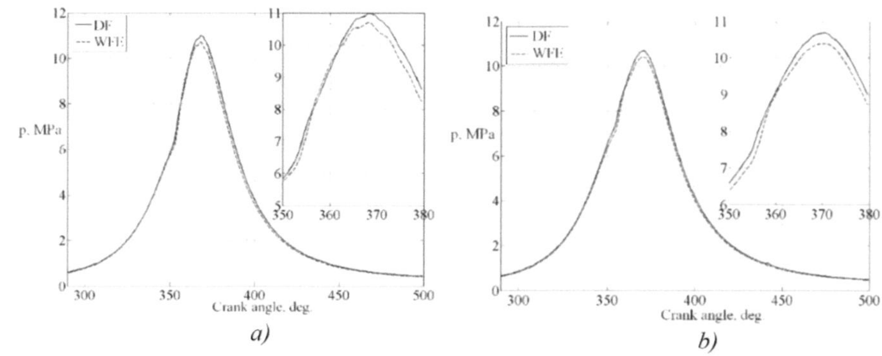

Fig. 5 Indicator diagrams of a diesel engine when operating on traditional diesel fuel and water-fuel emulsion: **a**—n = 1500 RPM, p_e = 1.076 MPa; **b**—n = 2000 RPM, p_e = 0.95 MPa

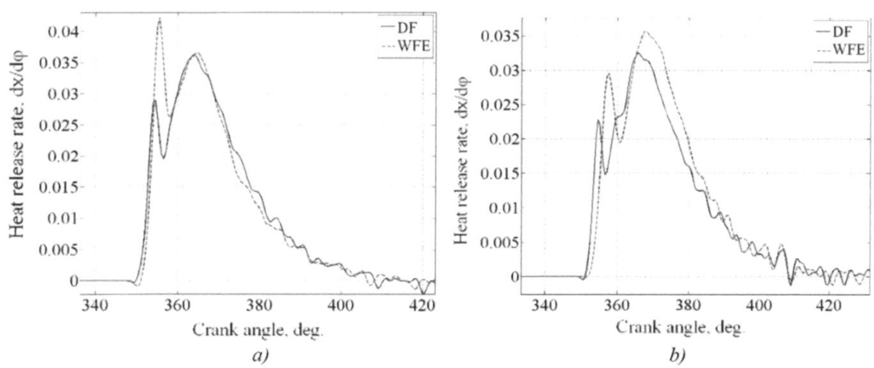

Fig. 6 Characteristics of differential heat release in a diesel engine cylinder when operating on traditional diesel fuel and on a water-fuel emulsion: **a**—n = 1500 RPM, p_e = 1.076 MPa; **b**—n = 2000 RPM, p_e = 0.95 MPa

water dissociation products on the fuel combustion process. Under high temperatures in the diesel combustion chamber, water vapor dissociates into hydrogen and oxygen, as well as hydrogen radicals and hydroxyl groups. The presence of hydrogen radicals in the combustion chamber contributes to the acceleration of chemical reactions and more complete combustion of hydrocarbons.

Another hypothesis explains the effect of WFE on the working process of a diesel engine by secondary atomization of fuel in the combustion chamber due to boiling of water ("micro-explosions" phenomenon) located inside a drop of water-fuel emulsion.

From the analysis of the characteristics of the differential heat release of the studied diesel engine, there is reason to believe that both mechanisms of influence occur during the combustion of WFE.

Thus, the introduction of the use of WFE in diesel engines allows simultaneously improving the economic and environmental characteristics of the diesel. However, taking into account the above differences in the working process of the diesel, in order to achieve maximum efficiency when operating diesel engines on WFE, it is advisable to recommend paying attention to finding rational diesel parameters when operating on this type of fuel [15].

4 Biofuel

To date, a large number of biofuels have been proposed, for example, vegetable oils in pure form and their mixtures with diesel fuel, methyl, ethyl, butyl esters of oils, etc. These fuels differ in physicochemical properties, which leads to changes in the course of processes in the engine cylinder [18].

The interest in using vegetable oils in diesel engines is related to their positive qualities. Rapeseed oil, for example, has a high cetane number, it contains practically no sulfur and polycyclic aromatic hydrocarbons. When burning vegetable oils, the amount of CO_2 that was absorbed by plants during photosynthesis is released into the environment, and thus the balance of "greenhouse" gases in the atmosphere is maintained. Plant residues and seed processing products are raw materials for animal feed and fertilizers. When such fuel falls on the ground, it does not cause environmental damage (unlike fuels of petroleum origin) [19, 20].

The work has studied the following most common biofuels, made on the basis of rapeseed oil. Today, the main ways of using vegetable oils in diesel engines of traditional design are their use in mixtures with diesel fuel. Therefore, the following fuels were selected for research:

- clean diesel fuel;
- mixtures of diesel fuel and rapeseed oil containing 25, 50 and 75% rapeseed oil by volume;
- pure rapeseed oil.

The physicochemical properties of traditional diesel fuel and rapeseed oil are given in Table 3. The engine characteristics when operating on various mixtures of alternative fuel with traditional diesel fuel are given in Fig. 7 [7].

The use of vegetable fuels leads to an increase in the specific effective fuel consumption, which is explained by a decrease in their lower calorific value compared to diesel. The change in the brake thermal efficiency of a diesel engine occurs to a much lesser extent and can both increase and decrease.

An increase in the rate of combustion during diffusion combustion phase and high quality of film mixture formation at high load conditions lead to an increase in brake thermal efficiency when using vegetable fuels.

At low loads, mixture formation in the combustion chamber volume deteriorates. In addition, the proportion of vegetable fuel burning in relatively cold wall zones increases, which explains the decrease in brake thermal efficiency at low loads.

Table 3 Physico-chemical properties of fuels

	Diesel fuel	Rapeseed oil
Approximate chemical formula	$S_{13}H_{24}$	$C_{57}H_{102}O_6$
Cetane number	45–55	44–52
C, %	85.8–87	77.6–77.9
H, %	12.6–14	11.5–11.9
Oh, %	0–0.4	10.2–10.9
Qn, MJ/kg (MJ/m^3)	40.2–43.5	36–37.3
Density, g/m^3 (20°C)	800–870	914–920
Dynamic viscosity, mm^2/s (20°C)	3–8	34–87
Heat of vaporization, kJ/kg	210–250	–
Autoignition temperature, °C	200–280	200

The work investigated the influence of the RPM and load on the output of carbon monoxide and nitrogen oxides NO_x and soot with diesel exhaust gases. The obtained data indicate that when using vegetable oil additives to diesel fuel, as well as when operating a diesel engine on rapeseed oil at high and medium load modes, there is a decrease in the formation of CO, soot and an increase in NO_x emissions. In most modes with low loads, CO and soot emissions increase slightly, and NO_x decreases (with the exception of rapeseed oil, for which NO_x emissions either do not change or decrease) [7].

By the content of CO and H_2 in EG, it is possible to estimate the completeness of hydrocarbon combustion in a diesel engine, since the CO yield characterizes the completeness of combustion, and the yield of H_2 characterizes the completeness of fuel cracking. If we assume that the amount of unburned hydrocarbons in the diesel combustion chamber is subject to the same laws as the formation of other products of incomplete combustion (CO, soot, H_2, etc.), then a simultaneous decrease in CO and H_2 with an increase in the proportion of rapeseed oil in diesel fuel for high loads will mean a decrease in emissions of unburned hydrocarbons from the combustion chamber. At low load modes, the yield of hydrocarbons will either not change or increase.

The change in the output of harmful substances from diesel exhaust gas when using vegetable fuels is explained by the peculiarities of the working process.

In most diesel engine operating modes, the combustion rate of vegetable fuel increases compared to diesel fuel during the diffusion combustion period. This leads to an increase in the average pressures and temperatures in the diesel cylinder. As a result, the yield of incomplete combustion products decreases and the yield of nitrogen oxides increases. The decrease in the yield of incomplete combustion products is also facilitated by an increase in the oxygen content and a decrease in the ratio of carbon and hydrogen atoms in the vegetable fuel molecule.

At low load conditions, when using vegetable fuels, the quality of volumetric mixture formation significantly deteriorates and the proportion of fuel that burns in relatively cold wall zones increases compared to diesel fuel. This is obviously the

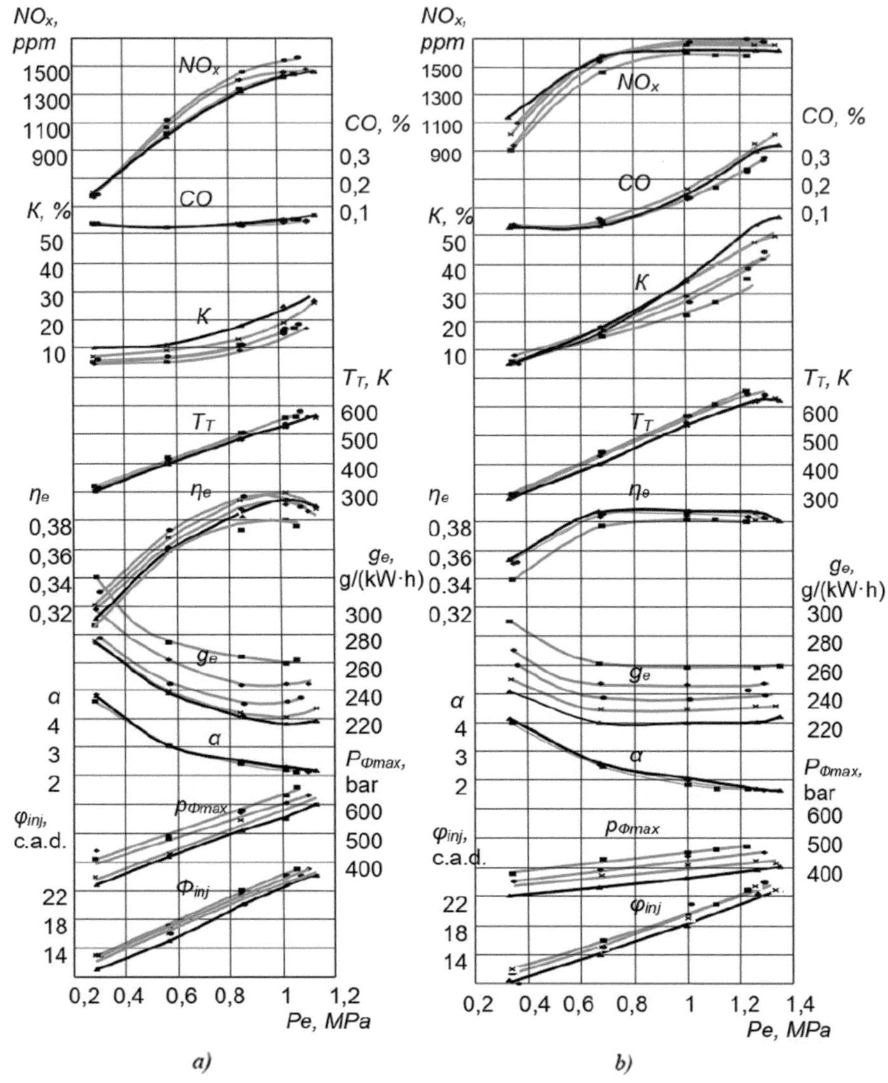

Fig. 7 The effect of load on diesel engine performance when operating on traditional diesel fuel and on various fuels: **a**—n = 2000 RPM; **b**—n = 1500 RPM; ▲—pure diesel fuel; ■—rapeseed oil; ×—25% rapeseed oil and 75% diesel fuel; ●—50% rapeseed oil and 50% diesel fuel; ◆—75% rapeseed oil and 25% diesel fuel

reason why the output of CO, C_nH_m and soot when using vegetable and diesel fuels is comparable, and NO_x emissions are reduced compared to diesel fuel when burning vegetable fuels [7].

5 Conclusions

Given the current trends in the aggravation of environmental problems in the operation of power plants with ICE, the relevance of tasks for the development and implementation of measures to improve the environmental characteristics of ICE is increasing. It is a generally recognized fact that the introduction of the use of alternative fuels is an effective measure to reduce the negative impact of ICE on the environment. This paper considers the potential for reducing the negative impact on the environment by using the following fairly common types of alternative fuels: low-calorie gas fuels, water-fuel emulsion, vegetable oils.

1. The use of low-calorie gas fuels is a promising direction because it allows to utilize by-products of industrial and agricultural production, significantly reduce the consumption of oil-based fuels, and in some cases reduce specific emissions of harmful substances by the engine. At the same time, engines on low-calorie gas fuels often require individual tuning by highly qualified specialists, taking into account the specific elemental composition (and therefore the properties) of the gas fuel. Also, low-calorie gas fuels are very difficult to use in vehicle engines, due to their low calorific value, which results in a small range of the vehicle.

The analysis of the current state of LCGF production and use showed that LCGF is advisable to use in stationary engines with prechamber-torch ignition of fuel–air mixtures, which are less demanding on the qualitative and quantitative composition of gas fuel compared to other ICE types.

The best LCGFs for usage in ICE has high methane number, lower calorific value, low content of inert components and wide ignition limits. However, the increased content of individual LCGF components has an ambiguous effect on fuel properties and engine performance.

The conducted studies have shown that when using most low-calorie gas fuels in a stationary 11GD100 engine at rated power, the efficiency indicators almost do not change compared to operation on natural gas. However, it should be noted an increase by 1.5–5.6 times in the volumetric flow rate of LCGF due to a significant decrease in the lower calorific value of these fuels.

2. The use of WFE allows to significantly reduce specific emissions of harmful substances with exhaust gases, primarily due to the positive effect of water particles on the dispersion and distribution of fuel in the volume of the combustion chamber (the phenomenon of "micro-explosions"). Also, in some cases, the use of water-fuel emulsion allows to reduce the specific consumption of petroleum fuel by 2…3% by the engine.
3. The use of vegetable oils in ICE can be considered promising due to the fact that such an energy source is renewable. The specific fuel consumption of the engine may increase slightly, but this is primarily due to the lower specific calorific value of such fuels, but the efficiency of the engine remains approximately at the same level, and may even increase slightly. The level of specific emissions of harmful

substances into the environment depends quite strongly on the design and control parameters of the engine. In the case of optimizing the engine parameters specifically for the use of fuel containing vegetable oil, environmental characteristics can be significantly better compared to traditional fuel.

Thus, the article provides an overview of the most common types of alternative fuels for ICE, performs a comparative analysis of them and shows their advantages and disadvantages. It is proven that the use of these alternative fuels will significantly reduce the negative impact of internal combustion engine power plants on the environment.

References

1. Kravchenko, S.S.: *Konvertatsiia statsionarnoho dvyhuna HD100 dlia roboty na nyz'kokaloriinykh hazovykh palyvakh* [Conversion of stationary HD100 engine to operate on low-calorific gas fuels]. PhD thesis. National Technical University “Kharkiv Polytechnic Institute”. Supervisor: Marchenko, A.P. (2015)
2. ECOTOWN: *Osnovni ekolohichni problemy Ukrainy u 2014 rotsi* [Main ecological problems of Ukraine in 2014] (2014). https://ecotown.com.ua/news/Osnovni-ekolohichni-problemy-Ukrayiny-u-2014-rotsi (Accessed: 8 December 2014)
3. Kanilo, P.M., Parsadanov, I.V.: ‘Analiz effektivnosti ispolzovaniya al’ternativnykh topliv v dvigatelyakh vnutrennego sgoraniya’ [Analysis of the efficiency of alternative fuels in internal combustion engines]. Dvigateli vnutrennego sgoraniya **1**, 8–13 (2009)
4. Marchenko, A.P., Osetrov, A.A., Kravchenko, S.S.: ‘Problemy i perspektivy proizvodstva sinteticheskikh gazovykh topliv (obzor)’ [Problems and prospects of synthetic gas fuel production (review)]. Energotekhnol. Resursosberezhenie **4**, 5–14 (2013)
5. NETL: *Worldwide Gasification Database Analysis*. National Energy Technology Laboratory (2010). http://www.netl.doe.gov/technologies/coalpower/gasification/worlddatabase/summary.html (Accessed: October 2010)
6. Pylev, V.A., Kravchenko, S.A., Prokhorenko, A.A., Zaslavskiy, E.G., Shpakovskiy, V.V.: ‘Shakhtnyy gaz – motornoe toplivo dlya DVS’ [Mine gas as a motor fuel for internal combustion engines]. Dvigateli vnutrennego sgoraniya **1**, 10–15 (2007)
7. Osetrov, A.A.: *Uluchshenie tekhniko-ekonomicheskikh pokazateley dizelya 4ChN12/14, rabotayushchego na biotoplivakh* [Improvement of technical and economic performance of 4ChN12/14 diesel operating on biofuels]. PhD thesis. National Technical University “Kharkiv Polytechnic Institute”. Supervisor: Marchenko, A.P. (2005)
8. Global Methane Initiative (2011) *Methane from coal mines: opportunities for emissions reduction, improvement of collection and utilization* [online]. GMI Report. Available at: https://www.globalmethane.org/documents/coal_fs_eng.pdf (Accessed: [May 21, 2025])
9. Przybyla, G., Postrzednik, S., Zmudka, Z.: Energy analysis of a small capacity SI engine fueled with lean air gas mixture. Powertrain and Transport **19**(2), 423–432 (2012)
10. Marchenko, A.P., Osetrov, A.A., Kravchenko, S.S., Khamza, O.A.: ‘Motornye svoystva nizkokaloriynykh gazovykh topliv i ikh vliyanie na pokazateli DVS’ [Motor properties of low-calorific gaseous fuels and their effect on engine performance]. Energotekhnol. Resursosberezhenie **5–6**, 3–12 (2014)
11. Marchenko, A.P., Osetrov, A.A., Kravchenko, S.S.: ‘Doslidzhennia ta matematychne modeliuvannia protsesu zghoriannia v dvyguni z forkamerno-fakel'nym zapaliuvanniam’ [Investigation and mathematical modeling of combustion in a prechamber-torch ignition engine]. Dvigateli vnutrennego sgoraniya **2**, 12–19 (2014)

12. Lebedevas, S., Raslavičius, L., Drazdauskas, M.: Comprehensive correlation for the prediction of the heat release characteristics of Diesel/CNG mixtures in a single-zone combustion model. Sustainability **15**(4), 3722 (2023). https://doi.org/10.3390/su15043722
13. Belousov, E., Marchenko, A., Gritsuk, I.V., Savchuk, V., Bulgakov, N., Mitienkova, V. and Pronin, S.: 'Research of the gas fuel supply process on the compression stroke in ship's low-speed gas-diesel engines. SAE Technical Papers, 2020–01–2107 (2020). https://doi.org/10.4271/2020-01-2107
14. Wang, Z., Lebedevas, S., Rapalis, P., Zaglinskis, J., Mickeviciene, R., Djackov, V., Liu, X.: Use of LNG cold potential in the cogeneration cycle of ship power plants. J. Mar. Sci. Eng. **8**(9), 720 (2020). https://doi.org/10.3390/jmse8090720
15. Savchenko, A.V.: *Vybir ta obgruntuvannia parametriv avtotraktornoho dyzelya pry yoho roboti na vodopalivniy emul'sii* [Selection and justification of tractor diesel parameters when operating on water-fuel emulsion]. PhD thesis. National Technical University "Kharkiv Polytechnic Institute". Supervisor: Marchenko, A.P. (2018)
16. Marchenko, A.P., Parsadanov, I.V., Prokhorenko, A.O., Savchenko, A.V., Osetrov, O.O., Meshkov, D.V.: 'Osoblyvosti protsesu zghoriannia v dyzeli pry roboti na vodopalivniy emul'sii' [Peculiarities of combustion in a diesel engine operating on water-fuel emulsion]. Dvigateli vnutrennego sgoraniya **1**, 3–10 (2016)
17. Marchenko, A.P., Prokhorenko, A.A., Savchenko, A.V.: 'Eksperimental'noe opredelenie energii aktivatsii vodotoplivnoy emul'sii' [Experimental determination of activation energy of water-fuel emulsion]. Dvigateli vnutrennego sgoraniya **2**, 9–14 (2016)
18. Othman, M., Adam, A., Mamat, R.: Green fuel as alternative fuel for diesel engine: a review. Renew. Sustain. Energy Rev. **80**, 694–709 (2017)
19. Geng, P., Cao, E., Tan, Q., Wei, L.: Effects of alternative fuels on the combustion characteristics and emission products from diesel engines: a review. Renew. Sustain. Energy Rev. **71**, 523–534 (2017)
20. Chakma, P., Cho, H.: Study on diesel engine characterization using alternative fuel and its blends. Int. J. Appl. Eng. Res. **12**, 5910–5915 (2017)

Application of Industry 4.0 Technologies for Marine Internal Combustion Engines Diagnostic Systems

Dmytro Minchev, Roman Varbanets, and Yurii Kucherenko

Abstract The specifics of marine internal combustion engines (ICEs) pose significant risks to the crew, cargo, and vessel. Malfunctions during voyages, port manoeuvres, or in environmental control areas (ECAs) can lead to violations of emission standards and financial penalties. Classification societies mandate regular monitoring and maintenance of main and auxiliary engines to ensure reliable operation. Shipowners invest in diagnostic systems to maintain engine health, while engine manufacturers equip modern ICEs with self-diagnostic and fault warning systems. Popular portable diagnostic tools like MarPRIME, IMES EPM-XP, DK-20, Premet X, LEMAG ECI, and DEPAS 4.0 Handy measure gas pressure patterns and sometimes include vibration sensors for valve and injection timing. These affordable systems improve engine efficiency but are typically delayed-action, requiring data analysis post-measurement. This limitation prevents realtime adjustments during engine operation. Future diagnostic systems will integrate Industry 4.0 technologies such as cloud computing, digital twins, wireless data transmission, and AI to enable real-time diagnostics, enhancing the scope and accuracy of engine condition monitoring.

1 Introduction

The specific operation of marine internal combustion engines (ICE) is characterized by the fact that their malfunctions and failures pose significant risks to the crew, cargo, and the vessel as a whole. These issues are undesirable both during voyages between ports and manoeuvring in port waters, as well as during the operation of the vessel in controlled environmental zones (ECA). For example, due to malfunctions in the fuel injection system, turbochargers, air supply systems, etc., the engine may fail to meet the actual requirements for toxic emissions in exhaust gases, resulting in substantial financial penalties when operating the vessel in specified zones.

D. Minchev (✉) · R. Varbanets · Y. Kucherenko
Department "Marine Power Plants and Technical Operation", Odesa National Maritime University, Odesa, Ukraine
e-mail: misaidima@gmail.com

© The Author(s), under exclusive license to Springer Nature Switzerland AG 2025
A. Marchenko and Z. Koruba (eds.), *Decarbonization of Transport Energy Installations in the Context of Sustainable Development Strategies*, Studies in Systems, Decision and Control 635, https://doi.org/10.1007/978-3-032-05884-3_14

This is why classification societies rules and vessel operation regulations mandate regular monitoring of the technical condition of main and auxiliary engines, along with timely implementation of planned maintenance procedures.

Shipowners have a vested interest in ensuring the proper condition of the ship's engines, thus they seek to apply and update the monitoring and diagnostic systems for engine technical conditions estimation. Internal combustion engine manufacturers, in turn, install on modern engines self-diagnostic, monitoring, and fault warning systems as well.

Portable diagnostic and analysing systems for monitoring the working processes of marine ICE, such as MarPRIME (Maridis Gmbh) [1], IMES EPM-XP [2], DK-20 (Icon Research) [3], Premet X (CM Technologies) [4], LEMAG ECI (Chris-Marine) [5], DEPAS 4.0 Handy (Depas Laboratory) [6], etc., are widely used and popular among senior mechanics of majority of shipping companies. These systems provide measuring the in-cylinder gas pressure diagrams (so called indicator diagrams), some additionally equipped with vibration sensors that capture the timing of gas distribution valves closure as well as the timing of the fuel injection. These systems are relatively inexpensive and significantly enhance the capabilities of the marine engine efficient operation.

Most of the existing portable diagnostic systems for marine engines on the market could be classified as the delayed-action systems. The engineer officer performs necessary measurements, storing data in the memory of the corresponding device. Visualization, processing, and analysis of data are carried out after testing on a personal computer. An obvious drawback of this approach is the inability to instantly determine the parameters of the working process in the investigated engine cylinder and to adjust the engine directly during its operation.

The further development of portable marine engine diagnostic systems involves integration of the technologies inherent in Industry 4.0. The most crucial among them include cloud computing, the use of digital twins, wireless data transmission, the application of artificial intelligence and providing real-time diagnosing process.

This paper explores the prospects for developing next-generation portable diagnostic systems for marine internal combustion diesel engines, which embody the mentioned technologies and enable an increase in the volume and quality of diagnostic information.

2 The Concept of the Next Generation of Portable Diagnostic Systems for Marine Internal Combustion Engines

The proposed concept of a next generation of portable marine engine diagnostic systems involves exploiting the capabilities of wireless data transmission from sensors to a computing device and internet connectivity. This enables periodic software updates, data storage, and subsequent "cloud" statistical processing. As a computing device, a smartphone, or another compatible device, is selected, aiming to reduce system costs while obtaining sufficiently high computational power characteristic of modern devices of this type.

Thus, the system functionally consists of the pick-up sensors, a sensor signals processing and transmission device (SPTD), and a computing device (smartphone, tablet, etc.) as it is shown in Fig. 1. Pressure sensor in the engine cylinder, a vibroacoustic sensor, and a temperature sensor form the set of sensors that pick-ups the corresponding physical parameters. Data transfer between the sensor signals processing and transmission device and the computing device is carried out through wireless communication (Bluetooth/Wi-Fi technology), and the computing device has internet access.

Key advantages of the proposed concept include the ability to monitor and analyse the in-cylinder processes in the near-real-time directly during measurements. The system's primary competitive advantages are realized through advanced software.

Some new capabilities provided by the implementation of real-time diagnostic system operation include:

- evaluation of engine operating parameters directly during engine operation, ensuring control over the registration of the necessary information;
- engine adjustment during operation with instant determination of the effect;
- assessment of the cyclic irregularity of the engine's operating process, fuel injection processes, and gas distribution mechanism to detect possible specific malfunctions.

Special criteria are proposed in [7] to provide an express assessment of the level of cyclic irregularity of the engine's operating process and the fuel delivery and gas distribution system.

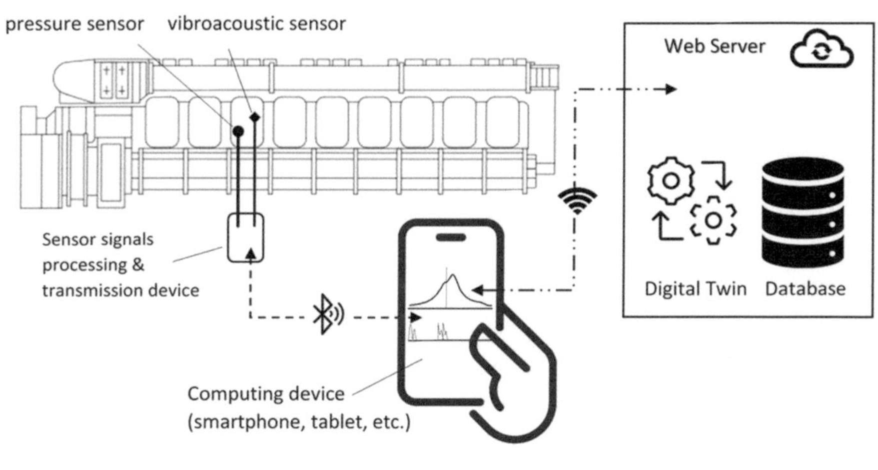

Fig. 1 Portable diagnostics system concept

3 Application of Digital Twin Technology

A digital twin is a virtual representation of a research object that replicates its behaviour within specified functional frameworks outlined by technical requirements. The widespread application of various types of digital twins across different industries has been provided in recent years. The concept of the digital twin, proposed initially by Grieves [8], has subsequently undergone rapid development and implementation in various fields of human activity. Digital twins are classified into types: digital twin instances (DTI), digital twin prototypes (DTP), and performance digital twins (PDT) [9–11].

In the context of diagnosing the current technical condition of marine internal combustion engines, a highly effective tool is a digital twin that reflects the operational processes in the engine and its systems. Such a digital twin can be implemented as software on a user device or be available online. The ability to synthesize engine processes using data from diagnostic system sensors provides broad possibilities for obtaining additional diagnostic information. This approach allows for a more accurate identification of the engine's deviations from normal operation, determination of parameters that are difficult or impossible to measure under specific conditions, and normalization of obtained diagnostic data considering operational conditions during measurements.

The concept of a digital twin as an integral element of the technical operation of a marine internal combustion engine is presented in Fig. 2. The sensors installed on the engine—either standard or additionally installed—transmit data on the measured parameters, which are used by the digital twin as input information. The digital twin must be pre-configured using data from the engine manufacturer as well as available experimental data.

Some of the data obtained from the engine sensors are used to identify the operating mode—these include crankshaft rotation frequency, position of fuel injection control elements (known as the fuel delivery index), etc. The set of such data may vary depending on the operational characteristics and available sensor set of a specific engine.

Other data are used to determine the correspondence of the actual engine operating process to the expected reference operating process defined by the digital twin. By comparing synthesized data with experimentally obtained data, deviations in engine operation from the reference can be identified.

If the real operating process significantly deviates from the reference, the functionality of the digital twin allows, firstly, to determine the consequences of engine operation with such deviation. Primarily analysed are changes in specific fuel consumption, emissions levels of harmful substances, changes in the temperature state of critical coolants and engine components. Thus, negative effects of engine operation are forecasted with a fixed deviation in the operating process.

Additionally, the digital twin can apply algorithms to search for potential causes of detected deviations in the operating process from the reference by synthesizing potential faults from the available database and known characteristics of the specific

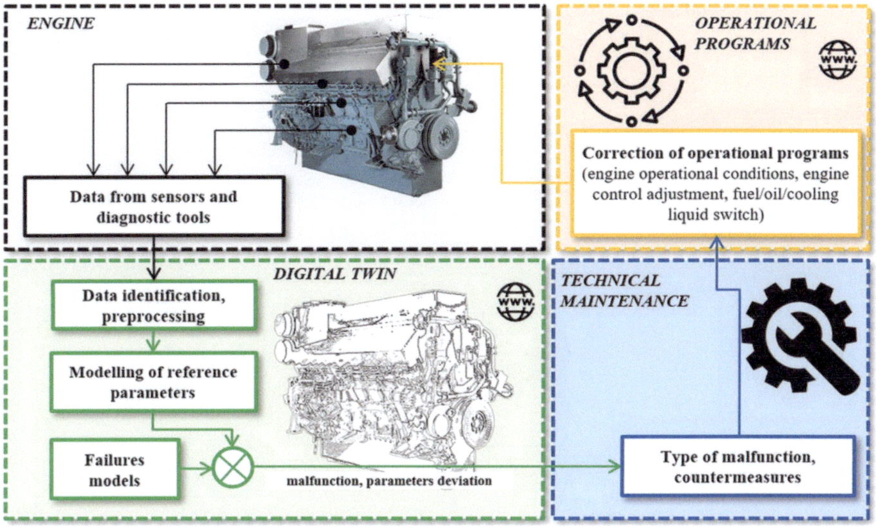

Fig. 2 Application of the performance type digital twin for marine internal combustion engines technical maintenance and diagnostics

engine. Examples of using mathematical simulation for diagnosing the operating process of marine ICEs are shown in [12].

The result of the digital twin's operation is information regarding the correspondence of the engine's operating process at a given mode to the expected reference, identification of potential faults, and recommendations for the need to intervene in the engine's operation process through appropriate technical maintenance.

4 Experimental Research

In the laboratory of the Department of Marine Power Plants and Technical Maintenance at ONMU, a research test-bench is being developed based on the Weichai WP4 marine diesel generator, which is equipped with necessary measuring instruments and controllers for investigating the possibilities of using a digital twin of the engine's working processes during its operation [13]. The diagram of the research test-bench is shown in Fig. 3, and a schematic diagram of the system for measuring the main engine parameters is given in Fig. 4. Measurements are provided for both high-speed parameters (cylinder pressure and fuel system components) and low-speed parameters characterizing the engine's operation (exhaust gases and intake air temperatures, air and fuel mass flows, etc.). The obtained data are used for analysis using the digital twin and obtaining diagnostic information regarding the current technical condition of the engine.

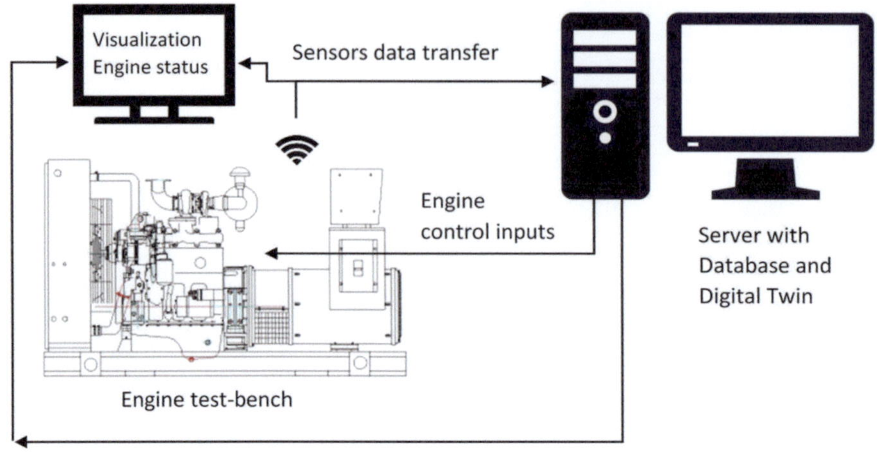

Fig. 3 Digital twin test-bench

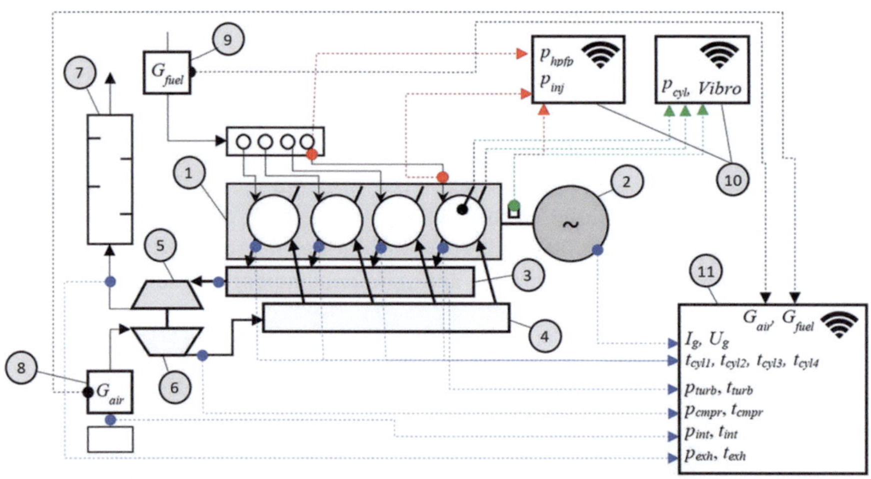

Fig. 4 Measurements diagram of the test-bench: 1—engine; 2—AC alternator; 3—exhaust manifold; 4—intake receiver; 5—turbine; 6—compressor; 7—silencer; 8—air mass flow sensor; 9—fuel mass flow sensor; 10—high-speed signals data transmission; 11—low-speed signals data transmission

For capturing high-speed parameters, a high-quality cylinder pressure sensor from IMES GmbH is installed, fuel pressure sensors at the inlet to the fuel injector and at the outlet from the high-pressure fuel pump manufactured by Bosch, a vibroacoustic sensor from Depas Lab, and an inductive crankshaft position sensor are used. The installation points of these sensors and the overall view of the research setup are shown in Fig. 5.

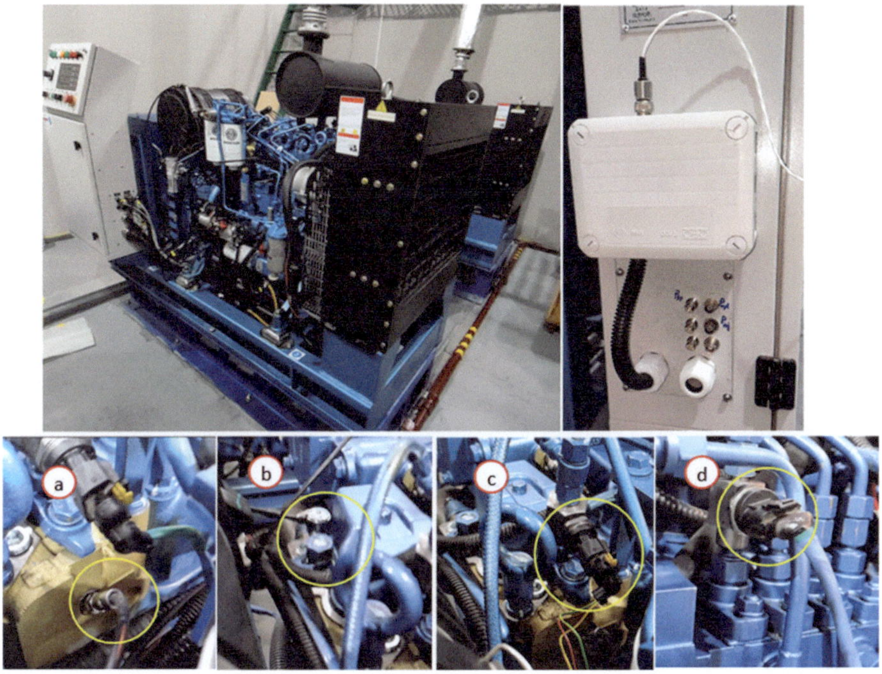

Fig. 5 Experimental test-bench view and sensors installation points: **a** cylinder pressure sensor; **b** vibroacoustic sensor; **c** fuel pressure sensor at injector inlet; **d** fuel pressure sensor at the fuel pump outlet

The test-bench is equipped with interfaces for direct signal output from the sensors to a digital oscilloscope, as well as signals processing and transmitting controller that transmits signals from the sensors to the network using Wi-Fi and Bluetooth interfaces.

The example of displaying of the high-speed parameters in the screen of the digital oscilloscope is presented in Fig. 6a. Up to 8 signals could be processed simultaneously providing excellent benchmark for the developing diagnostics system. In the Fig. 6b the prototype of the developing system software is shown, displaying the synchronous signals of the in-cylinder pressure sensor and the vibro-acoustic sensor. The data is transferred to the smartphone by Wi-Fi interface.

5 The Mathematical Model of the Digital Twin

The core of the digital twin is a mathematical model of the operational processes in the engine and its systems. This model provides the synthesis of operational processes for different types of engines with the required accuracy, performs fast calculations, and is user-friendly. One significant feature of the model is the use of the fuel combustion

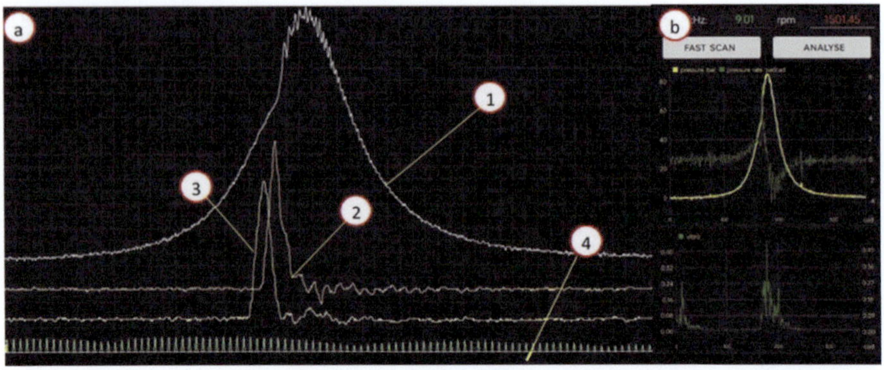

Fig. 6 Example of measuring of the high-speed parameters with the digital oscilloscope directly connected with the sensors **a** and with the smartphone coupled by Wi-Fi interface with the signals processing and transmitting device **b**: 1—cylinder pressure; 2—fuel pressure at injector inlet; 3—fuel pressure at the fuel pump outlet; 4—flywheel position

sub-model for diesel engines proposed by Razlejtsev, with important improvements. Modelling the parameters of supercharging units using the performance maps of the compressor and turbine of the turbocharger allows complete synthesis of the working process of a combined engine, considering such specific phenomena as turbocharger surging.

The composition diagram of the thermodynamic system "Engine" as it is considered for the mathematical model, is presented in Fig. 7. It is represented as a set of open thermodynamic subsystems interconnected by a common working matter that performs a closed working cycle. Each of these subsystems is an open thermodynamic system that exchanges energy with neighbouring systems in the form of enthalpy I_0, I_{cmpr}, $I_{res.in}$, $I_{res.out.i}$, $I_{int.i}$, $I_{exh.i}$, $I_{exh.in.i}$, I_{turb}, $I_{turb.out}$. Additionally, the subsystems exchange heat with engine cooling systems through heat flows Q_{CAC}, Q_{exh}, Q_{res}, $Q_{oil.i}$, $Q_{w.i}$, $Q_{oil.fr}$, and perform mechanical work $W_{cyl.i}$. Fuel injected to the engine puts heat and enthalpy into the respective subsystems, denoted as $Q_{f.i}$. The following thermodynamic subsystems are identified: "Cylinder"—working cylinder; "Receiver"—intake receiver; "Exhaust manifold"—exhaust manifold; "Compressor"—compressor; "Turbine"—turbine; "Intake pipe," "Exhaust pipe," "Compressor pipe"—intake, exhaust pipes, and compressor outlet pipe, respectively. It should be noted that depending on the configuration of the supercharging system, the number of subsystems for the system manifolds may vary.

For each of the interacting subsystems of the ICE "Engine" individual energy schemes and, in general, mathematical models of various types can be applied. Detailed features of the applied mathematical models are presented in sources [12, 14].

Recent developments aimed at enhancing the adequacy of engine cycle synthesis include improvements to the fuel evaporation and combustion models proposed by

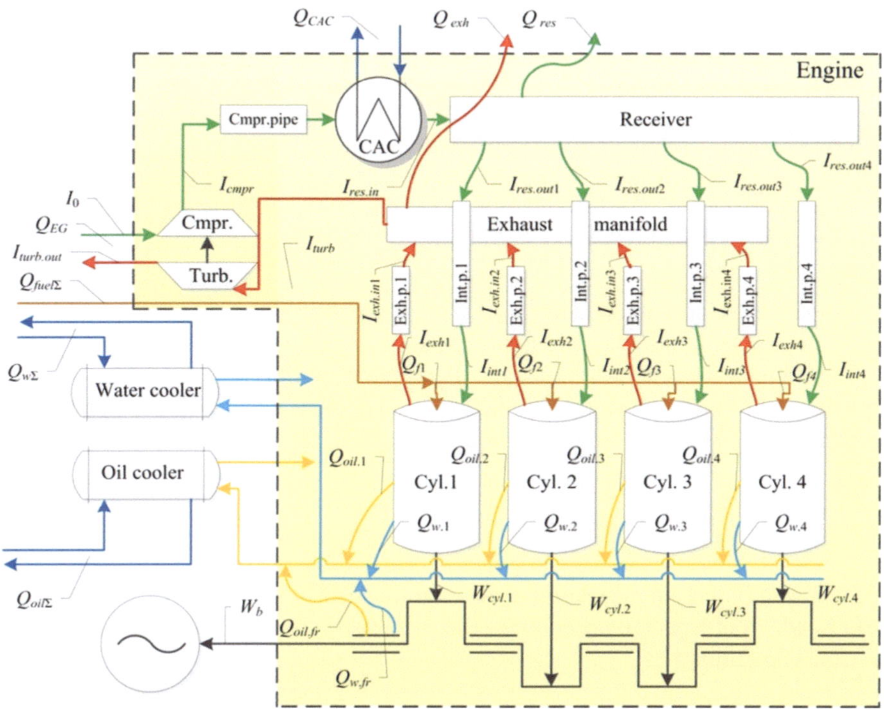

Fig. 7 Energy diagram of a multi-cylinder IC engine with a turbocharging system

Razletsev, as well as providing the capability to account for the unsteady operation of the centrifugal compressor of the turbocharger supercharging system.

6 Considering the Fuel Spray Variable Atomization During the Injection

According to the Razlejtsev fuel combustion model, the evaporation of the fuel sprays could be considered by applying the modification of the Sreznevsky law for the evaporation of an individual droplet. The fuel evaporation rate is calculated as the integral for all fuel portions injected up to given moment of time:

$$\frac{d\sigma_{ev}}{d\varphi} = \int\limits_{\varphi_{inj.start}}^{\varphi} \frac{3}{2}\chi_{wall}b_{ev}\left(1 - \chi_{wall}b_{ev}\frac{\varphi - \varphi_{inj.start}}{6n}\right)^{0.5}\left[\frac{d\sigma}{d\varphi}\right]d\varphi, \tag{1}$$

where σ_{ev} is evaporated portion of fuel; $d\sigma/d\varphi$ is relative injection rate; b_{ev} is evaporation constant; χ_{wall} is the reduction in the evaporation rate occurs when fuel interacts with the walls.

The fuel evaporation constant depends on the Sauter diameter of the fuel spray droplets d_{32} and can is calculated by:

$$b_{ev} = Y\frac{10^{12}}{d_{32}^2}\frac{1}{p_c} \tag{2}$$

where Y is the correction factor; p_c is the end pressure of the compression process; m_Y is the exponent of the evaporation function.

The original Razlejtsev approach assumes constant during the injection process both the fuel droplets diameter d_{32} and the fuel injection rate $d\sigma/d\varphi$ which results in a simplified equation:

$$\frac{d\sigma_{ev}}{d\varphi} = \frac{1}{\varphi_{inj}}\chi_{wall}\left(\left[1 - b_{ev}\left(\varphi - \varphi_{inj}\right)\right]^{3/2} - [1 - b_{ev}]3/2\right) \tag{3}$$

However, such a simplification causes significant errors in fuel evaporation calculations, which is especially crucial for the short fuel injection period at the engine's part-load operation [13].

The average Sauter diameter of fuel droplets is calculated as:

$$d_{32} = 10^6 E_c d_{inj.holes}\frac{M^{0.0733}}{(\overline{\rho} We)^{0.266}} \tag{4}$$

$$M = \frac{\mu_{fuel}^2}{d_{inj.holes}\rho_{fuel}\sigma_{fuel}};\ \overline{\rho} = \frac{\rho_c}{\rho_{fuel}};\ We = \frac{u_{inj}^2 d_{inj.holes}\rho_{fuel}}{\sigma_{fuel}},$$

Where τ, τ_c are time moment from injection start; μ_{fuel}, ρ_{fuel}, σ_{fuel} are dynamic viscosity, specific gravity and surface tension of the fuel; E_c is a coefficient; $d_{inj.holes}$ is the nozzle holes diameter; u_{inj}—injection velocity.

For simplified approach u_{inj} considered constant and averaged for whole injection period, and modified approach considers variable injection velocity from fuel injection diagrams.

As the illustration of importance of the consideration of the variable fuel atomization during fuel injection the results of comparative calculations for medium-speed marine diesel engine part-load operation are presented. The variation in fuel atomization size, the most significant at the beginning and at the end of injection process, results in different fuel evaporation rates diagram as the evaporation of the latest injected fuel portions lasts longer (Fig. 8a). This causes the longer and less efficient combustion at the final stage of combustion process (Fig. 8b). The smoke and NO_x development diagrams are also affected resulting in different values of engine's toxic emissions as it is shown in Fig. 9.

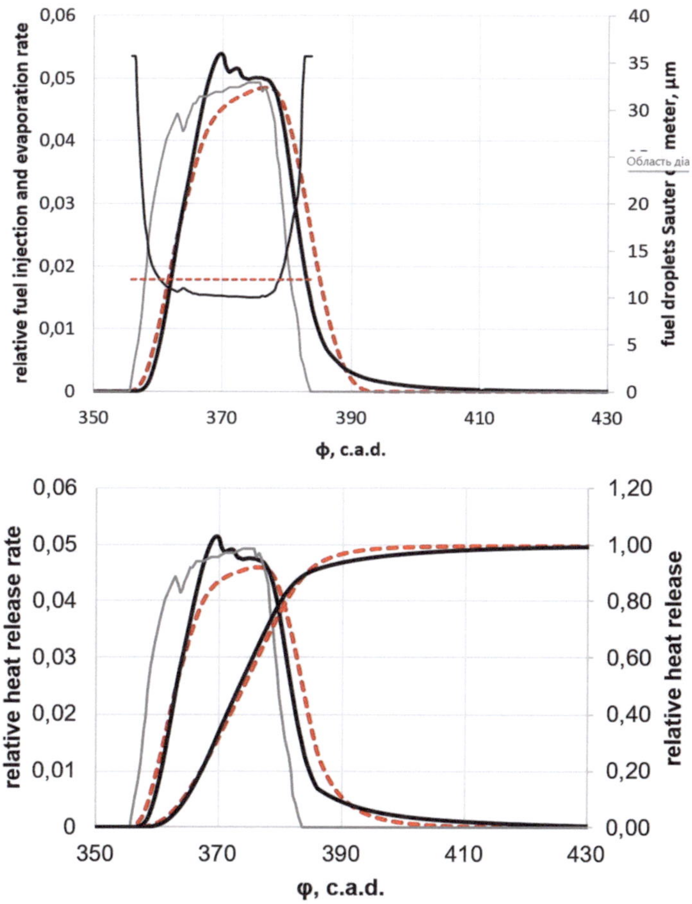

Fig. 8 Marine diesel engines simulated fuel evaporation and burning characteristics: solid lines—variable droplets size, dashed lines—constant droplets size

7 Ensuring the Calculation of the Engine's Operation in the Conditions of Turbocharger Surge

Centrifugal compressors are widely used for engine supercharging, both as part of a turbocharger and as mechanical superchargers. This type of compressor provides high efficiency combined with compact dimensions, but it also has some well-known drawbacks associated with matching its operation to the engine's piston part. Firstly, centrifugal compressors require a control system to achieve the desired level of boost pressure over a wide range of engine operating conditions, and secondly, they have a range of operating modes characterized by unstable air compression system performance. Unstable compressor operation is distinguished as rotating stall and

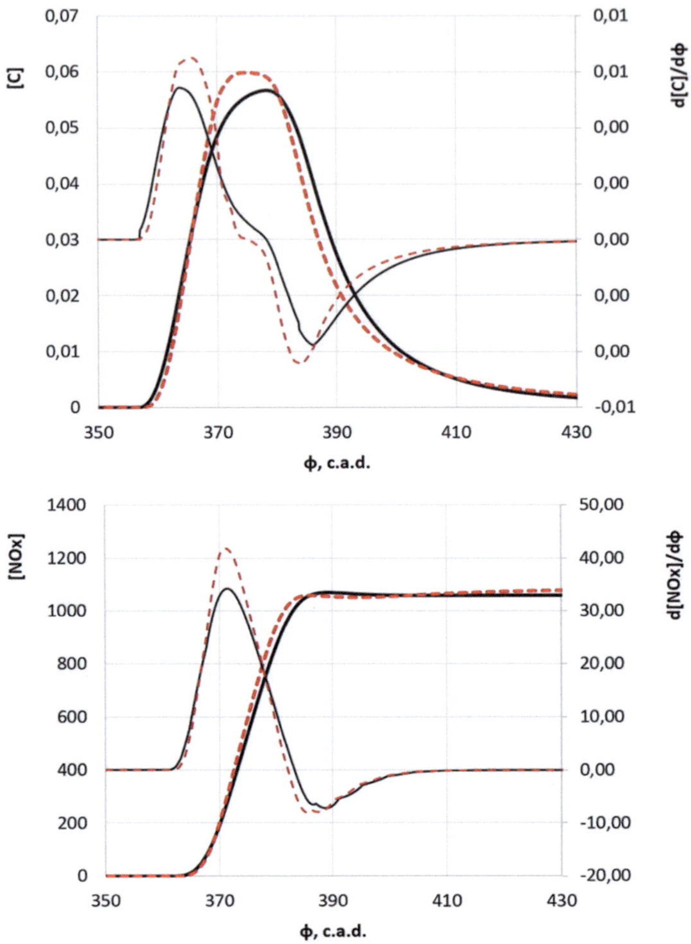

Fig. 9 Marine diesel engines simulated in-cylinder toxic emissions development: solid lines—variable droplets size, dashed lines—constant droplets size

surge mode. Both of these modes can reduce the lifespan of the turbocharging unit and the engine as a whole, and can even lead to engine shutdown due to failure.

From the standpoint of mathematical synthesis of the engine's working cycle, it is important to ensure the calculation of all possible compressor operating modes, including those in the unstable operating zone. This is a challenging task because surge is caused by a complex of gas dynamic phenomena, including vortex formation in compressor flow path elements and their complex interaction and development. However, the operation of the system in surge mode depends not only on the compressor's characteristics but also on the parameters of the entire compression system, which includes collectors, pipes, and valves.

When calculating the steady-state operation of the engine, tasks usually include calculating the static operational parameters of the engine, predicting the engine's thermal balance, studying combustion processes and heat transfer, and so on. Among the important tasks is the computational matching of the turbocharging unit with the engine, determining rational control methods of the turbocharging system. Therefore, when calculating the steady-state operating mode, it is important to be able to predict the possibility of the compressor unit's unstable operation, the type of this unstable operation, and its main parameters. It is also necessary to ensure quick and stable execution of the calculation.

Consider the operation of the compressor with a constant rotational speed.

Figure 10b shows the diagram of pressure change in the engine intake receiver under conditions of deep surge, with arrows indicating the direction of pressure change. The direction of change in the instantaneous working point of the compressor is shown by black arrows. The deep surge mode occurs if the system resistance exceeds a certain level at which the pressure in the receiver reaches point *B* (the extremum). With further reduction in air flow through the compressor, the air pressure at the compressor outlet begins to decrease rapidly, causing a slowdown in the air flow in the intake pipe. The pressure in the engine's receiver, due to its significant volume and short time during which the pressure drop at the compressor intake occurs, remains almost constant. As a result of the decrease in compressor intake pressure, the air flow reverses, moving from the receiver to the compressor intake, which is observed simultaneously with the pressure drop in the engine's receiver. The compressor parameters are determined by the *CD* section. When the pressure in the receiver reaches the pressure corresponding to the compressor intake pressure at zero air flow (point *C*), an instantaneous stop of air movement in the intake pipe occurs, and then it starts moving towards the receiver, accelerating due to the increase in compressor intake pressure, as the compressor operation corresponds to line *CA*. The pressure in the receiver changes insignificantly during this process due to the short duration for which it occurs. The compressor again enters the stable operation section *AB*, and the pressure in the receiver begins to increase, and when its value reaches point *B*, the process repeats.

Detailed principles of simulation the operation of a centrifugal compressor under surge and surge conditions are presented in [15]. Figures 11 and 12 show the results of calculating the operating mode of a marine medium-speed diesel engine with the compressor operating under conditions of so called mild surge. The engine operates at partial load according to the propeller curve at a crankshaft speed of 632.6 rpm and a power of 2212 kW (60% of the rated power). As can be seen, the engine's operation under compressor surge is characterized by both pressure fluctuations in the air receiver and pulsations in air flow through the compressor with periodic reversal of its direction. In this case, an example of mild surge is shown, under which the engine is still able to operate although for a limited time. Notably, for this operating mode, the frequency of pressure pulsations in the receiver due to surge synchronized with the frequency of the engine's crankshaft and amounted to 10.54 Hz.

To ensure the ability to alter the stability of the compressor operation in simulations, it is proposed to use the multiplier μG_{surge}. This multiplier allows scaling

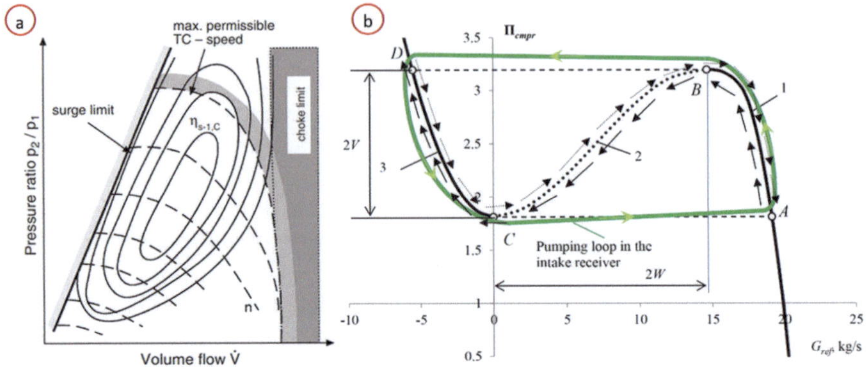

Fig. 10 Centrifugal compressor operational range **a** [16] and constant-speed line treatment for "deep" surge simulation **b**: 1—compressor stable operation line *AB*; 2—compressor unstable operation region *BC*; 3—backflow part of the constant-speed line *CD*

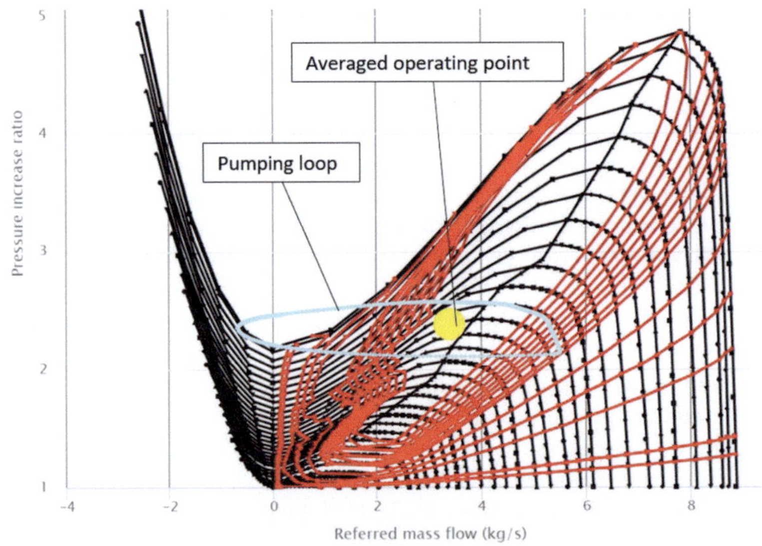

Fig. 11 Simulation of the compressor surge. Medium-speed marine diesel engine, 60% MCR

the unstable part of the compressor's characteristic as illustrated in Fig. 13, line 2. The values of the multiplier obbey to the condition $0 < \mu G_{surge} < 1$, and it is used as follows:

$$\prod_{cmpr}^{\prime} = \prod_{cmpr}^{B} - \mu G_{surge} \left(\prod_{cmpr}^{B} - \prod_{cmpr} \right),$$

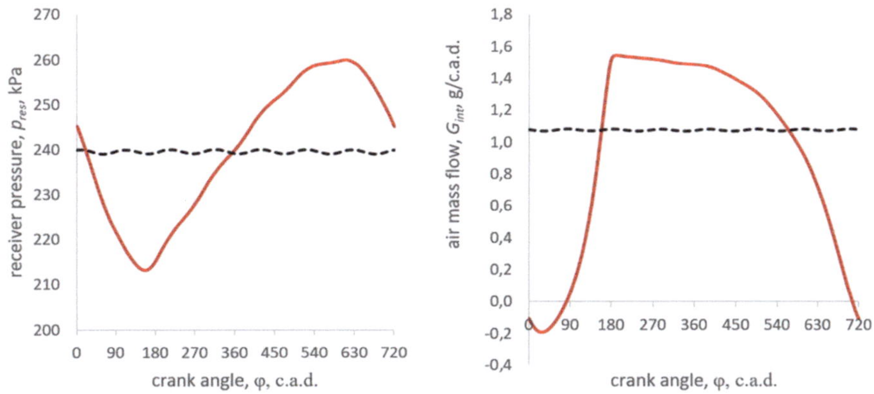

Fig. 12 Comparison of the intake receiver parameters if the turbocharger operates with mild surge (solid lines) and at stable operating point (dashed lines)

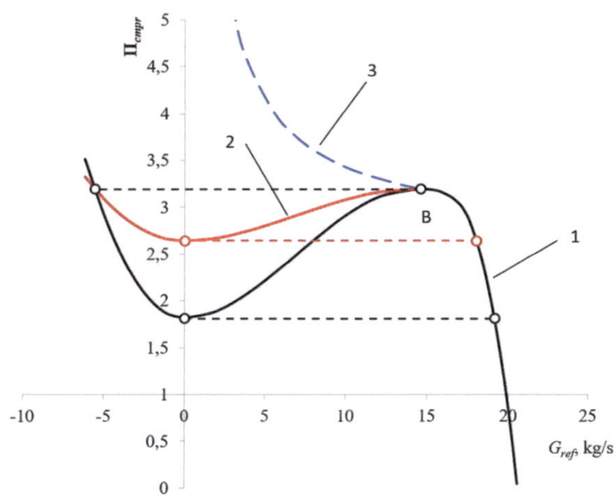

Fig. 13 Example of constant-speed line treatment in the range of instable operation of the compressor to modify the surge behavior foe different multiplier μG_{surge}: 1 $\mu G_{surge} = 1$; 2 $\mu G_{surge} = 0.4$; 3 $\mu G_{surge} = 0$

where $\prod_{cmpr}^{B}$–the value of compressor's pressure increase ratio on the critical point (point B).

It is evident that by using μG_{surge}, one can alter the shape of the surge loop and adjust it according to available experimental data. For rapid and stable calculations, it is recommended to apply the hyperbolic form of the unstable part of the compressor characteristic if the user selects $\mu G_{surge} = 0$.

8 Application of Advanced Analysis of Engine Vibroacoustic Signals

An important feature of the proposed system is the use of a vibroacoustic sensor synchronously with a pressure sensor in the engine cylinder [17]. This technology enables the capture the gas distribution and fuel injection timing directly during engine operation [18]. In addition, the proposed methods allow diagnosing the technical state of the turbocharger, bearings, injector lubrication oil supply to the engine cylinder liner by analysing vibroacoustic signals [19].

The application of advanced high-efficiency algorithms for determining the synchronization of the engine cylinder pressure diagram and the crankshaft rotation angle without using the crank position sensor, applying a vibroacoustic sensor for real-time engine diagnostics, allows the development of new diagnostic methods and criteria, including cyclical irregularities in engine operation, which represents an important prospect for further research (Fig. 14).

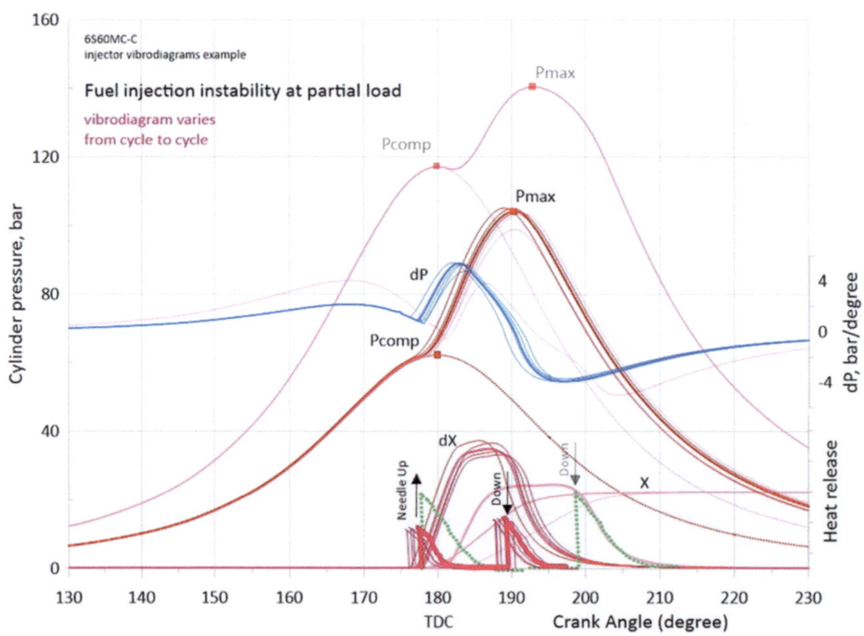

Fig. 14 Example of cyclic irregularities in the operation of a marine diesel engine under partial load due to a defect in the fuel supply system

9 Conclusion

Portable diagnostic systems for internal combustion marine engines based on the parameters of the working process are in high demand and are widely used. Further development of such systems should be based on the application of key Industry 4.0 technologies.

The use of wireless data transmission technologies and modern smartphones as computing devices allows for cost reduction, increased mobility, and convenience of the system, as well as regular and convenient software updates. However, the key advantage is the possibility of implementing real-time system operation. This ensures the ability to determine the technical condition of the engine directly during its operation and develop specific diagnostic methods that allow, among other things, to consider cyclic irregularities in engine operation.

The use of cloud computing technologies allows for the involvement of methods related to the use of digital twins of internal combustion engines in the diagnostic process, reflecting their working processes. Digital twins enable obtaining more diagnostic information from sensors data, identifying potential causes of malfunctions, and have the potential to provide automated diagnosis of the engine's technical condition.

The effectiveness of using digital twins depends on the quality of the mathematical model of the working processes in the engine cylinder, the fuel system, and the air supply system, as well as on special algorithms that ensure their use based on data obtained from cylinder pressure sensors, vibroacoustic sensors, etc.

References

1. MARIDIS: Maritime diagnosis & service. https://www.maridis.de/en/mobile-devices/marprime-marprime-ultra.html. Last accessed 03 2024
2. IMES|: The cylinder pressure people. https://www.imes.de/epm-xp.html. Last accessed 03 2024
3. ICON Reasearch: https://iconresearch.co.uk/diesel-engine-analysis/doctor-portable/. Last accessed 03 2024
4. CMT: Electronic Innovation [Online]. https://www.cmtechnologies.de/de/produkte/leistungs-und-effizienzueberwachung/premet-x.html. Last accessed 03 2024
5. Chris-Marine: https://chris-marine.com/products/monitoring-of-diesel-and-gas-engines/electronic-cylinder-pressure-indicator-eci/. Last accessed 03 2024
6. Depas Laboratory: http://depas.od.ua/products.html. Last accessed 03 2024
7. Varbanets, R., Minchev, D., Kucherenko, Y., Zalozh, V., Kyrylash, O.: Real-time diagnostics of marine diesels: algorythms and methods. Polish Maritime Res. (2024)
8. Grieves, M., Vickers, J.: Digital Twin: Mitigating Unpredictable, Undesirable Emergent Behavior in Complex Systems, Transdisciplinary Perspectives on Complex Systems: New Findings and Approaches. pp. 85–113. https://www.researchgate.net/profile/Michael-Grieves/publication/306223791_Digital_Twin_Mitigating_Unpredictable_Undesirable_Emergent_Behavior_in_Complex_Systems/links/5aa54e1ea6fdccd544bc386f/Digital-Twin-Mitigating-Unpredictable-Undesirable-Emerge

9. Botín-Sanabria, D., Mihaita, A.-S., Peimbert-García, R., Ramírez-Moreno, M., Ramírez-Mendoza, R., Lozoya-Santos, J.: Digital twin technology challenges and applications: a comprehensive review. Remote Sens. **14**(6), 1335 (2022). https://doi.org/10.3390/rs14061335
10. Li, L., Aslam, S., Wileman, A., Perinpanayagam, S.: Digital twin in aerospace industry: a gentle introduction. IEEE Access **10**, 9543–9562 (2022). https://doi.org/10.1109/ACCESS.2021.3136458
11. Singh, M., Fuenmayor, E., Hinchy, E., Qiao, Y., Murray, N., Devine, D.: Digital twin: origin to future. Appl. Syst. Innov. **4**(2) (2021). https://doi.org/10.3390/asi4020036
12. Minchev, D., Varbanets, R., Alexandrovskaya, N., Pisintsaly, L.: Marine diesel engines operating cycle simulation for diagnostics issues. Acta Polytechn. **61**(3), 428–440. https://doi.org/10.14311/ap.2021.61.0435
13. Minchev, D., Varbanets, R., Shumylo, O., Zalozh, V., Aleksandrovska, N., Bratchenko, P., Truong, T.H.: Digital twin test-bench performance for marine diesel engine applications. Polish Maritime Res. **30**(4), 81–91 (2023)
14. Minchev, D.S.: (2020) Blitz-PRO. User's Manual. http://blitzpro.zeddmalam.com/extra/Tutorial/Help.pdf (2020)
15. Minchev, D., Gogorenko, O., Varbanets, R., Moshentsev, Y., Píštěk, V., Kučera, P., Shumylo, O., Kyrnats, V.: Prediction of centrifugal compressor instabilities for internal combustion engines operating cycle simulation. Proc. Instit. Mech. Eng. Part D J. Automobile Eng. **237**(2–3), 572–584 (2023). https://doi.org/10.1177/09544070221075419
16. Prenninger, H.H., Prenninger, P.: Charging the Internal Combustion Engine. Springer-Verlag, Wien (2007)
17. Neumann, S., Varbanets, R., Minchev, D., Malchevsky, V., Zalozh, V.: Vibrodiagnostics of marine diesel engines in IMES GmbH systems. Ships Offshore Struct. (2022). https://doi.org/10.1080/17445302.2022.2128558
18. Varbanets, R., Shumylo, O., Marchenko, A., Minchev, D., Kyrnats, V., Zalozh, V., Aleksandrovska, N., Brusnyk, R., Volovyk, K.: Concept of vibroacoustic diagnostics of the fuel injection and electronic cylinder lubrication systems of marine diesel engines. Polish Maritime Res. **29**(4), 88–96 (2022). https://doi.org/10.2478/pomr-2022-0046
19. Varbanets, R., Fomin, O., Píštěk, V., Klymenko, V., Minchev, D., Khrulev, A., Zalozh, V., Kučera, P.: Acoustic method for estimation of marine low-speed engine turbocharger parameters. J. Mar. Sci. Eng. **3**(9). https://doi.org/10.3390/jmse9030321

Main Directions and Results of Scientific Research Ivchenko-Progress JSC on Reducing Negative Impact on Humans and the Environment When Using Aircraft Engines

Volodymyr Zhuravlov, Sergiy Smirnov, Dmytro Kozel, Igor Kravchenko, Mykhailo Mitrakhovych, and Sergiy Dmytriyev

Abstract This article summarizes the research efforts of Ivchenko-Progress JSC in mitigating the environmental impact of aircraft engines. Key areas include reducing acoustic emissions and harmful exhaust gases. By advancing technologies in engine design and operation, the enterprise aims to enhance environmental performance while maintaining high efficiency. The findings underline the importance of innovation in addressing global challenges associated with aviation's environmental footprint, supporting international goals for sustainable air transport.

1 Introduction

The signing of the Association Agreement between the European Union and Ukraine in 2014 obliged Ukraine to adopt some of the obligations and rules in force in the European Union, including in the field of aviation.

The main coordinating center for the development of aviation industry policy in the European Union is the Advisory Council for Aeronautics Research and Innovation (ACARE).

The ACARE-developed Flightpath 2050 program includes environmental goals that recognize the need for aviation to accelerate its efforts to reduce emissions affecting climate change, noise impact, and air quality to ensure sustainable growth of transport.

ACARE estimates that aviation produces about 2% of all anthropogenic CO_2 emissions and proposes to reduce them by increasing fuel efficiency through technology,

V. Zhuravlov · S. Smirnov · D. Kozel · I. Kravchenko · M. Mitrakhovych · S. Dmytriyev (✉)
Zaporizhzhia Machine-Building Design Bureau Progress Joint Stock Company named after, Academician O.H. Ivchenko (Ivchenko-Progress JSC), Zaporizhzhia, Ukraine
e-mail: s.dmytriyev@ivchenko-progress.com

© The Author(s), under exclusive license to Springer Nature Switzerland AG 2025
A. Marchenko and Z. Koruba (eds.), *Decarbonization of Transport Energy Installations in the Context of Sustainable Development Strategies*, Studies in Systems, Decision and Control 635, https://doi.org/10.1007/978-3-032-05884-3_15

operations, and infrastructure, as well as by developing sustainable low-carbon aviation fuels.

The program proposes to reduce the level of NOx and particulate emissions, reduce the negative impact of noise on human health, which requires measures to improve the design of vehicles and engines, operational and infrastructure facilities, and a deeper study of the impact of acoustic radiation on human health.

It is known that civil aviation emissions account for about 2–2.5% of global greenhouse gas emissions. Given the constant growth of air transport (approximately 4…5% per year), ACARE proposes to consider measures. Regarding the potential reduction of harmful emissions from aviation activities, primarily through the development of modern technologies, aircraft engine design, and their energy sources. The Ivchenko-Progress JSC actively participates in the implementation of the Flightpath 2050 program regarding the achievement of Goal 3—"Environmental protection and energy conservation". Achieving the 2050 goals of the Flightpath 2050 program requires radical changes in the methodology of designing and operating aircraft and the use of alternative sources. This section of the monograph presents the main results of the research of the Ivchenko-Progress JSC, aimed at achieving the goals of the Flightpath 2050 program to reduce the negative impact of aircraft engine operation on the environment in the following areas.

2 Main Directions and Results of the Research of the Ivchenko-Progress JSC on Reducing the Level of Acoustic Radiation of Aircraft Power Units

It is known [1] that the main sources of noise of dual-circuit engines with high and ultra-high bypass ratio in the front hemisphere are the fan and compressor, in the rear hemisphere—the fan, jet stream, turbine, and combustion chamber. The reduction of engine noise level of leading aircraft engine companies in the world [2–7] is ensured by increasing the bypass ratio of the engine, reducing the fan rotation speed, optimizing the parameters of the working cycle, using rotors with opposite rotation, using sound-absorbing structures, and reducing the level of turbulence at the fan inlet. The development of an optimized guide vane to reduce the level of turbulence at the engine inlet significantly reduces discrete and broadband noise, and the use of an inlet device of a special design reduces the level of vortex noise at the fan inlet [1]. The use of sound-absorbing structures and materials, improving the aerodynamic characteristics of turbine blades, and reducing the fan rotation speed allows reducing the level of tonal noise of the low-pressure turbine. The goal of the Flightpath 2050 ACARE program [8, 9] is defined that the reduction of acoustic emission of the aircraft can be achieved by developing technologies aimed at:—researching optimal fan reducer designs;—improving modeling methods;—acoustic and aerodynamic improvement of fan blade shape;—researching boundary layer control;—development of composite material applications;—development of sound-absorbing structures;—further acoustic perfection of engines;—optimization of aircraft and power unit integration with the objective function of minimizing the acoustic emission of the aircraft;—researching active control of jet acoustic characteristics using microjets and combining active noise control and the use of chevron nozzles;—reducing noise from internal sources—compressor and air bypass valves, combustion chamber, and turbine.

This section presents the results of using the newly created modern open acoustic testing center to facilitate the research of the Ivchenko-Progress JSC in improving the D-18 T series 3 turbofan engine to series 3 M, in line with several directions of the Flightpath 2050 ACARE program, specifically assessing the impact of:

- reducing the rotation frequency of the fan rotor;
- application of sound-absorbing structures;
- acoustic improvement of the fan turbine;
- application of sound-absorbing structures of the internal contour nozzle.

The general view of the modern open acoustic testing center of the Ivchenko-Progress JSC is shown in Fig. 1.

The open stand is located on a concrete pad with measurement points marked in the far field according to the national industry standard.

The stand area is free from structures and obstacles to the sound field. The microphone installation scheme of the measuring belt on the stand is shown in Fig. 2.

Microphones are installed at a distance of 50 m from the acoustic radiation object (ARO).

Fig. 1 Open acoustic testing center of the Ivchenko-Progress JSC

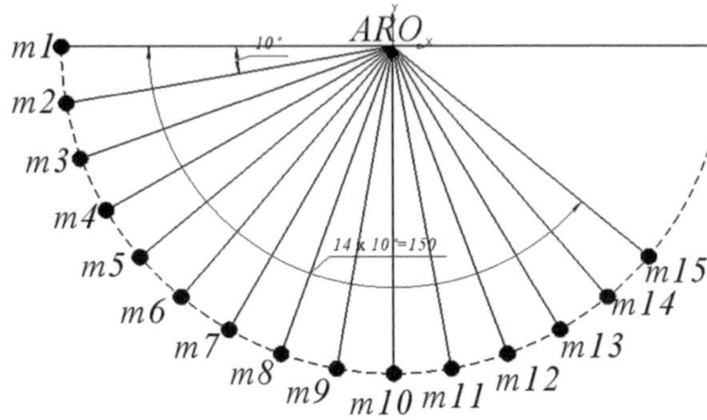

Fig. 2 Microphone placement scheme on the open test stand

During testing, sound pressure levels are measured in the direction of course angles from 0 to 150°.

The open stand is equipped with a set of electroacoustic equipment produced by "Bruel & Kjaer" and a MIC-455 computing complex.

Measurements of sound pressure levels were carried out in the far acoustic field according to the approved enterprise program and methodology at specified modes.

The MIC-455 computing complex processes the test results and brings them to standard atmospheric conditions and a relative air humidity of 70%.

The object of study chosen is the serial D18T series 3 turbofan engine, which is improved to series 3 M by:

- reducing the rotation frequency of the fan rotor by 5%;
- application of new two-mode tubular sound-absorbing structures;
- acoustic improvement of the fan turbine;
- application of sound-absorbing structures of the internal contour nozzle.

Comparative results of testing the D18T series 3 and series 3 M turbofan engines are shown in Fig. 3.

The assessment of the total effectiveness of the proposed measures on the D-18 T series 3 M engine at 3 control points for noise certification on the ground shows that the implementation of these measures allows reducing the level of acoustic radiation by 5.5 dB.

An important area of research for the Ivchenko-Progress JSC in reducing the impact of aviation noise on the environment is work aimed at reducing the acoustic emissions of propellers. The noise of an aircraft with a propeller during takeoff and altitude gain is primarily determined by emissions at the harmonics of the blade rotation frequency, as well as emissions at the combination harmonics. The noise of the leading edge of the blade is caused by the unsteady component of the aerodynamic force due to turbulence and inhomogeneity of the incoming flow, or instability of the rotation frequency. The noise of the trailing edge is caused by pressure pulsations on the trailing edge of the propeller blade due to flow separation, as well as pressure pulsations in the wake [10]. Considering the vast number of factors and parameters of airflows interacting with the propeller, it is essentially a single-mode aerodynamic

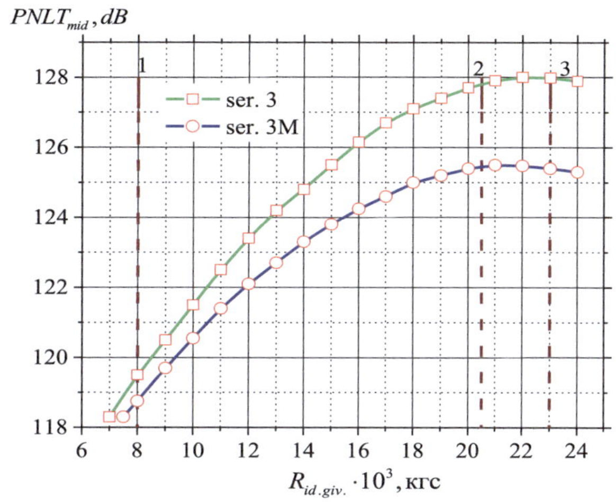

Fig. 3 Comparison dependencies of average perceived noise levels, corrected for tonality *PNLTcp* of D-18 T series 3 and D-18 T series 3 M engines

machine that converts engine rotation energy into thrust. Therefore, the power plant of the aircraft will have high efficiency only if the propeller is designed taking into account the actual conditions of use. Ivchenko-Progress has experience in comprehensively considering the task of designing a propeller for a specific aircraft as part of its power plant. When setting the task of designing a propeller, the following initial data are determined: conditions of use of the aircraft, its aerodynamic characteristics, main data of the power plant engine. The result is a propeller that will have maximum efficiency at the calculated flight modes of this aircraft as part of a power plant with a specific engine. When designing a propeller, the possibility of optimizing the layout of the power plant and the geometry of the propeller blades is studied in order to minimize the interrelated negative interaction of blades, aircraft surfaces, and airflows that flow around them. The scientific and methodological apparatus created at the enterprise allows for consistent mathematical modeling of the propeller flow using vortex and impulse theories. To determine a compromise option between the most favorable factors affecting the aerodynamic, strength, and acoustic characteristics of the propeller, a sequential verification aerodynamic calculation of each variant of the propeller geometry is performed [11]. The task of reducing the level of acoustic emission of the propeller is complex and requires numerous computational and experimental studies. When designing and optimizing the parameters and characteristics of the propeller, maximum reduction of the impact of sources of acoustic emission from the propeller is ensured, as well as emissions caused by its aerodynamic interference with the wing and fuselage, provided that the thrust and efficiency are not reduced. The use of modern, highly efficient aerodynamic profiles, composite materials, careful optimization of the geometry of the propeller blades and the layout of the power plant allows to reduce the blade's flow area, their mass, and the intensity of free vortex shedding from the blade tips, while ensuring a high efficiency of the air propeller at the calculated operating mode.

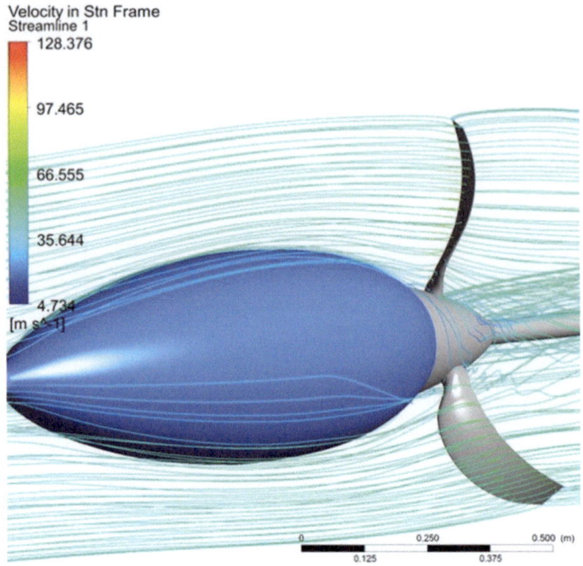

Fig. 4 Visualization of the results of mathematical modeling of a thrust air propeller

During the execution of verification aerodynamic calculations, modern software packages (ANSYS CFX, ANSYS FLUENT, etc.) are used.

An example of visualizing the results of mathematical modeling of an air propeller is shown in Fig. 4.

A practical result of the activities of the Ivchenko-Progress JSC in this direction is the creation of a highly efficient five-blade air propeller AI-R500V5C, with a power of 400 kW, adjustable pitch, and blades made of composite material (Fig. 5).

The five-blade air propeller AI-R500V5C has an efficiency of up to 93% in the calculated mode, a mass of 40 kg, which is 20% less than similar samples.

An important advantage of the created air propeller is a 10% reduction in noise level compared to existing analogs.

The results of the research of the Ivchenko-Progress JSC on the improvement of air propellers with high technical characteristics allow to increase the fuel efficiency

Fig. 5 Five-blade air propeller AI-R500V5C

of the power plant, increase the useful load, due to the reduction of fuel reserves, and as a result, reduce the total emissions of harmful substances and the level of acoustic radiation.

3 Main Directions and Results of Research on Reducing the Level of Harmful Emissions of Aircraft Engines

In conditions of annual growth of air transportation volumes and energy generation using gas turbine drives, the issue of reducing emissions of harmful substances contained in the exhaust gases of combustion products of gas turbine engines is very important due to their negative impact on the environment, climate, and human health [12].

The main components of combustion products in the exhaust gases of gas turbine engines can be conditionally divided into two groups [13, 14]:

- harmful substances: carbon monoxide (CO), unburned hydrocarbons (UCH), sulfur oxides (SOx), nitrogen oxides (NOx) and solid soot particles (non-volatile solid particles), smoke;
- non-toxic substances: carbon dioxide (CO_2) and water vapor.

Since carbon dioxide CO_2 is a greenhouse gas, its increased concentration in the atmosphere leads to a negative impact on the climate, in the form of a "greenhouse effect", causing global warming.

Moreover, an increased content of CO_2 in the air has a toxic effect on humans.

Emissions of water vapor are also dangerous due to the appearance of a "greenhouse effect" in the atmosphere.

Carbon monoxide CO is a toxic gas, which causes slowing down the transport of oxygen by tissues, leading to severe poisoning.

Unburned hydrocarbon emissions into the atmosphere are also harmful and cause a carcinogenic effect on living organisms.

In addition, sulfur oxides are toxic and can lead to the formation of sulfuric acid droplets in the atmosphere. Due to the oxidation during combustion of almost all sulfur contained in the fuel, the only effective method to reduce their emissions into the atmosphere is to purify the fuel from sulfur by refining the fuel.

Of all the nitrogen oxides, nitric oxide and nitrogen dioxide NO_2 are the most dangerous. They are several times more toxic than CO and when burning hydrocarbon fuels are released in large quantities.

Nitrogen oxide emissions are also dangerous due to their destructive action on the ozone layer in the atmosphere, leading to an increase in ultraviolet radiation on the surface of the earth. Therefore, special attention is currently paid to NOx emissions.

Emissions of solid non-volatile particles and soot (smoke) are also harmful, as they contain many toxic substances.

It should be noted that CO_2 emissions practically do not depend on the technology of fuel combustion in the combustion chamber, but are determined by the amount of burned fuel.

CO and unburned UCH hydrocarbon emissions are the result of incomplete combustion and not complete burning of fuel, while NOx emissions are the result of oxidation of inert nitrogen, which is in the air, in combustion zones with high temperature.

The level of smoke and non-volatile solid particles is mainly determined by the presence of small fuel-rich zones in the flame.

All this leads to the fact that with the increase in the engine operating mode or the increase in the level of air parameters in the combustion chamber, NOx emissions, smoke and non-volatile solid particles usually increase, while CO and unburned UCH hydrocarbons decrease.

Figure 6 shows the dependencies of the relative value of the emission index of the gaseous component CO—$EI(CO)_{rel}$, the relative value of the emission index of the gaseous component UHC—$EI(UHC)_{rel}$, the relative value of the emission index of the gaseous component NOx—EI(NOx) and the amount of smoke SN from the relative thrust of the engine—R_{rel}.

It should be noted that CO_2 emissions practically do not depend on the technology of fuel combustion in the combustion chamber, but are determined by the amount of burned fuel. Emissions CO and unburned hydrocarbons UCH are the result of incomplete combustion and not complete burning of fuel, while emissions NOx are the result of oxidation of inert nitrogen, which is in the air, in combustion zones with high temperature. The level of smoke and non-volatile solid particles is mainly determined by the presence of fuel-rich zones in the flame. of small fuel-enriched

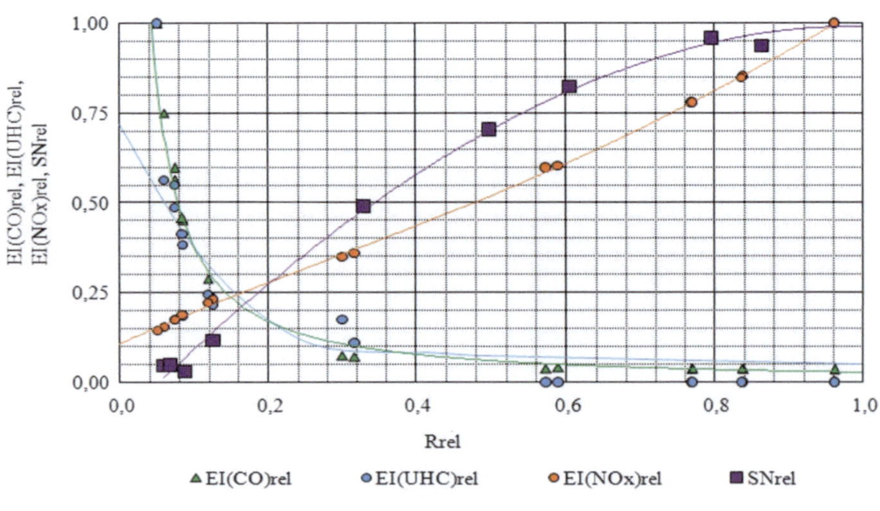

Fig. 6 Dependence of relative values of emission indices and the amount of smoke from the relative thrust of the engine

zones. All of this leads to the fact that with an increase in the engine operating mode or an increase in the level of air parameters in the combustion chamber, emissions of NOx, smoke and non-volatile solid particles usually increase, while CO and unburned hydrocarbons UCH decrease.

The application of different schemes and technologies for fuel combustion in combustion chambers also leads to a divergent trend of change for different components of harmful emissions. Therefore, changes in engine operating modes and flight conditions, the application of different design solutions and fuel combustion technologies for combustion chambers lead to a decrease in some emission components and an increase in others. Considering the fact that different components of harmful emissions have different impacts on the climate as a whole, or on local air quality, or directly on the health of living organisms, three main areas of research can be identified in the issues of harmful emissions from gas turbine engines [15].

The first direction is the development and design of low-emission combustion chambers and engines.

The second direction of research involves studying the impact of harmful emissions on human health and the environment.

The third direction, as a result of seeking a compromise solution and a rational balance of the results of the first two directions, is the development of effective measures to limit and regulate standards for harmful substance emissions.

This section discusses the development and design of low-emission combustion chambers, describes the main methods and approaches used in the design and calculation of combustion chambers. Also, in the context of discussing issues of compliance with existing regulatory restrictions for harmful emissions, the requirements of regulatory standards are briefly described.

3.1 Mechanisms of Harmful Emissions Formation and Design Solutions for Their Reduction

To determine the factors that influence the amount of harmful emissions, it is necessary to understand the mechanisms and principles by which emissions are formed during fuel combustion in gas turbine engine combustion chambers.

Let's consider in more detail the mechanisms of harmful substance emission formation.

The experience of studying gas turbine engine combustion chambers shows that in practice the amount of CO emissions significantly exceed thermodynamic equilibrium values, and the maximum emission values are realized at low engine operating modes, when the gas temperature in the combustion chamber is not high and the fuel–air mixture is lean.

This fact indicates that incomplete fuel combustion contributes significantly to the thermodynamic equilibrium emission. Indeed, the conversion of intermediate oxidation reactions of carbon-containing fuel CO into CO_2 is almost entirely determined by the chemical reaction:

$$CO + OH \rightarrow CO_2 + H \tag{1}$$

Therefore, the task of reducing CO emissions comes down to ensuring the completion of its oxidation reaction, and the reasons for the increased emission of CO can be caused by:

- low combustion speed in the combustion chamber due to low temperature and air pressure at the inlet, lack of fuel in lean operation mode;
- local “freezing” of combustion products in the combustion chamber by cold air used in the cooling system of the furnace tube walls;
- inhomogeneity of the fuel–air mixture due to insufficient mixing, leading to the presence of local areas with a lean mixture and low combustion completeness;
- insufficient residence time of the fuel–air mixture in the combustion chamber due to the small volume of the furnace tube;
- poor quality of liquid fuel atomization and, as a result, a long evaporation time of liquid fuel droplets, leading to incomplete oxidation.

Unburned hydrocarbons, like CO, are a result of incomplete fuel combustion. Therefore, the reasons for increased emissions of UCH are the same as for CO. Similarly to emissions, the maximum values of UCH emissions occur at low engine operating modes, and with an increase in the engine operating mode, the emissions of UCH decrease.

NOx are formed in high-temperature zones of the combustion chamber due to nitrogen oxidation. Therefore, in contrast to CO and UCH, with an increase in the engine operating mode, NOx emissions increase.

There are three main mechanisms for the formation of NOx emissions:

- thermal (Zeldovich mechanism), where direct oxidation of atmospheric nitrogen occurs in high-temperature zones of the combustion chamber;
- “fast” mechanism of nitrogen oxides formation due to rapid reactions of nitrogen with hydrocarbon radicals at the flame front;
- fuel, where nitrogen oxides are formed as a result of nitrogen oxidation contained in the fuel.

In combustion chambers with a diffusion combustion mode, the main part of NOx emissions (approximately 90%) is formed by the Zeldovich mechanism.

The process of nitrogen oxide formation by this chain mechanism is endothermic and proceeds at a noticeable rate only at temperatures above 1800 K, depending on the flame temperature—exponentially.

Therefore, the reasons for high nitrogen oxide emissions can be high flame temperature values in the combustion zone and a relatively long residence time of the fuel–air mixture in these zones.

Regarding the smoke level, it can be noted that it is associated with the formation of dispersed soot particles in areas with a high concentration of fuel.

In these areas of the primary zone of the combustion chamber, soot can be formed in large quantities, burning out below the flow in zones with relatively high temperature.

Thus, the smoke level in the combustion products of the engine exhaust jet is the result of the amount of soot formation in the combustion zone and the fraction of combustion of this amount of soot in the subsequent zones of the combustion chamber. Overall, the level of non-volatile solid particle emissions correlates with the level of smoke.

Thus, the level of smoke generally increases with the increase in engine operating mode due to the predominance of soot generation in fuel-rich areas of the combustion zone, and the main reasons for the increased level of smoke and non-volatile solid particle emissions are:

- high fuel concentration in the combustion zone of the combustion chamber;
- uneven mixing of fuel with air, leading to the appearance of fuel-rich areas in the combustion zone;
- small volume or low temperature of gases in the intermediate zones of the combustion chamber, leading to insufficient combustion of the soot formed in the combustion zone.

Based on the described mechanisms of harmful emissions formation, the following main factors determining the level of harmful emissions from gas turbine engines can be identified, which should be taken into account when choosing design solutions for low-emission combustion chambers [16, 17], namely:

- the degree of homogeneity of the combustion process in the primary zone of the combustion chamber;
- temperature and excess air coefficient in the primary zone;
- residence time of the fuel–air mixture and combustion products in the primary and intermediate zones;
- "freezing" of combustion by cooling air near the walls of the combustion tube.

The degree of homogeneity of the combustion process in the primary zone of the combustion chamber is mainly determined by the design of the front device of the combustion chamber, which should provide high intensity and efficiency of fuel atomization, its evaporation and mixing with air.

The temperature and excess air coefficient in the primary zone depend on the distribution of air by elements and zones of the combustion chamber, which should be rationally chosen considering the range of engine operating modes and ensuring such characteristics as reliable ignition (including high-altitude ignition), stable combustion, and acceptable level of gas temperature field unevenness at the outlet of the combustion chamber.

The residence time of the fuel–air mixture and combustion products in the primary and intermediate zones depends on the choice of the combustion tube volume.

Therefore, the decisive factor in designing low-emission combustion chambers is the choice and experimental verification of both the design scheme of the front device and the balance of fuel concentration distribution (air distribution) by zones of the combustion chamber.

Given the developed and experimentally verified solutions for the front device and fuel concentration distribution by zones of the chamber, the task of ensuring low emissions of harmful substances is reduced to the task of choosing the volume of the combustion tube, its relative length, and rational choice of the cooling system scheme.

Also, from the above, it can be noted that the regularities of the formation of different components of harmful emissions are opposite, as a result of which many ways to reduce harmful emissions by improving The combustion process in combustion chambers with a single-zone diffusion combustion mechanism is a compromise between the amount of CO and UCH emissions, on one hand, and the amount of NOx and smoke emissions, on the other.

3.2 Emission Standards for Aviation

Research results dedicated to the problematic issues of global energy show that the total global power of gas turbine engines (GTE) and gas turbine units (GTU) is approximately 20 times less than the total power of the world's internal combustion engine (ICE) fleet. Moreover, due to the peculiarities of the working process, GTEs are less toxic compared to ICEs. The main source of atmospheric pollution is automotive transport (approximately 60%), while aviation emissions account for about 2%. However, in the area of airports, due to significant specific air consumption through GTEs, aviation emissions can lead to very significant local air pollution. To limit the emissions of harmful substances from aviation engines operating in the airport area, the International Civil Aviation Organization (ICAO) has introduced emission standards for UCH, CO, NOx, and smoke emissions from turbojet and turbofan engines of civil aircraft in the airport area since 1980. Based on the analysis and forecasting of the annual growth of air transport volumes and the associated increase in the amount of fuel burned, the ICAO is implementing a policy of tightening standards for NOx emissions—as the most harmful air pollutant [18] (see Fig. 7). For UCH and CO components, the ICAO does not yet foresee tightening of standards.

For the standardization of harmful emissions, the emission parameter is used, which equals the total mass of harmful emissions during the standard "takeoff–landing" cycle, related to the takeoff thrust. Thanks to such an approach to standardization, this parameter reflects both the quality of the working process in the combustion chamber and the perfection of the engine as a whole (its fuel efficiency).

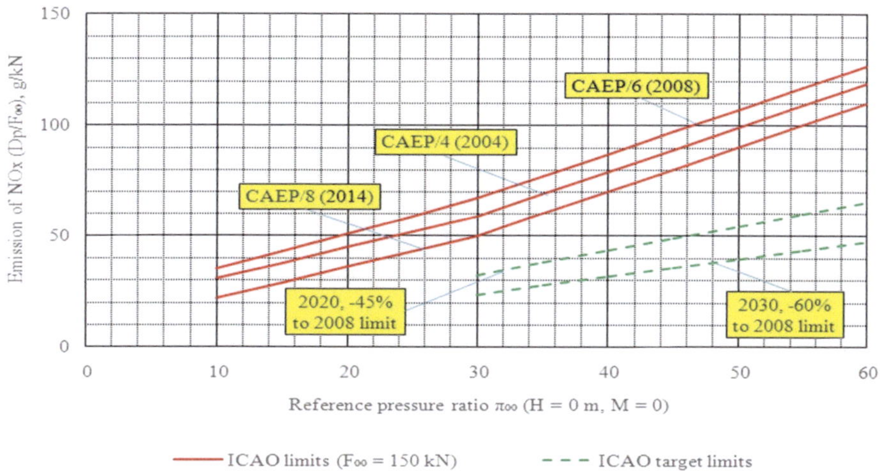

Fig. 7 ICAO standards for NOx emissions

3.3 Experience in Designing and Modernizing Combustion Chambers of GTEs, Ensuring Compliance with Environmental Standards

The task of ensuring the compliance of the engines being developed with the ICAO standard requires research to reduce the level of harmful emissions. At theIvchenko-Progress JSC, significant experience has been accumulated and scientific and technical achievements have been made in the design of combustion chambers with low emissions of harmful substances. Based on these works, the main directions of research and reduction of harmful emissions have been identified, as well as measures for the combustion chambers being developed, which are as follows:

- The application of a front device with double-tier swirlers, which allows for reliable operability of the combustion chamber with the necessary degree of fuel–air mixture lean, its homogeneity, intensification of preparation and combustion processes of the fuel–air mixture;
- The selection of the main air distribution in the elements of the combustion tube, which provides fuel concentrations close to optimal in its cavity at main operating modes;
- The application of a highly efficient, optimized by length and density of holes, cooling system for the combustion tube walls;
- The development of a method for calculating the volume of the combustion tube and ensuring a rational value of residence time.

As noted above, the emission characteristics of harmful substances, and even the start-up and stable operation of the combustion chamber directly depend on the design scheme of the front device. The design solution with double-tier air swirlers

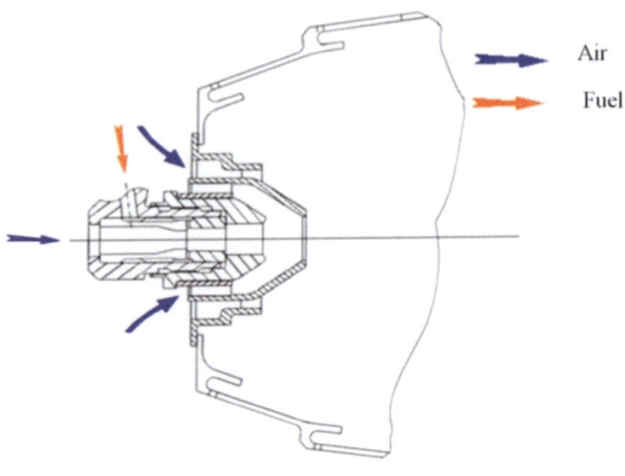

Fig. 8 Design solution for the front device with double-tier air swirlers and single-circuit fuel injectors with air atomization

and single-circuit fuel injectors with air atomization (see Fig. 8) developed at the Ivchenko-Progress JSC and tested by long-term operation allowed to provide a rational compromise for the entire set of combustion chamber characteristics.

Based on the results of experimental studies and a set of calculations, it was established that for aviation combustion chambers with a single-zone diffusion combustion mechanism and a uniform design solution for the front device, emissions are best described by functional dependencies: regarding the emission index of the gaseous component CO

$$EI(CO) = f_1(\Omega), \tag{2}$$

where $\Omega = \frac{W_{31}}{V_{31}\cdot(p_{t3})^{1,8}\cdot\exp\left(\frac{T_{t3}}{300}\right)}, \frac{kg}{s\cdot m^3\cdot(MPa)^{1,8}\cdot\exp\left(\frac{K}{300}\right)}$–load parameter;

W_{31} air flow in the combustion chamber, kg/s;
V_{31} combustion chamber volume, m^3;
p_{t3} total pressure at the combustion chamber inlet, *MPa*;
T_{t3} total temperature at the combustion chamber inlet, *K*;
V_{31} combustion tube volume;

regarding the emission index of the gaseous component *UHC*

$$EI(UHC) = f_2(\Omega); \tag{3}$$

regarding the emission index of the gaseous component *NO* x

$$EI(NOx) = f_3(\sigma), \tag{4}$$

where $\sigma = \frac{W_{31}\cdot\sqrt{T_{t3}}}{V_{31}\cdot(p_{t3})^{1,45}\cdot\exp\left(\frac{T_{t3}}{230}\right)}, \frac{kg}{s\cdot m^3\cdot(MPa)^{1,45}\cdot\exp\left(\frac{K}{230}\right)}$—boosting criterion.

To predict the level of NOx emissions from the combustion chamber, which is designed according to the principle described above, a method for calculating the level of NOx emissions has been developed.

This method is based on the Zeldovich kinetic equation, which links the concentration of nitrogen oxides formed during the combustion of a homogeneous mixture with the parameters in the combustion chamber. The calculated value of the emission index NOx—$EI(NOx)_{calk}$ is determined by the equation:

$$EI(NOx)_{calk} = \frac{\exp\left(-\frac{45000}{T_{tpz}}\right)\cdot\left(\frac{p_{t3}}{T_{tpz}}\right)\cdot V_{31}\cdot p_{t3}\cdot A_{NOx}}{T_{t3}\cdot W_{31}}, \quad (5)$$

where T_{tpz} is the total temperature in the primary zone of the combustion chamber; T_{t3} is the total temperature at the entrance to the combustion chamber; p_{t3} is the total pressure at the entrance to the combustion chamber; A_{NOx} is the coefficient. Figures 9 and 10 show the dependence of the emission index of the gaseous component CO and UHC on the load parameter of the combustion chamber for the D-18 T series 3 engine with a thrust in takeoff mode of 23,430 kgf and the D-436–148 engine with a thrust in takeoff mode of 7007 kgf.

Figure 11 shows the dependence of the emission index of the gaseous component NOx on the combustion chamber boosting criterion for the D-18 T series 3 engine with a thrust in takeoff mode of 23,430 kgf and the D-436–148 engine with a thrust in takeoff mode of 7007 kgf.

The structural diagrams of the combustion chambers of these D-18 T series 3 and D-436–148 engines are shown in Figs. 12 and 13, respectively.

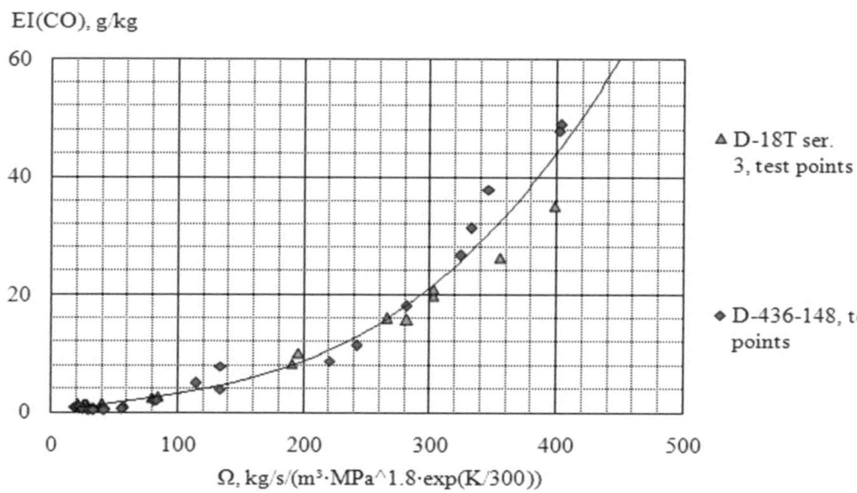

Fig. 9 Dependence of the emission index of the gaseous component CO on the load parameter of the combustion chamber

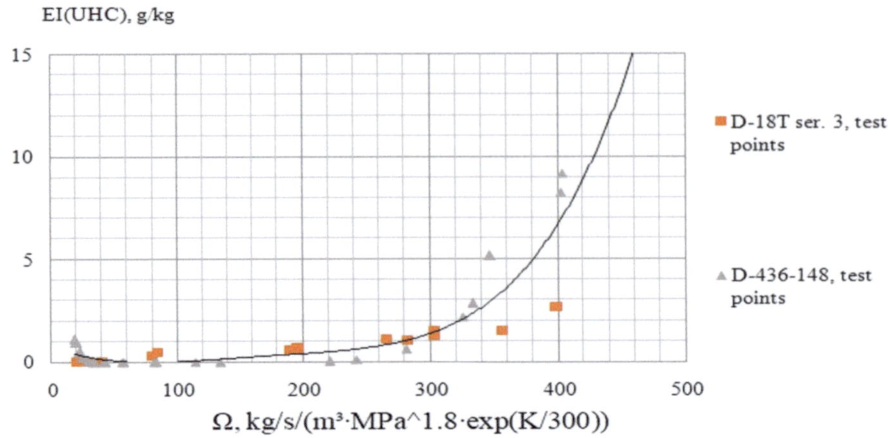

Fig. 10 Dependence of the emission index of the gaseous component UHC on the load parameter of the combustion chamber

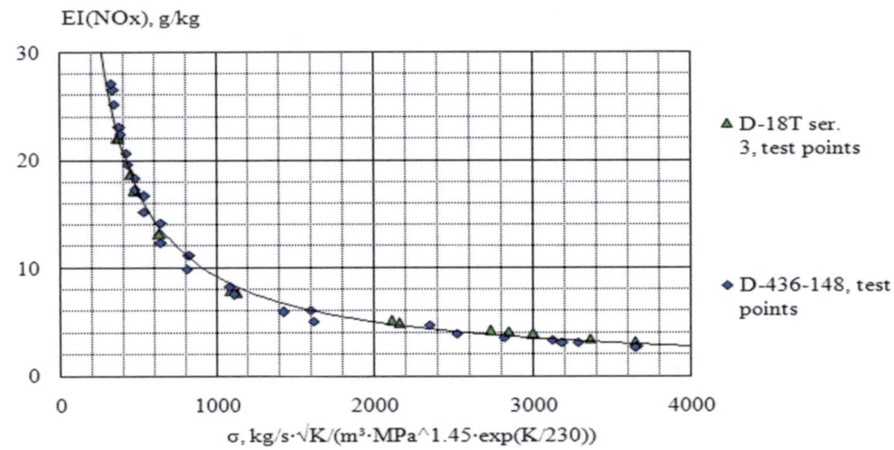

Fig. 11 Dependence of the emission index of the gaseous component NOx on the combustion chamber boosting criterion

Fig. 12 Structural diagram of the combustion chambers of the D-18 T series 3 engines

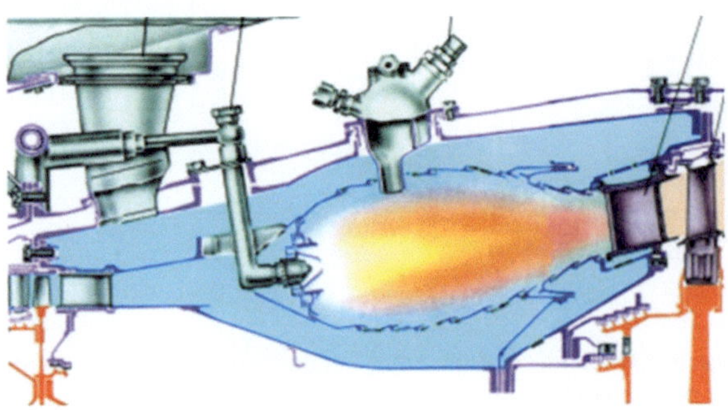

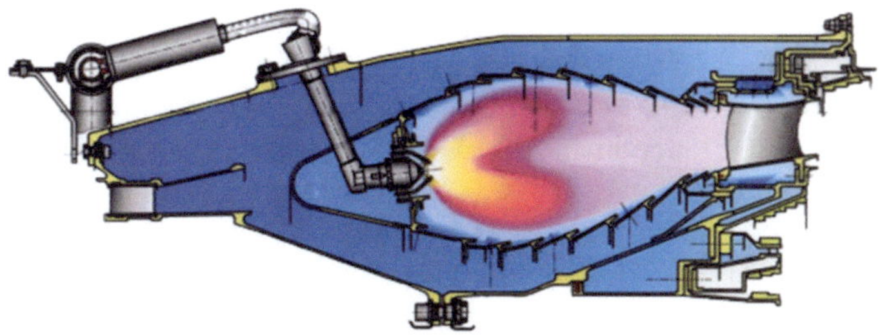

Fig. 13 Structural diagram of the combustion chamber of the D-436–148 engine

These combustion chambers are designed taking into account the results of studies on the selection of the front device, the choice of distribution of the main air through the elements of the fire tube, and the application of a highly efficient cooling system for the fire tube walls. At the same time, the similarity of the structural schemes of these combustion chambers is preserved, the same approaches to the selection of the front device, cooling system, and distribution of the main air are used. Despite the fact that these engines are representatives of different classes of engines in terms of thrust and size, the application of the principle of similarity in the design of the combustion chambers of these engines ensures the stability of functional dependencies regardless of the type and size of the engine, as can be seen from Figs. 9, 10 and 11. The comparison of the results of the physical experiment with the results calculations of emission indices NOx are shown in Fig. 14.

The results shown in Fig. 14 suggest that the proposed method for calculating emission indices NOx provides satisfactory accuracy.

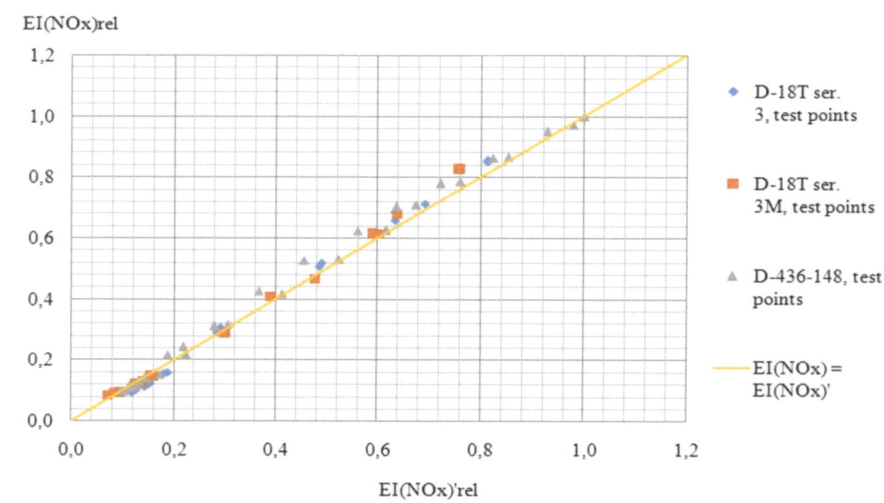

Fig. 14 Comparison of physical experiment results with the calculations of emission indices *NO*x

The influence of engine operating parameters at the combustion chamber inlet can be analyzed based on the dependencies presented in Figs. 9, 10 and 11.

Increasing the pressure and temperature of the air entering the combustion chamber leads to a reduction in CO, UCH emissions and an increase in NOx emissions.

Conversely, increasing the air flow rate (reducing residence time) leads to an increase in CO, UCH and a decrease in NOx emissions.

These results confirm the conclusions about the influence of operating parameters on the level of harmful emissions, made earlier when considering their formation mechanisms.

As for the influence of combustion chamber design features on the amount of harmful emissions, the dependencies presented in Figs. 9, 10 and 11 suggest that for structurally similar combustion chambers using an effective front device and air distribution by flame tube elements, reducing the volume of the flame tube (forcing the combustion chamber by residence time) leads to a decrease in NO x emissions and an increase in CO and UCH emissions.

Since NOx emissions are the most harmful and ICAO standards for them are being tightened (while maintaining standards for UCH and CO), reducing the volume of the flame tube while maintaining the design scheme of the developed optimal front device and maintaining a rational choice of air distribution by flame tube elements is an effective way to reduce NO x and ensure prospective ICAO standards for combustion chambers of future engines and when upgrading existing ones.

This approach allows reducing the level of NOx emissions with minimal increase in UCH and CO emissions while ensuring the entire set of combustion chamber characteristics, including startup, stable operation, and gas temperature field at the outlet.

During the development of new engine modifications D-18 T ser. 3 M and D 436 148FM to ensure their compliance with updated and stricter standards, the described experience was used, the flame tubes of these engines were modernized by reducing their combustion volume. As a result, a reduction in NOx emissions with an increase in UCH and CO was achieved. For example, Fig. 15 shows a comparison of the design schemes of flame tubes of D 18 T ser. 3 M and D-18 T ser. 3 combustion chambers, and Figs. 16, 17 and 18 show the results of tests for measuring harmful emissions. Emissions and their significance in the modes of the standard LTO cycle.

For combustion chambers of prospective engines, the fuel/air ratio increases, which is due to the optimal thermodynamic cycle conditions for modern engines. In this case, the application of the described technical and design solutions for combustion chambers with low emissions may be insufficient, as the increase in fuel concentration requires even better mixing and homogenization, as well as the search for solutions for more efficient combustion of fuel that enters the cold wall zones. An original scheme of air flow in the flame tube with a transverse slot and a directed cooling air flow was applied to solve such a task using the combustion chamber of the D-27 engine with high thermodynamic cycle parameters. The application of such a

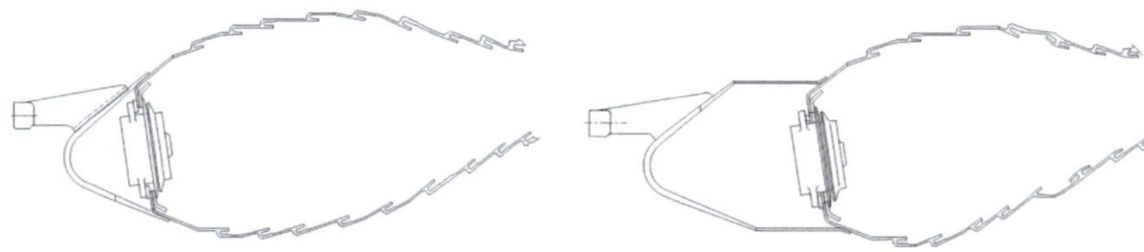

Fig. 15 Comparison of the combustion chamber flame tube design schemes for D-18 T ser. 3 and D-18 T ser. 3 M engines

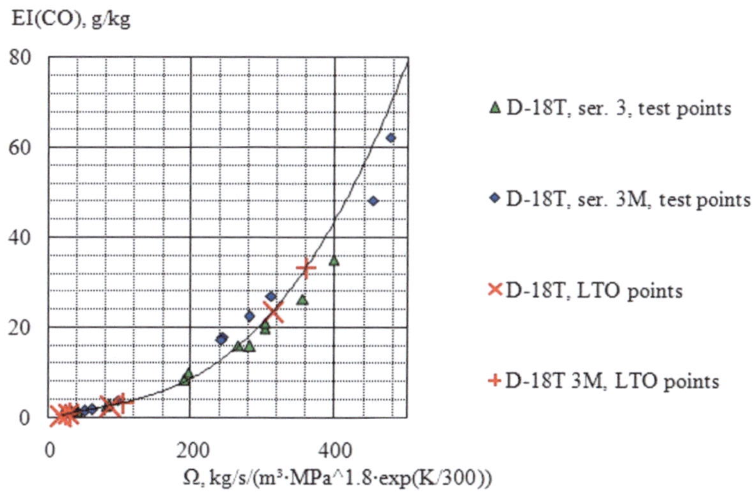

Fig. 16 Dependence of the emission index of the gaseous component CO on the load parameter with corresponding values in the modes of the standard LTO cycle

scheme allowed to use the previously developed frontal device, ensure better homogenization of the fuel–air mixture in the primary zone, and ensure efficient combustion of fuel due to the return of cold wall-rich fuel zones back to the combustion zone.

The design scheme of the D-27 engine combustion chamber is shown in Fig. 19.

The obtained characteristics of the environmental perfection of the D-27 combustion chamber, operating at high values of the fuel/air ratio, are at least not worse than the characteristics of the D-18 T and D-436–148 engine families, and the high level of engine fuel efficiency ensures a high level of environmental perfection of the engine as a whole.

The use of an effective frontal device based on the concept of a two-tier air swirl, in which the air twisted by the blades is divided into two streams, ensures the division of air and preliminary mixing with its portion. The use in conjunction with a fuel injector that employs the principle of aerodynamic atomization not only allows for a wider range of combustion chamber operation when lean fuel–air mixture is achieved

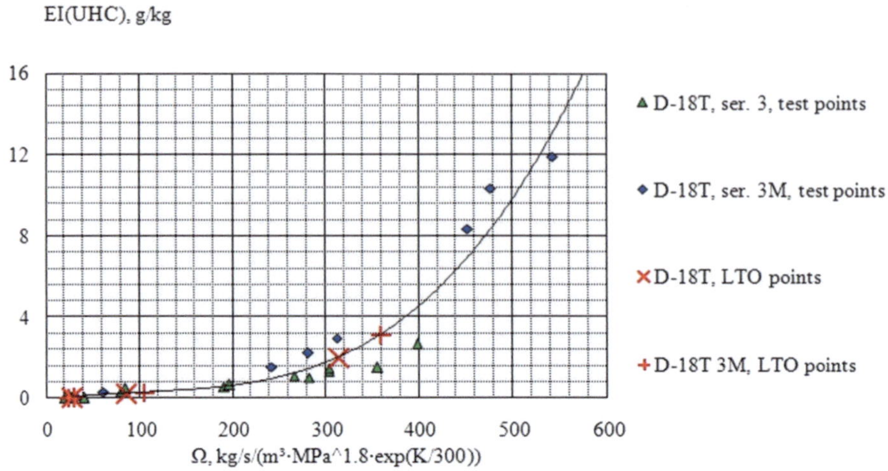

Fig. 17 Dependence of the emission index of the gaseous component UHC on the load parameter with corresponding values in the modes of the standard LTO cycle

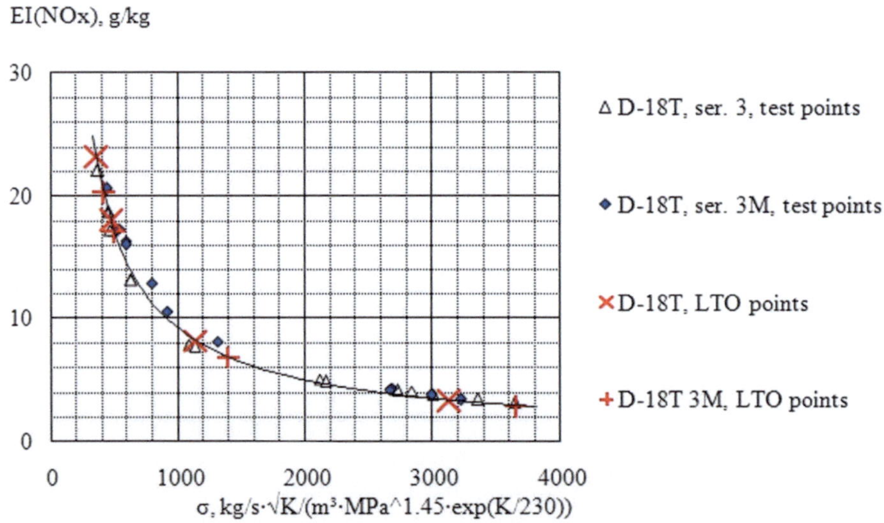

Fig. 18 Dependence of the emission index of the gaseous component NOx on the boosting criterion with corresponding values in the modes of the standard LTO cycle

through fuel aeration and proper fuel/air ratio selection, but also creates conditions for smokeless combustion with low emissions of harmful substances in all modes.

Figure 20 shows the results of smoke level measurements when changing the excess air coefficient in the primary zone of the combustion chamber.

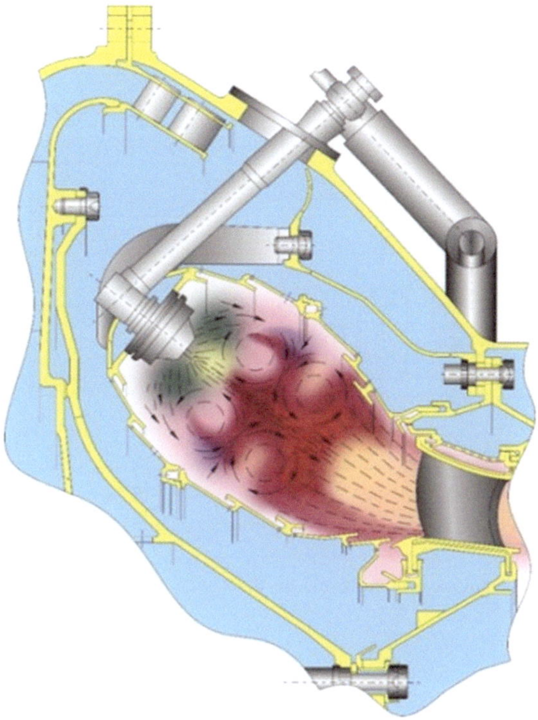

Fig. 19 Design scheme of the D-27 engine combustion chamber

As we can see, reducing the volume of the fire tube for the modification of the D-18 T ser. 3 M engine combustion chamber led to an increase in smoke levels at the same fuel concentrations in the primary zone compared to the original combustion chamber.

This result can be explained by the reduction in soot residence time in areas below the flow from the primary zone with a relatively high temperature, where soot combustion occurs. Therefore, in order to reduce the maximum smoke number *SN* to the level of the original combustion chamber, the air flow was redistributed to lean the primary zone. As a result, for the engine modification, a reduction in *NO* x emissions was achieved while maintaining the maximum value of $SN_{\max}$ at approximately the same level (the change in $SN_{\max}$ is comparable to the measurement error).

Based on the results of the conducted studies, the smoke parameter of engines with combustion chambers—A_{SN}, on which the described type of front device is applied, is best described by the functional dependence

$$A_{SN} = \frac{p_{t3}^2}{\alpha_{pz} \cdot W_{pz} \cdot Tt_{pz}} \cdot ((1 - B \cdot V_{31} \cdot \alpha_{sz} \cdot \exp(0,0011 \cdot Tt_{sz})), \quad (6)$$

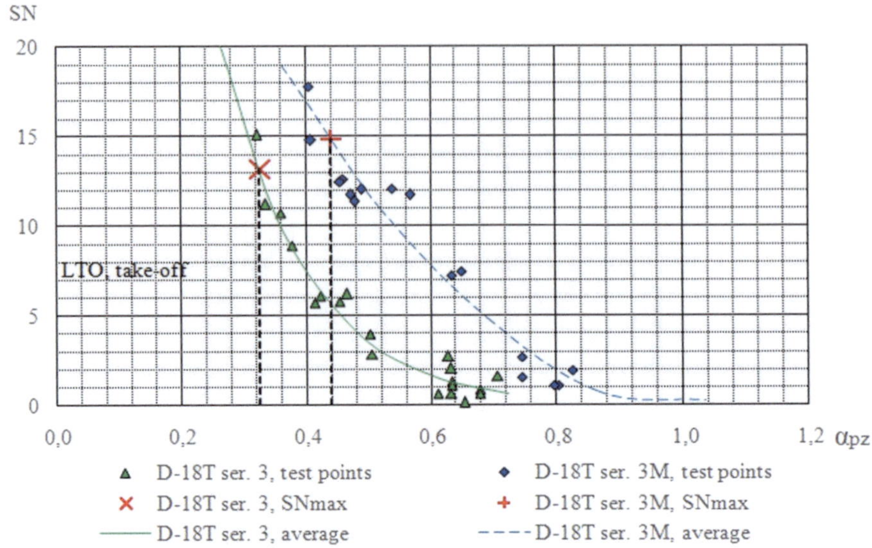

Fig. 20 Dependence of engine smoke levels on the concentration of the fuel–air mixture in the primary zone

where

α_{pz} excess air coefficient in the primary zone of the combustion chamber;
α_{sz} excess air coefficient in the secondary zone of the combustion chamber;
p_{t3} total pressure of the working medium at the entrance to the combustion chamber;
Tt_{pz} total temperature of the working medium in the primary zone of the combustion chamber;
W_{pz} air flow rate in the primary zone of the combustion chamber;
B coefficient.

Figure 21 shows the dependence of the smoke level on the smoke parameter.

Thus, when designing and modernizing combustion chambers, to assess the level of smoke and analyze the impact of parameters and characteristics of the combustion chamber on the level of smoke, it is sufficient to use the parameter A_{SN}.

Figure 22 shows the relative level of emissions from engines developed by Ivchenko-Progress JSC and demonstrates the compliance of these engines with existing standards.

The results of research aimed at finding new ways to solve the problem of reducing harmful emissions based on a compromise using extensive experimental studies, the application of the most effective means of ensuring most characteristics of the high-temperature combustion chamber (intensification of fuel–air mixture preparation and combustion processes, its dilution and homogenization, especially in the primary zone), optimization of air distribution and residence time of the fuel–air mixture in

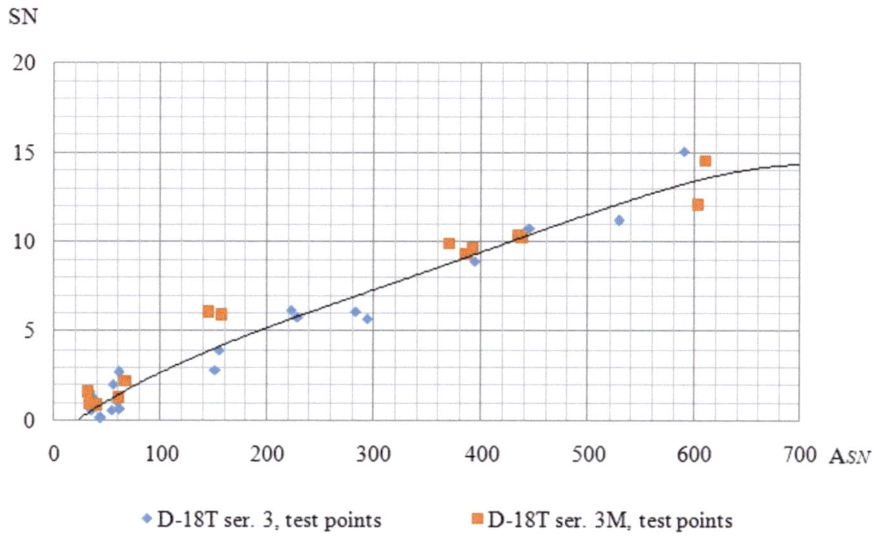

Fig. 21 Dependence of engine smoke levels on the parameter A_{SN}

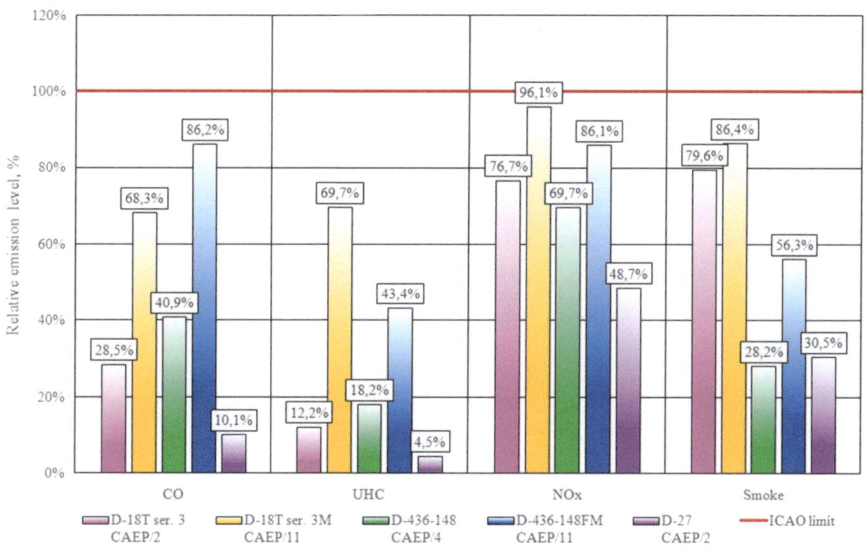

Fig. 22 Relative emission levels of engines developed by Ivchenko-Progress JSC

combustion and mixing zones, have allowed the creation of engines that meet current and prospective ICAO standards.

3.4 Ways to Further Reduce the Level of Harmful Emissions

Research aimed at developing prospective combustion chambers for gas turbine engines has revealed combustion chamber schemes that can meet prospective standards for harmful emission levels—these are combustion chambers with multi-zone combustion of diffusive and/or homogeneous mixtures, or with single-zone combustion of a homogeneous mixture, implementing the LPP (Lean, Premixed and Prevaporised) concept.

The LPP concept is based on low-temperature ($1600 \le Tt_{pz} \le 1900$ K) combustion of a pre-mixed lean fuel–air mixture. By excluding droplet combustion and creating a homogeneous lean mixture in the primary zone, it is possible to ensure a low reaction temperature, eliminate local hot zones in the combustion area, and thus significantly reduce NOx emissions. To maintain complete combustion while operating on a lean homogeneous mixture, it is necessary to increase its residence time in the combustion zone compared to a diffusive mixture. If the mixture is lean and well mixed, the residence time has little effect on the creation of NOx, and this allows to "break" the negative link between NOx and CO (see Fig. 23) for classical diffusive combustion chambers [16].

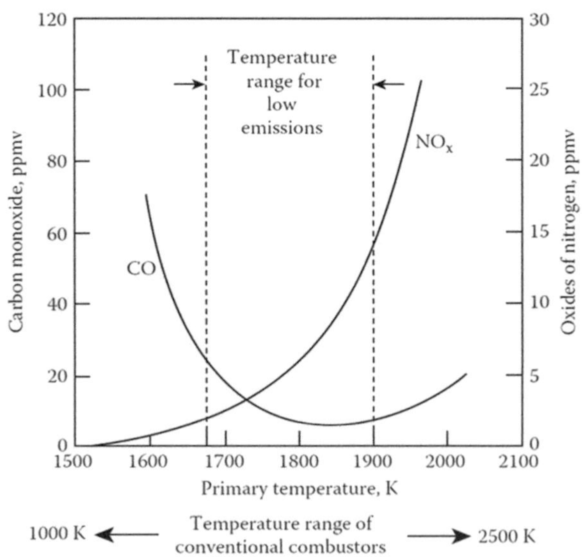

Fig. 23 Dependence of CO and NOx emission levels on temperature in the combustion zone [16]

Fig. 24 Combustion chamber developed by Ivchenko-Progress JSC with a two-zone fuel combustion scheme

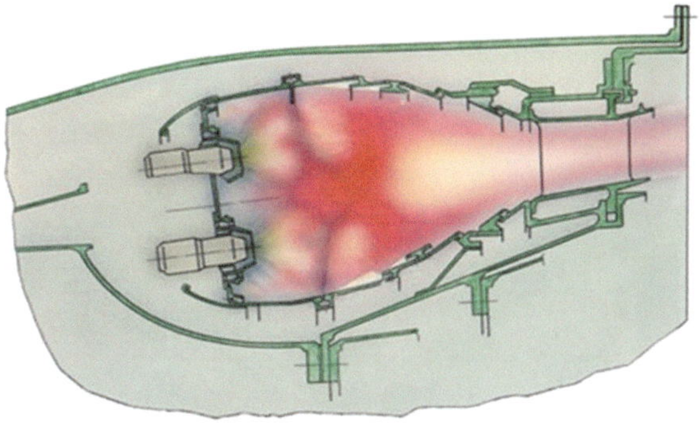

The LPP concept gives the maximum effect in combustion chambers with variable geometry, which allows controlling the concentration and temperature of the flame in the combustion zone when changing the operating mode, ensuring optimal values.

The development of low-toxic aviation combustion chambers with a two-zone fuel combustion scheme (see Fig. 24) showed that, despite the application of a new combustion concept, new design solutions for organizing the working process with complex and costly fuel supply and regulation automation, it was possible to reduce the NOx emission level by only 30 … 35%.

Ivchenko-Progress JSC has experience in the development, manufacture, and operation of combustion chambers using the LPP concept for ground application gas turbine engines (see Fig. 25).

The results obtained for NOx emissions reduction using the LPP concept compared to a combustion chamber with a classic diffusion fuel combustion scheme with original and reduced volume of the combustion tube are shown in Fig. 26.

Structural schemes of combustion chambers with a classic diffusion fuel combustion scheme are shown in Fig. 27.

The presented results were obtained using natural gas as fuel. From the presented data, it can be seen that the use of the LPP concept allowed to reduce the maximum

Fig. 25 Combustion chamber developed using the LPP concept for GTE AI-336

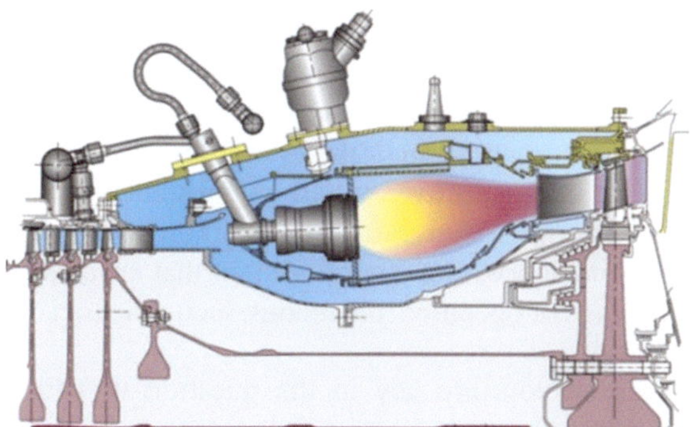

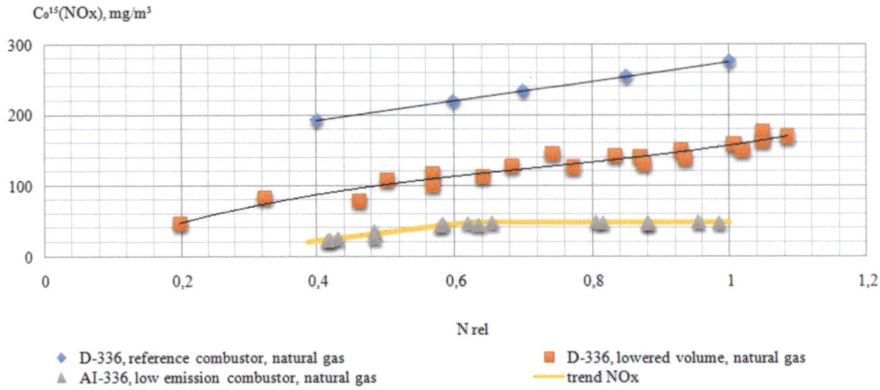

Fig. 26 Dependence of the concentration of NOx, reduced to 15% oxygen content from the relative power of the gas turbine drive for diffusion combustion chambers and combustion chambers developed using the LPP concept

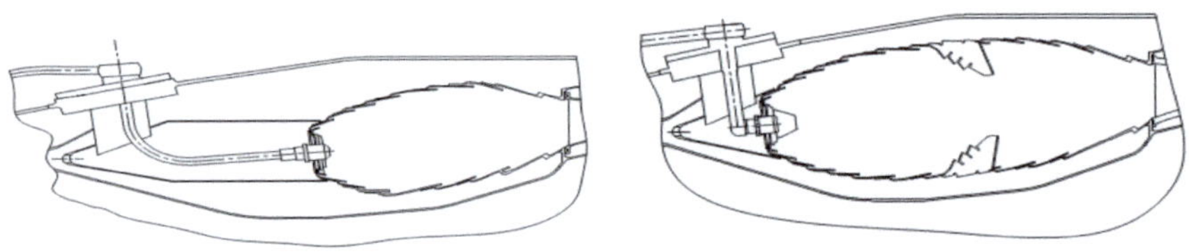

Fig. 27 Structural schemes of GTE D-336 combustion chambers with classic diffusion fuel combustion

NOx emissions approximately five times compared to the original design, and three times compared to the diffusion combustion chamber with a reduced volume of the combustion tube.

The results of harmful emissions level research for LPP combustion chambers, operating on gas and natural gas, are presented in Figs. 28 and 29.

When using gas instead of natural gas for this combustion chamber, NOx emissions increased**, CO**—decreased.

Thus, the application of the LPP concept for combustion chambers with variable geometry in prospective aircraft engines allows to significantly reduce NOx emissions. However, for aircraft GTE, the implementation of the design of such chambers In combination with a control system that ensures stable operation throughout the entire range of operation and modes, including transitional modes, it is very complex and costly.

These factors are key in the question of widespread and mass use of LPP combustion chambers in aircraft engines.

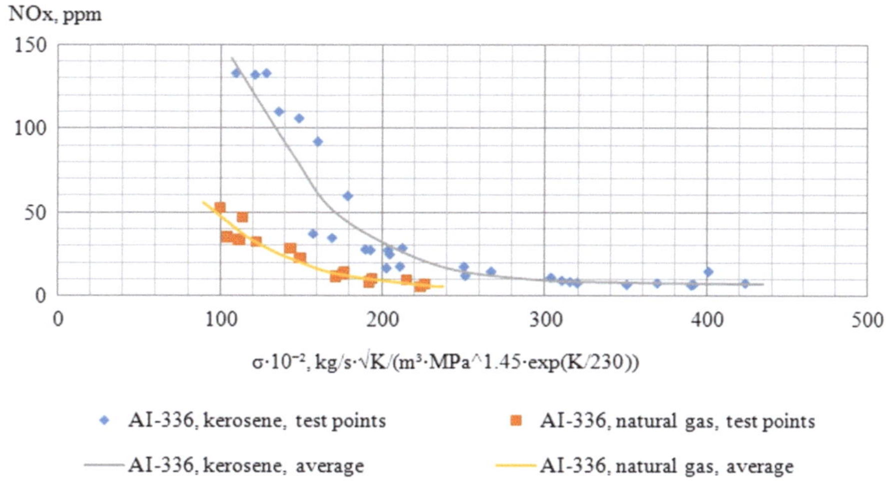

Fig. 28 Results of harmful emissions NOx for LPP combustion chambers operating on gas and natural gas

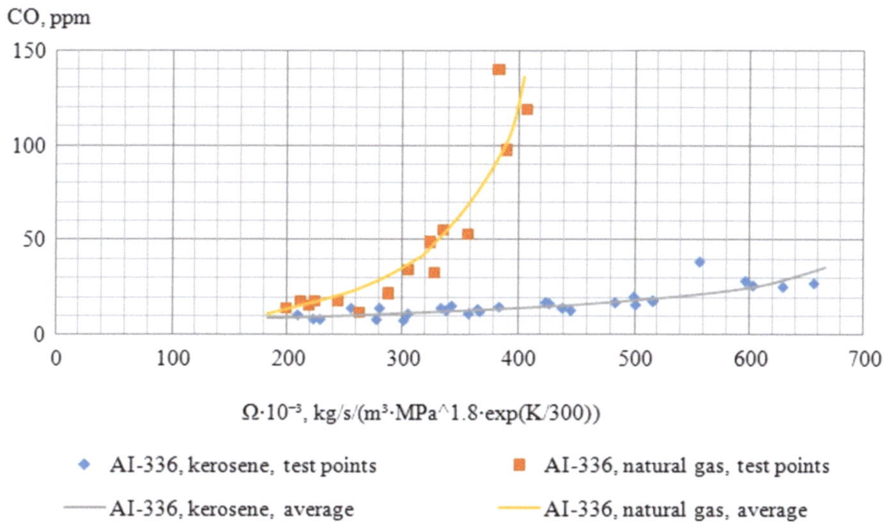

Fig. 29 Results of harmful emissions *CO* for LPP combustion chambers operating on gas and natural gas

References

1. Kröger, G., Schnell, R., Humphreys, N.D.: Optimised aerodynamic design of an OGV with reduced blade count for low noise aircraft engines. In: Proceedings of the ASME Turbo Expo 2012: Turbine Technical Conference and Exposition. Vol. 8: Turbomachinery, Parts A, B, and C. ASME, Copenhagen, Denmark, pp. 407–418. (2012). https://doi.org/10.1115/GT2012-69459
2. Beply, M.: Innovations in the field of nacelles. SAFRAN J. **17**, 34–35 (2015)
3. http://gereports.ru/post/103189404800/primenenie-kompozitnyh-polimernyh-materialov
4. http://www.ciam.ru/research/nodes/compressors/
5. http://www.tsagi.ru/research/aeroacoustics/
6. Environmental Protection: Aircraft Noise, vol. 1. Appendix 16 to the Convention on International Civil Aviation. ICAO, Moscow (2014)
7. Doc 9501 AN/929: Environmental Technical Manual, vol. 1. Aircraft Noise Certification Methodologies. ICAO, Moscow (2014)
8. http://ncp.tsagi.ru/cleansky/main
9. http://ncp.tsagi.ru/cleansky2/main
10. Moshkov, P.A., Samokhin, V.F., Yakovlev, A.A.: Problems of Reducing noise created on the ground by aircraft with turbofan engines. Aviat. Technol. **4**, 126–128. Izv Universities (2018)
11. Smirnov S.A. (2021) Sketch design of an air screw, Practical guide. Sketch design of an air screw, Zaporozhye.
12. Seminozhenko, V.P., Kanilo, P.M., Ostapchuk, V.N., Rovensky, A.I.: Energy. Ecology. Future, Kharkov: Flag, 464. (2003)
13. Kanilo, P.M.: oxicity of GTE and prospects for the use of hydrogen. Sci. Thought 140 (1982)
14. Khristich V.A., Tumanovsky A.G.: Gas turbine engines and environmental protection. K.: Technika 144 (1983)
15. Lieuwen, T.C., Yang, V.: Gas turbine emissions, p. 368. Cambridge aerospace series; 38 (2013)
16. Lefebvre, A.N., Ballal, D.R.: Gas Turbine Combustion: Alternative Fuels and Emissions, 3rd edn., p. 557. Taylor & Francis (2010)
17. Gupta, A., Lilly, D., Sayred, N.: (1987) Swirling flows, M.: world. p. 588
18. Benezekein, M.: Propulsion strategy for the 21 century: a vision into the future. In: 15th International Symposium on Airbreathing Engines (XV ISABE), Bangalore, India, p. 11 (September 2–7, 2001). (ISABE-2001-1005)

Main Directions and Results of Scientific Research Ivchenko-Progress JSC on Improving Parameters of the Working Process and Schemes of Gas Turbine and Hybrid Engines, Use of New Fuels to Reduce Negative Impact on Humans and the Environment

Igor Kravchenko, Mykhailo Mitrakhovych, Sergiy Smirnov, Sergiy Dmytriyev, Maksym Kirichkov, and Andriy Popuga

Abstract The paper details the initiatives of Ivchenko-Progress JSC to improve gas turbine and hybrid engines for reduced environmental impact. Research focuses on increasing efficiency through higher bypass ratios, adopting hybrid and electric power systems, and exploring hydrogen and fuel cell technologies. The use of new fuels and production methods aligns with global decarbonization trends, aiming to enhance performance while minimizing emissions. These efforts contribute to advancing sustainable aviation technologies and supporting broader environmental goals.

1 Introduction

Global warming and climate change have negative effects on the planet and are increasingly concerning humanity. One of the measures to address this problem is the UN Framework Convention on Climate Change in the form of an international climate agreement on joint efforts to combat the consequences of dangerous human interference in the environment, signed in 1992, which Ukraine joined and in 2020 committed to not exceed 60% of the 1990 greenhouse gas emissions level by 2030 and confirmed its intention to achieve the goals of the 2015 Paris Agreement, namely to keep the increase in global average temperature below 2 °C and to make efforts to limit it to 1.5 °C.

I. Kravchenko · M. Mitrakhovych · S. Smirnov · S. Dmytriyev (✉) · M. Kirichkov · A. Popuga
Zaporizhzhia Machine-Building Design Bureau Progress Joint Stock Company named after, Academician O.H. Ivchenko (Ivchenko-Progress JSC), Zaporizhzhia, Ukraine
e-mail: s.dmytriyev@ivchenko-progress.com

© The Author(s), under exclusive license to Springer Nature Switzerland AG 2025
A. Marchenko and Z. Koruba (eds.), *Decarbonization of Transport Energy Installations in the Context of Sustainable Development Strategies*, Studies in Systems, Decision and Control 635, https://doi.org/10.1007/978-3-032-05884-3_16

In the context of constantly increasing requirements for the environmental performance of aircraft power units, leading aircraft manufacturing companies in the world have set goals to reduce by 2025, compared to existing engines, the level of CO_2 emissions by 40 ... 50%, NOx–by 75...80%, the level of acoustic radiation by 2 times, compared to the requirements of Chap. 3 of ICAO [1].

Achieving such indicators is planned through the implementation of the following measures:

- increasing the efficiency of engines as a "heat engine" that has high values of internal efficiency, through the application of new technical solutions and design schemes, new materials, improving the basic parameters of the gas turbine engine (GTE) working process (gas temperature at the combustion chamber outlet, degree of pressure increase in the compressor) and improving the efficiency of all the main elements and systems of the power unit;
- improving the efficiency of engines as "propulsors" that have high values of flight efficiency, by increasing the degree of bypass, applying turbofan engine schemes with a rear-mounted unducted fan of the "open rotor" type.

The monograph discusses the results of research by Ivchenko-Progress JSC on ways to improve the efficiency of bypass engines by increasing the degree of bypass, researching ways and technologies for applying hybrid and developing electric power units, problems of using hydrogen and its production technologies, problems of using fuel cells for hybrid power units, aimed at improving their environmental friendliness.

2 Justification of Ways to Ensure the Increase in Efficiency of Bypass Engines by Increasing the Degree of Bypass

In the life cycle of an aircraft, fuel costs account for a significant portion (over 50%) of direct operating costs, which is explained by relatively high global prices for hydrocarbon fuel and their constant growth, forcing global aviation engine manufacturing to seek new technologies to improve the fuel efficiency of aircraft power units. On the other hand, despite the fact that aviation transport accounts for the use of just a little more than one percent of oil fuel, a number of factors lead to the fact that the level of harmful substance emissions from GTEs is significant. The presence of intensive air traffic creates an additional negative contribution to the environmental situation.

Environmental requirements for aircraft and their power units are constantly being strengthened, so the fuel efficiency of the aircraft and its characteristics regarding the emission of harmful substances and acoustic radiation are interconnected factors. There is a continuous strengthening of environmental standards, which is caused by various factors: the growing role of aviation transport, the level of urbanization in many countries, the achievements of scientific and technological progress, the increasing influence of means and public organizations, and others.

It can be assumed that this is a steady trend, so the growth of environmental requirements for aviation transport will continue in the future.

The application of turbojet engines with an ultra-high bypass ratio can provide a reduction in specific fuel consumption by 10...15% lower than the similar indicator of the best existing turbojet engines, which will lead to a corresponding reduction in the emission of harmful substances.

The criteria for optimizing the parameters and characteristics of GTEs are ratios that allow assessing the efficiency of the engine as part of an aircraft, taking into account its purpose and range of use.

For subsonic passenger and transport aircraft, which have short altitude gain sections during takeoff, the main criterion for efficiency is the theoretical flight range is determined under the assumption that all fuel reserves are consumed during the cruise flight.

For passenger and transport aircraft with a significant flight range and a defined payload, the main criterion for selecting optimal power plant parameters is the practical flight range of the aircraft, which is determined taking into account takeoff, altitude gain, cruise flight, and landing.

In some cases, when optimizing the parameters of power plants for such aircraft, the flight range is not determined. In this case, the flight time and the total fuel mass in the aircraft's fuel system are determined at a known flight speed. For most aircraft, the relative mass of fuel and the power plant accounts for about 50% of the aircraft's takeoff weight. This determines the importance of the power plant's efficiency on the aircraft's effectiveness.

There is a dependence of the power plant efficiency optimization criterion on the total specific mass of fuel and the mass of the power plant.

Therefore, the dependence of the total specific mass of fuel and the mass of the power plant on the main parameters of the engine's working process (total degree of pressure increase in the compressor, degree of pressure increase in the fan, gas temperature at the outlet from the combustion chamber, and degree of bypass) is the basis for optimizing the power plant parameters according to the defined criterion of aircraft efficiency.

The most important criterion for optimizing aircraft efficiency is the specific mass of fuel and the power plant, as this criterion depends on the effective values of specific fuel consumption and specific mass of the power plant.

Thus, the efficiency of the turbojet engine is significantly influenced by its working process parameters, such as the degree of pressure increase in the compressor and the gas temperature at the outlet from the combustion chamber. The increase in the degree of pressure increase in the compressor is limited by the complexity of the design and its mass at acceptable efficiencies, and the increase in the gas temperature at the outlet from the combustion chamber is limited by the limit on *No* x, which for ordinary hydrocarbon fuels is less than the stoichiometric combustion temperature.

In addition, the fuel efficiency of the turbojet engine is also significantly influenced by the ratio of air consumption through the external circuit to air consumption through the internal circuit, which is the degree of bypass of the engine.

With the optimal distribution of turbojet engine cycle energy with separate output of flows between its circuits under constant external conditions and working process parameters, an increase in the degree of bypass leads to a decrease in the working body speed behind the engine and an increase in flight efficiency, thanks to the distribution of cycle energy to a larger mass of the working body. However, an increase in the degree of bypass leads to an increase in the influence of hydraulic losses in the external circuit on overall efficiency. The opposite effect of these factors leads to the fact that the engine efficiency has a maximum for the degree of bypass (i.e., specific fuel consumption has a minimum).

At the same time, the value of the optimal degree of bypass of the engine often significantly exceeds the value of the degree of bypass of existing and prospective turbojet engines. This is explained by the fact that with an increase in the degree of In a dual-circuit system, there is a continuous decrease in specific thrust, and to maintain the necessary thrust, it is necessary to increase the air consumption through the engine. This leads to an increase in the size of the engine nacelle and, as a result, an increase in external aerodynamic drag. When considering the increase in losses from the external resistance of the power unit, the value of the optimal degree of dual-circuitry, from the point of view of fuel efficiency, significantly decreases (Fig. 1).

The optimal degree of dual-circuitry depends on the specific work of the cycle, losses in the external circuit, aerodynamic perfection of the power unit elements, and the typical flight cycle.

Figure 2 presents an estimate of the required relative total mass of fuel and the mass of the power unit for a range of 2000 and 3000 km of a virtual aircraft with a turbojet engine. From the dependencies shown in Fig. 2, it can be seen that as the flight range increases, the optimal value of the degree of dual-circuitry increases.

Thus, when developing measures for the selection of optimization criteria for the efficiency of turbojet engines, the use of turbojet engines with a high degree

Fig. 1 Dependence of the relative specific fuel consumption of a turbojet engine on the degree of dual-circuitry: 1—for an engine without considering the needs of the aircraft, losses in the intake device, and resistance of the nacelle; 2—considering the needs of the aircraft and losses in the intake device; 3—considering the resistance of the nacelle

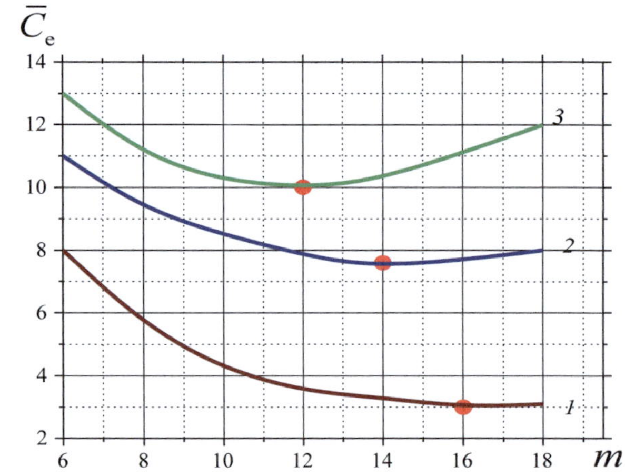

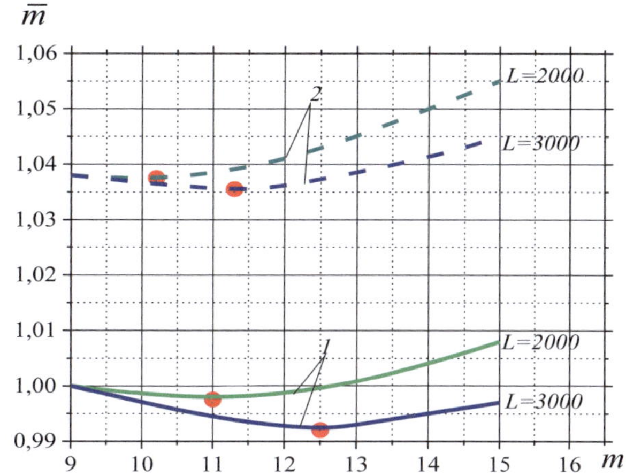

Fig. 2 Dependencies of the total relative mass of the power unit and fuel on the degree of dual-circuitry: 1—without considering the resistance of the nacelle; 2—considering the resistance of the nacelle; at a flight range of the aircraft of 2000 and 3000 km

of dual-circuitry can provide a reduction in specific fuel consumption by 10…15% lower than the similar indicator of the best existing turbojet engines, which leads to a corresponding reduction in harmful emissions.

3 Research of Ways and Technologies of Application of Hybrid and Development of Electric Power Units

In order to reduce the impact of aviation on the environment, engine-building companies around the world are introducing new methods and ways of creating hybrid power units based on:

- the use of battery and turboelectric schemes (hybrid technologies);
- direct combustion of hydrogen or hydrogen mixtures in the combustion chamber;
- the use of hydrogen fuel cells that power electric motors.

The analysis of the directions of development of hybrid power units shows that the creation of hybrid turboelectric power units involves several ways:

1. The use of gas turbine engines on hydrogen, which is burned in the combustion chamber. This is the simplest schematic solution that uses an existing gas turbine engine with improved with a combustion chamber and new fuel lines. However, significant changes to the aircraft's design and fuel storage and distribution systems are necessary.
2. Use of a gas turbine engine (GTE) and an electric motor, connected through a common reducer to drive the propeller. In this scheme, it is advisable to rationally use the stored energy of the battery on certain flight segments. The use of electrical energy depends on the operating conditions of the GTE, flight conditions, and profile. This approach requires the use of an onboard energy distribution unit.

3. Use of a GTE and a separate electric motor powered by a battery or fuel cells. This scheme is used on certain flight segments of the aircraft.

Analysis of research shows that hydrogen technologies can reduce climate impact by 50–75%, the use of fuel cells—by 75–90%. The most modest climate benefits are provided by synthetic fuel: 30–60%.

Traditional power units running on hydrocarbon fuel have a negative impact on the environment [2, 3] and therefore require improvement.

Hybrid and electric power units potentially give the aircraft advantages capable of solving a host of technological problems, primarily affecting the increase in fuel efficiency and environmental friendliness.

By "hybridity" we mean a scheme that allows combining in the power unit the conversion into thrust of engines of different types.

3.1 Aviation Hybrid Power Units

An important trend in modern engine development is the creation of hybrid power units, which combine a heat engine (piston or gas turbine) and an electric one, providing power to the propeller.

This combination significantly increases the fuel efficiency of the aircraft and its environmental friendliness (reduces the level of harmful emissions).

In aircraft construction at present, traditional power unit schemes with internal combustion engines (ICE) are widely used, the scheme of which is shown in Fig. 3 and consists of a propeller—1, ICE–2, fuel reserve—3 and other support systems, as well as power units with a gas turbine engine (GTE), which is shown in Fig. 4 and consists of a propeller—1, GTE–2, fuel reserve—3 and corresponding support systems.

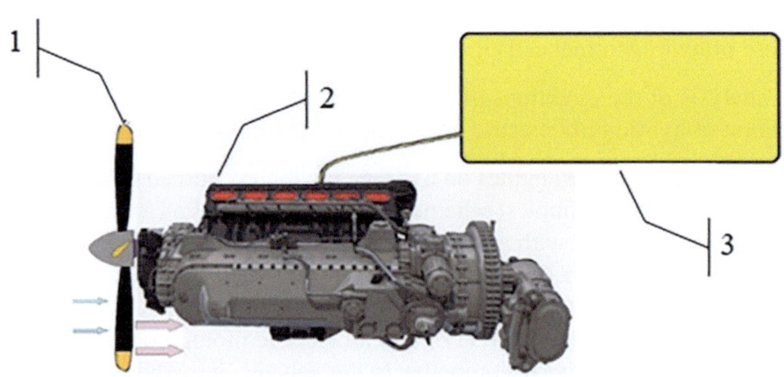

Fig. 3 Scheme of a traditional power unit with internal combustion engines

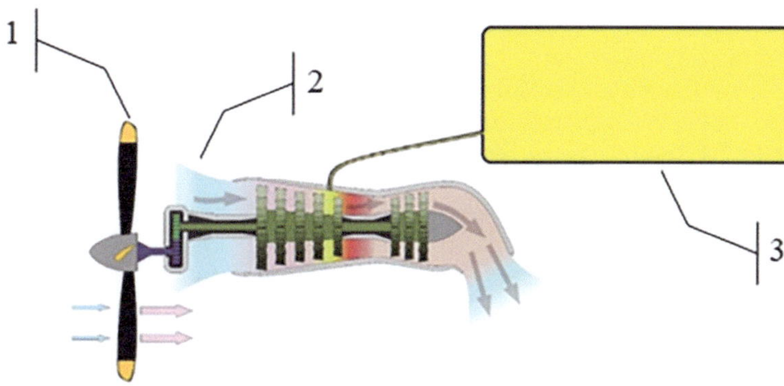

Fig. 4 Scheme of a traditional power unit with a gas turbine engine

Based on the results of an information search regarding general concepts of creating aviation hybrid power units, problems of supplying energy to the hybrid part using hydrogen, fuel cells, and/or batteries (generators), Ivchenko-Progress JSC and Antonov have identified the main trends and problematic issues that Additional research is required to develop modern, efficient hybrid power units [4].

Ivchenko-Progress JSC, within the framework of the European project EFACA (Environmentally Friendly Aviation for All Aircraft Classes) under the Horizon Europe program, has assessed the possibility of creating a commercially attractive and environmentally safe regional propeller passenger aircraft with a hybrid turbo-electric power unit. The aim is to develop highly efficient hybrid power unit schemes, study their efficiency in aircraft during development and testing of technology demonstrators.

It is known [4] that there are several projects for the design of hybrid power units of an aircraft, which are divided into two types: with sequential and parallel arrangement of engines.

In the works [5, 6], an assessment of the possibility of achieving regulatory indicators of reliability of hybrid power units of sequential and parallel schemes has been carried out.

An important issue in designing hybrid power units of a sequential scheme is the increased requirements for reliability, fault tolerance of individual elements and their redundancy, which leads to an increase in the total mass and complexity of the design.

Among the various types of hybrid power unit designs, schemes with sequential and parallel engine arrangements [6] deserve special attention.

Figure 5 shows a scheme of a hybrid power unit with a sequential arrangement of engines, consisting of a turboprop engine (TPE)–1, transmission—2 (if necessary), generator—3, engine controller—4, air propeller—5, electric motor—6, battery—7, fuel reserve—8.

The operation of a hybrid power unit with a sequential arrangement of engines (Fig. 5) is that the turboprop engine—2 rotates the generator—3 and provides energy

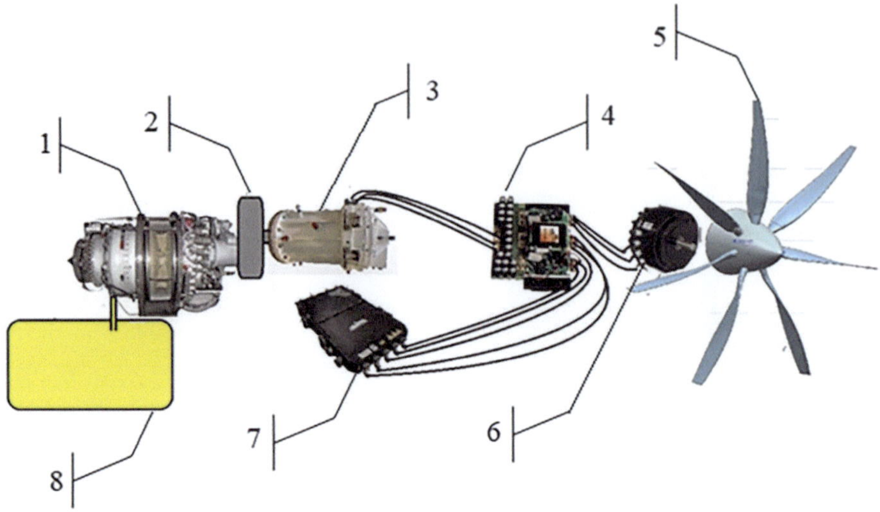

Fig. 5 Scheme of a hybrid power unit with a sequential arrangement of engines

to the electric motor—6, which in turn rotates the air propeller—5, creating the thrust of the power unit.

Figure 6 shows a scheme of a hybrid power unit with a parallel arrangement of engines, consisting of a turboprop engine (TPE)—1, transmission—2 (if necessary), electric motor—3, engine controller—4, air propeller—5, battery—6, fuel reserve—7.

Within the framework of research on the problems of creating a hybrid turboelectric power unit for regional aircraft, in collaboration with Antonov, it was found that the most optimal is a hybrid unit with a 20% degree of hybridization, the internal combustion engine provides 80% of the power and 20% is provided by the electric motor. In such a scheme, the internal combustion engine operates at modes no higher than cruising, which positively affects its resource, significantly reduces the necessary fuel reserve. fuels for flight execution and, accordingly, total emissions of harmful substances decrease [7].

3.2 Investigation of the Problems of Using Electrodynamics Machines in Aviation Hybrid Power Units

The application of electrodynamics machines (electric generators and electric motors) in aviation hybrid power units requires solving a number of important design and technological problems, the use of new materials, and a number of others.

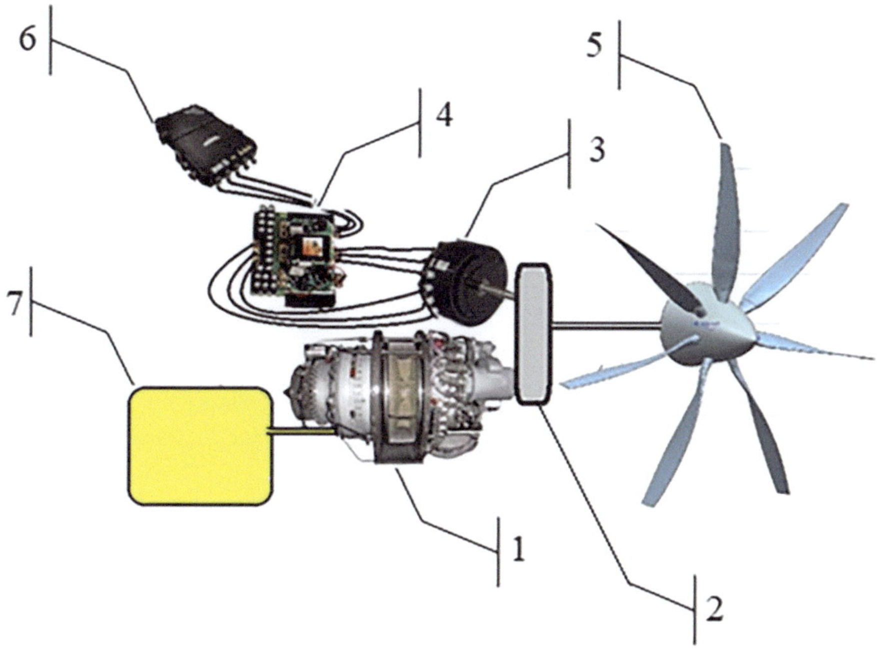

Fig. 6 Scheme of a hybrid power unit with a parallel arrangement of engines

An important criterion for the perfection of electrodynamics machines is the specific power density, which is determined by the ratio of the power of the electrodynamics machine to the total mass of the device.

The task of increasing the power density of electrodynamics machines used in hybrid and electric power units of aircraft is being solved by the Ivchenko-Progress JSC in cooperation with the Institute of Electrodynamics of the National Academy of Ukraine.

Two schemes of electrodynamics machines with excitation from permanent magnets, with axial and radial direction of magnetic flux (Fig. 7), are being studied.

The use of modern construction materials with low magneto-electric losses in the study, by mathematical modeling using modern calculation software complexes, determines possible ways to increase the energy efficiency of high-speed electric motors and generators for aviation hybrid power units.

The rotation frequency of such electrodynamics machines is chosen equal to the rotation frequency of the power turbine of the gas turbine engine, which allows reducing the mass of the electric machine and the transmission of the hybrid power unit.

An analogue of such a unit is the turbo-electric power unit of Honeywell with a total power of 400 kW with two electric generators and a specific power density of the unit of 0.57 kW/kg (Fig. 8).

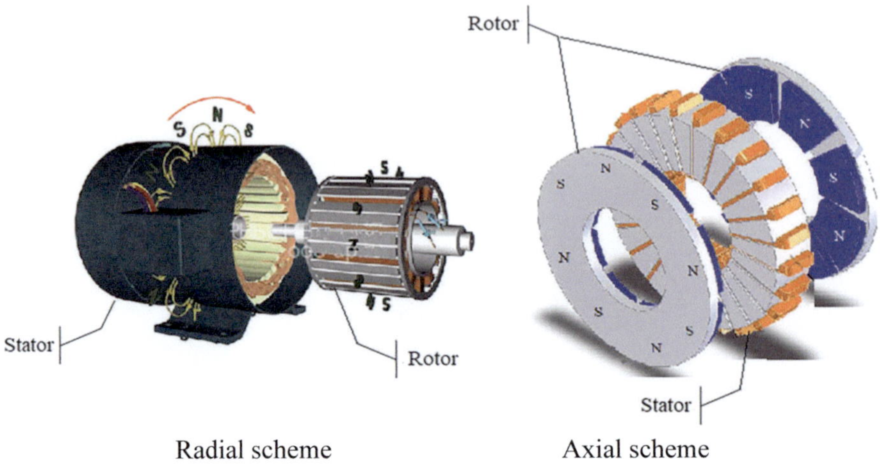

Fig. 7 Schemes of electrodynamics machines with excitation from permanent magnets, with axial and radial direction of magnetic flux

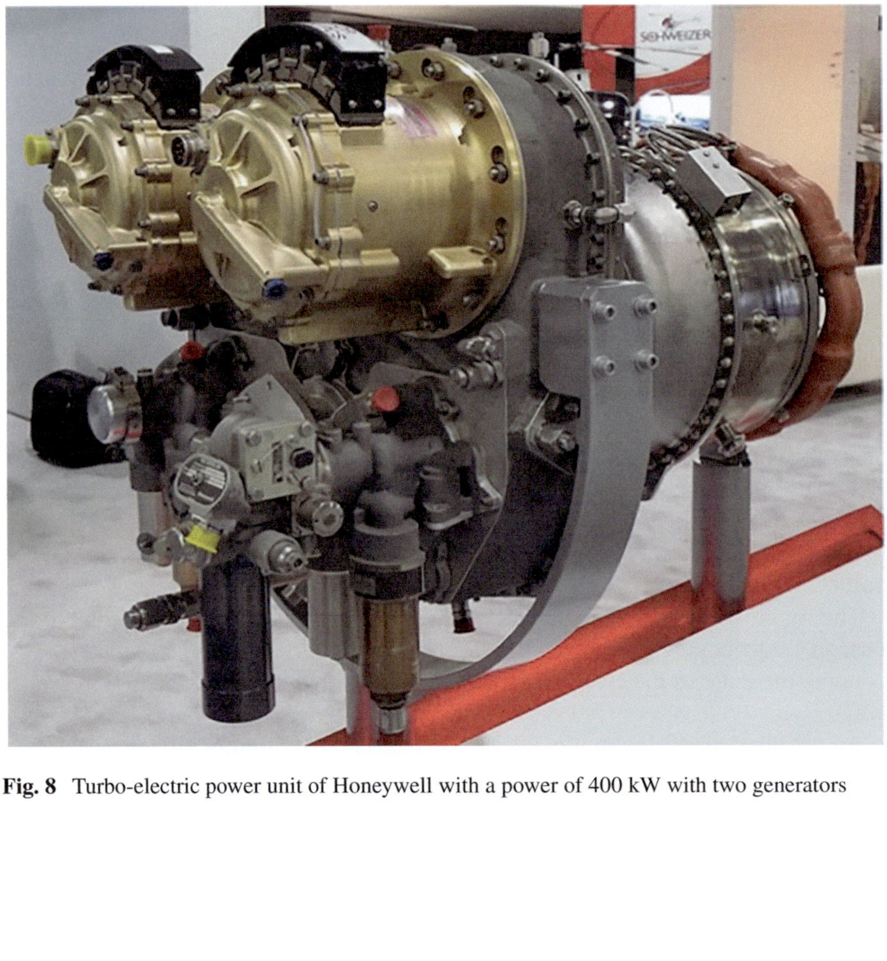

Fig. 8 Turbo-electric power unit of Honeywell with a power of 400 kW with two generators

The experience of the Belgian startup Magnax shows that when using electric motors with an axial magnetic field and a rotation frequency of 2000 … 3000 rpm, for direct drive of the air screw, it is possible to obtain a larger torque at a smaller mass and have high efficiency, characterized by a specific power density up to 3.0 kW/kg (Fig. 9).

An important element of hybrid power units is electronic frequency Converters (controllers) used to manage the operation of electric motors.

Figure 10 shows an example of an industrial controller from MGM COMPRO used for managing electric motors in cars and aircraft, including UAVs, with a specific power density of 17.2 kW/kg.

Fig. 9 Electric motor with axial magnetic flux produced by the Belgian startup Magnax

Fig. 10 Example of an industrial controller from MGM COMPRO

Thus, electromagnetic machines are very attractive for use in electric and hybrid power units not only for their energy efficiency, but also for their reduced impact on the environment.

3.3 Analysis of Problems and Directions for Improving Battery Technology

It is known that the technology of generating electricity in batteries is based on catalytic chemical reactions, methods of increasing their specific power density per unit mass in the following directions:

- Research into the chemical properties of materials manufactured using nanotechnology;
- Application of thin plates or membranes;
- Use of active and catalytic materials with a highly developed surface.

Current aviation lithium-polymer batteries are currently insufficiently effective for use in electric and hybrid power units, primarily due to their specific power density, which is no more than 0.30 kW/kg.

The lifespan of such batteries, despite the use of high-quality separation and corrosion-resistant alloys, is small, 150…200 cycles.

Figure 11 shows the general appearance of a lithium-polymer battery.

The production and disposal of batteries at the current technological level requires significant efforts to minimize harmful environmental impact, but when onboard power sources achieve a specific power density at the level of a power unit with a turboprop engine (0.6 kW/kg), such sources of electrical energy will be able to compete in aviation transport with internal combustion engines on hydrocarbon fuel and will allow a significant step in preserving the environment.

Fig. 11 General appearance of a lithium-polymer battery

3.4 Comparative Analysis of the Characteristics of Power Units of Traditional and Hybrid Schemes

A comparative assessment of the efficiency of power units of different schemes is considered using the example of their application for a light regional aircraft, the main characteristics of which are given in Table 1.

A comparative assessment of the efficiency of four types of power unit schemes: (1) with a piston engine; (2) with a turboprop engine; (3) hybrid power unit with sequential arrangement of engines (GTE + electric motor); (4) hybrid power unit with parallel arrangement of engines (GTE + electric motor) of a light regional aircraft, are given in Table 1.

The main criteria for comparing the efficiency of power units of selected schemes are the specific power density (the ratio of the power unit's power to its mass) and the ratio of the mass of the useful load to the mass of the power unit with fuel.

The diagram in Fig. 12 shows the relationship between the payload weights of a light regional aircraft and the power unit with fuel for four types of selected power units.

From the diagram (Fig. 12), it can be seen that the requirements of the characteristics of a light regional aircraft can be met by the relationship between the payload

Table 1 Comparison of four types of power unit schemes

Main characteristics	Type of Power Unit			
	1	2	3	4
Power, kW	700	700	700	700
Engine power, kW	700	700	560	560
Engine mass, kg	650	230	230	190
Electric motor power, kW	–	–	700	140
Electric motor mass, kg	–	–	273	35
Generator power, kW	–	–	560	–
Generator mass, kg	–	–	98	–
Air screw mass, kg	70	70	70	70
Fuel reserve, kg	640	800	670	670
Battery mass, kg	–	–	370	370
Power unit mass, kg	1360	1100	1711	1335
Specific power density, kW/kg	0.51	0.64	0.41	0.52
Useful load mass, kg	1900	2160	1549	1925
Glider mass, kg	1740	1740	1740	1740
Maximum aircraft weight, kg	5000	5000	5000	5000
Payload weight, kg	1900	2160	1549	1925
Payload deficit, %	−5		−23	−3.75
Estimated potential for reducing harmful emissions, %	–	–	−20	−20

Fig. 12 Relationship between the payload weights of a light regional aircraft and the power unit with fuel for four types of selected power units

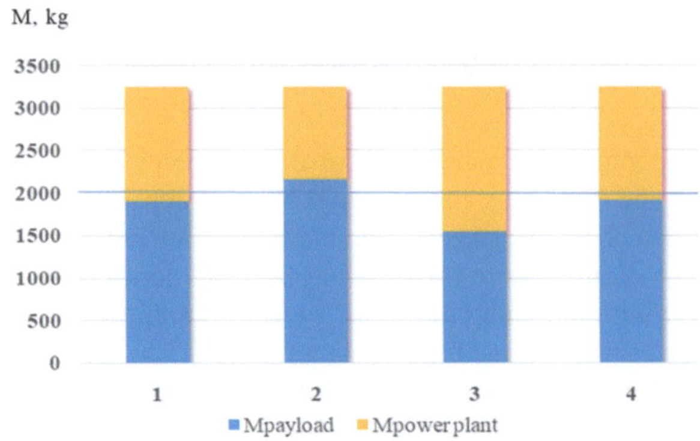

weights of a light regional aircraft and the power unit with fuel when using a power unit with a turboprop engine (option 2).

Figure 13 shows the diagram of the specific density level of four types of selected power units for a light regional aircraft.

From the analysis of the diagram (Fig. 13), it can be seen that only the traditional power unit with a turboprop engine (option 2) meets the requirements of the characteristics of a light regional aircraft.

Thus, at the current level of technical development, the low specific power density of battery packs and especially fuel cells significantly reduces the characteristics of aviation power units.

For commercial hybrid aviation power units, it is necessary to continue work on increasing the specific power density of energy sources to a level of more than 0.6 kW/kg (see Fig. 13).

The increase in weight, complexity of the structure, increased technical risks of implementing projects with hybrid power units, and their high price compared to traditional power units currently restrain their use in commercial aviation.

Fig. 13 Specific density level of four types of selected power plants of a light regional aircraft

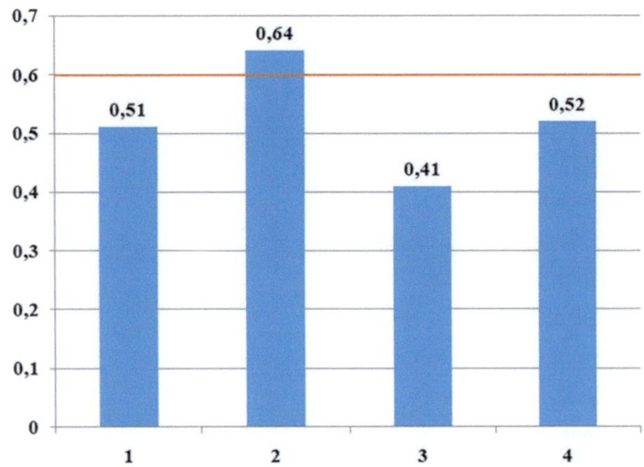

When using hybrid power units (options 3 and 4) for this type of aircraft, with a reduced payload, it is possible to reduce harmful emissions by 20%.

At present, until more efficient hybrid power units appear, from the point of view of reducing the harmful impact of the power unit's operation on the environment, traditional power units are promising. with low-emission combustion chambers and the use of environmentally neutral fuel.

4 Analysis of the Problems of Hydrogen Application and Its Production Technologies

With each passing year, the issue of integrating hydrogen into human economic activity becomes increasingly relevant. Hydrogen and the hydrogen-based energy sector are promising means of decarbonizing existing energy, industry, and transport.

Hydrogen energy is the most efficient way to use existing energy sources, a method of accumulating and storing energy, or a way to gain significant advantages. Hydrogen in nature exists only in a bound state, so it needs to be produced, consuming energy in the process. At present, there is an unofficial color gradation of hydrogen production, depending on the method of its production, as shown in Table 2 Currently, the bulk of hydrogen is produced from fossil fuels, and "green" hydrogen accounts for about 4% of total production. One of the reasons for this distribution is the fact that a large part of the electricity for electrolysis is produced from fossil fuels with an efficiency of about 30%, and the efficiency of electrolysis is 70%, which means that in the full cycle of production, the efficiency of hydrogen energy use is 20%. Thus, obtaining hydrogen by electrolysis, using fossil energy resources, is a pure energy loss.

The efficiency of wind energy is 30–40%, which allows achieving an efficiency of hydrogen production in the cycle of 25%. The best industrial solar batteries have an efficiency of 10% or it is necessary to spend 9 units of solar energy to get 1 unit of hydrogen energy.

Ivchenko-Progress JSC actively participates in the implementation of European programs that have environmental goals, recognizing the need for aviation to accelerate its efforts to reduce emissions that affect climate change, noise impact, and air quality. In particular, the company participates in the scientific and technical project EFACA (Clean Aviation for All Aircraft Classes), funded by the European Commission under the Horizon Europe program, under the "Climate, Energy, and Mobility" cluster.

One of the research topics of the project is to assess the possibilities of creating a commercially attractive and environmentally safe regional propeller passenger aircraft with a hybrid turboelectric power unit based on a gas turbine engine and hydrogen fuel cells, as well as forming higher-level requirements for its power unit.

At the initial stage of this research, a review of over sources of scientific and technical information on general concepts of creating aviation hybrid power units

Table 2 "Color" gradation of hydrogen production

"Color" gradation	Brown	Blue	Gray	Turquoise	Yellow	Green
Technology	Steam conversion of coal	Steam conversion of methane+ CCS 2	Steam conversion of methane	Pyrolysis of methane	Water electrolysis	Water electrolysis
Energy source	TPP	TPP	TPP	TPP	NPP	RES
Emission of gaseous carbon in the production process, CO_2 eq/kg[1]	18–20	0.4–4.5	9–11	Solid carbon (soot)	0	0
Cost, $/kg	≈ 2–3	up to 9	≈ 2–3	≈ 5	≈ 9–10	≈ 9–10

Note (1) CO_2 eq/kg—carbon dioxide equivalent per kilogram of produced hydrogen;
(2) CCS (carbon capture and storage)—technology for capturing and utilizing gaseous carbon oxide.

with energy supply of the hybrid part using hydrogen (fuel cells) and/or batteries (generators) was conducted.

Based on the analysis of this scientific and technical information, a number of general conclusions were obtained that determine the main trends in the study of problems of hydrogen application in aviation power units:

1. In commercial aviation, hydrogen or hydrogen-based fuel mixtures have been chosen as an alternative fuel for existing and prospective aircraft. In In any case, hydrogen will be on board the aircraft and will be burned in the combustion chamber. However, hydrogen can be stored and used in two forms: in cryogenic form or as synthetic fuel.
2. All ways to improve the characteristics of existing aircraft are based on the use of hybrid systems with the application of hydrogen technologies.

It should be noted that the transition to the use of hydrogen in the working process of gas turbine engines has a number of features:

- Engines running on hydrogen fuel, due to the absence of carbon formation in the gas-air path, can have a lifespan approximately 25% longer than engines running on aviation kerosene;
- The calorific value of liquid hydrogen is three times greater than that of kerosene, which means the higher cost of hydrogen fuel can be offset by reducing its specific consumption;
- The use of gas turbine engines on hydrogen fuel is impossible without a detailed calculation of the gas-dynamic parameters of the hydrogen engine. A controlled

process of hydrogen supply and combustion must be ensured to prevent ignition and explosion.

The main benefit from hydrogen technologies is achieved when using "green" hydrogen, as its production uses renewable energy with the possibility of energy storage and transportation.

Hydrogen infrastructure includes three main parts: production, transportation, and storage.

Hydrogen production, especially "green", is carried out by water electrolysis—a chemical process carried out in a special device—an electrolyzer, in which water under the influence of electric current decomposes into hydrogen and oxygen.

At present, there are several promising schemes of electrolyzers, which are presented in Table 3 One of the main problems of hydrogen energy is the high cost of transporting and storing hydrogen. The problem is primarily associated with the extremely low density of gaseous hydrogen (at atmospheric pressure and room temperature, 1 kg of hydrogen has a volume of 11 m ^{3}). Therefore, the issue of its compact storage is very acute. The most commonly used and promising methods of hydrogen storage are presented in Table 4 [8–10].

Hydrogen can be transported by road, sea, and rail, as well as by pipelines.

Transporting hydrogen in cryogenic tanks by road and rail can be considered both the simplest and the least profitable method and should be used when the delivery distance is small or there are no other alternatives [11].

Sea transportation, analogous to the delivery of liquefied natural gas, is still in the stage of technology development and requires not only the refinement of ship systems but also port infrastructure.

Table 3 Comparison of main electrolyzer technologies

Type	Commercial status	Advantages and disadvantages
Alkaline	Mass-produced	• simplicity of design; • slow dynamic response: poorly suited for operation with variable generation of renewable energy; platinum and iridium are required for operation;
Proton exchange membrane (PEM)	Rapidly developing	• fast dynamic response: suitable for operation with variable generation of renewable energy and regulated voltage
Solid oxide (SOEC)	Technology demonstrator	• process is not regulated (no boosting and power shedding); • well suited for hydrogen generation at a fixed productivity
Anion exchange membrane (AEM)	Limited application	• precious metals are not used in this process; • membrane is cheaper than in PEM

Table 4 Hydrogen storage methods

Storage methods	Advantages	Disadvantages
• cylinders with compressed hydrogen (T = 300 K, P ≤ 200 bar)	• technology is well established and available; cost is relatively low;	• low volumetric content ($7.7\ kg/m^3$ at a pressure of 100 bar);
• tanks with liquid hydrogen (T = 20.4 K)	• technology is available, high density (71 kg/m^3);	• high energy consumption for liquefaction; • hydrogen loss due to evaporation; • need for superinsulation; • high cost;
• cryoadsorption: activated carbon (T = 155 K)	• technology is simple and well-developed	• low volumetric content (0.5…20 kg/m^3); • need for cooling and compression;
• carbon nanostructures: nanotubes, fullerenes;	• technologies can provide high hydrogen storage density (30…100 kg/m3);	• production of carbon structures is not sufficiently developed; • hydrogen retention results are not reproducible;
• metal hydrides, alloys, intermetallic compounds, and composites	• convenience and safety of storage in solid phase (in a bound state); • several technologies are quite well-developed	• insufficient capacity; • need for heating; • degradation over time; • relatively high cost

A special place is occupied by transportation using pipelines. There are two directions in solving this issue: the construction of specialized pipelines and the use of existing gas pipelines.

Experimental studies of the possibility of transporting hydrogen Studies on gas pipelines [12] have shown that the loss of technological hydrogen is 3–3.5 times higher than the loss of gas. However, due to the high heat of combustion of hydrogen, the total energy losses in the pipeline are approximately equal to the losses during gas transportation.

The first main hydrogen pipeline was put into operation in 1938 in Germany and still operates without critical accidents [13]. The number of such facilities is increasing, but the rapid growth in the length of pipelines is hindered by large capital investments and the still insufficient volume of hydrogen transportation.

Within the scope of hydrogen energy, transport is considered one of the main consumers of hydrogen. The use of hydrogen in aviation as a fuel is of particular interest and relevance and is considered in several directions: hydrogen as an additive to hydrocarbon fuel, hydrogen as the main fuel [14], the use of hydrogen fuel cells in hybrid and electric power plants.

Due to its unique physicochemical properties (wide concentration range of occupancy), hydrogen as an additive to the main fuel allows for the combustion of lean fuel–air mixtures (α_1 >1,5), thereby reducing emissions of carbon oxides and unburned hydrocarbons. Also, working on lean mixtures positively affects the lifespan of combustion tubes by reducing radiant heat flux from flame tips. Operating an engine on hydrocarbon fuel with such an excess air coefficient in the combustion zone is challenging due to ignition reliability issues, the possibility of lean flameout on variable modes, and reduced fuel combustion completeness.

Using hydrogen as the main fuel for gas turbine engines will completely eliminate carbon emissions and unburned hydrocarbons.

Also, hydrogen, being a cryogenic fuel, has a significant cooling resource, which can be used in hypersonic aviation.

The use of hydrogen in aviation is associated with certain problems, caused by the properties of hydrogen itself:

- Complications and enlargement of the aircraft's fuel tanks, hydrogen diffusion into the metal and subsequent structural failure (use of composites, metal composites, high-density polymers for hydrogen storage and transportation elements);
- Fuel delivery from the tank to the combustion chamber's burn zone (a fuel system with a minimum of detachable connections and pipeline insulation is required, replacement of existing fuel pumps, for example, with a turbopump unit of a liquid rocket engine is necessary);
- Ensuring operational safety (installation of hydrogen leak detectors, engine tract purging with inert gas in case of unsuccessful launch);
- Change in the aggregate state of fuel during transportation by the fuel system (hydrogen is stored in the tank in liquid form for compactness considerations, but as experimental experience shows, it is necessary to supply hydrogen to the combustion chamber in gaseous form to stop its boiling in the injectors, further generation of low-frequency oscillations leading to fluctuations in the turbocharger rotation frequency, for this a gasifier heat exchanger is introduced into the fuel system).

Aviation hybrid power units—similar to automotive ones, i.e., a gas turbine engine is used as a drive for the generator and creating additional thrust during the aircraft's takeoff. The main advantage of this concept is the limitation of the gas turbine engine's operation, especially at low modes in the airport area.

Aviation hybrid power units on fuel cells allow abandoning the heat engine and switching to a fully electric aircraft. This is facilitated by the rapidly developing fuel cell market, with a fairly wide range of powers from 10 kW to 5 MW.

Also, the advantages of this concept include high environmental friendliness of fuel cells and increased efficiency compared to heat engines.

However, there are several factors that restrain this direction:

- high cost of fuel cells—precious metals are used in their production;
- the need for very chemically pure fuel and catalyst;
- low, at present, reliability for ensuring flight safety;
- most fuel cells with high efficiency are high-temperature, so they require heat dissipation systems.

Thus, it can be noted that at present, hydrogen energy has a complex of global problems that do not allow its widespread use:

- cost—the cost of hydrogen, especially "green", is higher than hydrocarbon fuel;
- technology readiness—some technologies in the hydrogen creation chain are still at the technology demonstrator level and require refinement for large-scale production;
- efficiency—the hydrogen production chain is still associated with quite large energy losses at each stage of the life cycle;
- sufficient amount of renewable energy—by 2050, the production of "green" hydrogen may consume about 21,000 TWh—almost as much electricity as is produced in the world today [15]. And through the trend towards greater electrification of industry and transport, a lack of sufficient renewable energy could become a bottleneck for "green" hydrogen;
- political and regulatory uncertainty—although more than 140 countries have committed to achieving zero emissions in the coming decades, the speed at which these goals will be achieved remains uncertain;
- standards and certification—in countries, there are no institutional mechanisms for tracking the production and consumption of any "color" of hydrogen and determining its characteristics (for example, origin and emissions throughout the life cycle) [16] is not taken into account in the official statistics of total final energy consumption, and the economic value of the contribution of clean hydrogen to emission reduction is not recognized;
- the problem of demand/supply—without demand, investments in hydrogen energy remain too risky for large-scale production, which could reduce the final cost of hydrogen, and without the effect of scaling, technologies remain too high cost.

5 Analysis of the Problems of Using Fuel Cells for Hybrid Power Plants

An important issue for further scientific and technical research on the characteristics of hybrid power plants is the rational use of thermodynamic properties of the process of obtaining electricity from hydrogen for fuel cells.

To implement hydrogen technologies in aviation, it is necessary to solve a number of technical tasks, it is about increasing the overall efficiency of technology implementation by means of:

- lighter tanks for liquid hydrogen (12 kW·h/kg, gravimetric indicator 35%);
- fuel cell systems (5 kW/kg, including auxiliary systems: cooling, supply, heat recovery);
- systems for distributing liquid hydrogen inside the aircraft;
- application of combustion chambers and turbines capable of burning hydrogen with a low level of NOx emissions, etc.

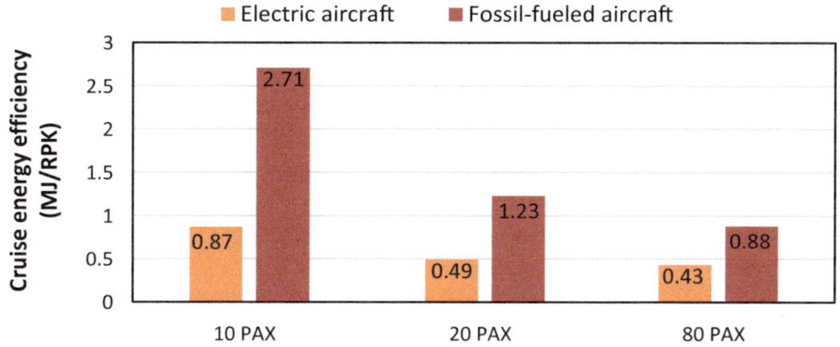

Fig. 14 Comparison of energy efficiency of electric aircraft and aircraft operating on fossil fuels during the cruise phase of flight

There are forecasts that such solutions will likely be created within 5–10 years.

Thus, the analysis of scientific and technical research shows that for small aviation and general-purpose aviation, the use of hydrogen as fuel instead of aviation gas (or additives to the combustion chamber process) is not advisable.

In addition, the use of hydrogen on board such aircraft significantly increases the weight of the structure and equipment (storage tanks, hydrogen supercharging and supply).

It is necessary to take into account in the design that different methods and capacities of hydrogen storage have different rates of its evaporation.

Therefore, for such aircraft, it is advisable to use power plants with battery and turboelectric schemes (hybrid technologies).

Comparison of the cruise energy efficiency of electric aircraft (assuming that the power capacity of the electric motor is 10 kW/kg, i.e., twice greater than the current possible 5 kW/kg) and aircraft operating on fossil fuels with varying numbers of passengers (Fig. 14).

Analysis of the diagram (Fig. 14) allows the following conclusions to be drawn:

- an aircraft operating on fossil fuels and carrying 10 passengers is 3.1 times more efficient than an electric one;
- an aircraft operating on fossil fuels and carrying 20 passengers is 2.5 times more efficient than an electric one;
- an aircraft operating on fossil fuels and carrying 80 passengers is 2.0 times more efficient than an electric one.

There is a forecast that power units for regional aircraft (from 50 to 90 passengers) will use fuel cells and battery packs, and by 2030 they could switch to hydrogen.

Such power units using fuel cells and battery packs are the most energy-efficient, environmentally friendly, and economical option for suburban and regional aircraft.

Compared to conventional aircraft, their operating costs are approximately 10% higher per passenger. Such aircraft may appear in the period 2030–2035.

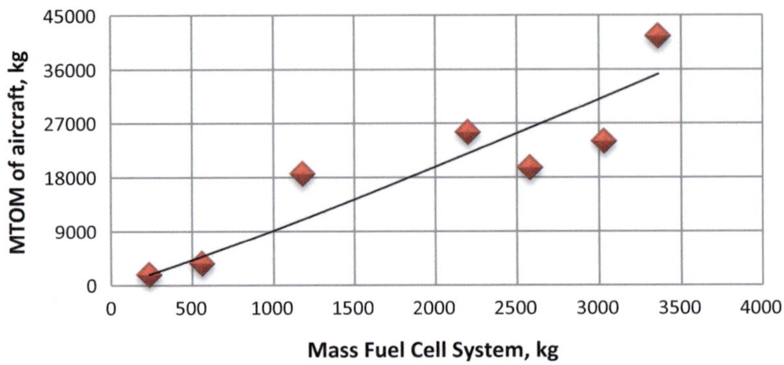

Fig. 15 Dependence of the mass of the energy system with fuel cells on board the aircraft on the mass of fuel

Based on the analysis of modern layout schemes of aircraft with gas turbine engines and electric motors using fuel cells as a power source, dependencies of the mass of the energy system with fuel cells on board the aircraft on the mass of fuel have been constructed (Fig. 15).

Analysis of the research results (Fig. 15) shows that on short-range missions the weight of the energy converter, i.e., the fuel cell or jet engine, is more dominant.

This leads to disadvantages of fuel cell systems compared to jet engines. For long-haul flights, the gravimetric energy density of the energy storage is decisive.

For near-mainline aircraft (capacity 90–120 passengers) the best fit is a hybrid power unit (combustion of hydrogen and fuel cells), which increases the cost per passenger by 20–30%.

Aviation engines for near-mainline aircraft can also operate on environmentally friendly aviation fuel (SAF), and approximately from 2050 – on hydrogen.

Within the framework of joint research by the Ivchenko-Progress JSC and the State Enterprise "Antonov" on the creation of a hybrid turboelectric power unit based on a gas turbine engine and fuel cells for regional aircraft, the possibility of using energy hydrogen fuel cells on polymer membranes Polymer Electrolyte Membrane Fuel Cell [7] (Fig. 16) was considered.

It was found that at the current level of technical development, fuel cells with systems to ensure their operation have a specific power density of no more than 0.2…0.25 kW/kg.

Hydrogen fuel cells of the membrane type consume up to 30% for the operation of their own auxiliary devices, thus the actual specific power density decreases to 0.17 kW/kg.

Therefore, at present, fuel cells are not attractive for use in the design of hybrid power units, despite the absence of greenhouse gases during their operation.

General trends regarding the directions of improving systems that use fuel cells are given in the work [17].

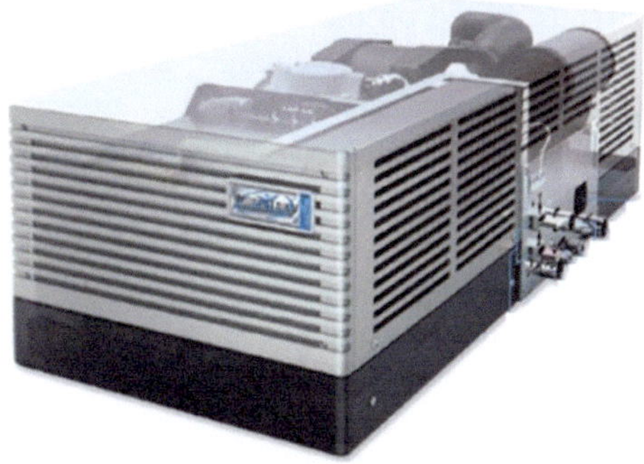

Fig. 16 Hydrogen fuel cell on polymer membranes

It has been found that for short-distance flights, only a future more advanced and efficient system with fuel cells will be able to compete with the current jet engine that operates on aviation fuel. During long flights, the jet engine also has a higher specific power density and specific energy density than current fuel cells with modern tank technology. The mass of the energy system together with hydrogen storage technology directly affects the range of the aircraft. With the improvement of hydrogen storage technology, the use of fuel cells becomes more efficient than the use of a jet engine that operates on aviation fuel.

Medium-range aircraft with a passenger capacity from 220 to 350 require a significant expansion of the fuselage to ensure the storage of liquid hydrogen and it is predicted that it will consume 25% more energy than a regular aircraft. In addition, the cost per passenger increases by 30–40%.

For medium-haul aircraft, it is advisable to apply power with hydrogen technologies. The weight fraction of the structure and the hydrogen itself is acceptable, in terms of ensuring specific weight. However, for this class of aircraft, it is necessary to have on board from 2000 to 5000 kg of hydrogen, which is quite difficult to place on board. For example, to implement the Airbus ZEROe project, the company chose direct combustion of hydrogen in the engine combustion chamber. Research has shown this process to be more technological and profitable than the use of fuel cell technology [18].

There are also known long-haul aircraft using hydrogen (Airbus A380). However, such projects are currently too expensive. A more effective option for such aircraft is the use of synthetic fuel.

Despite the significant advantages of hydrogen fuel cells, there are disadvantages and technical problems that need to be solved for future decarbonization of the energy system, namely:

hydrogen production requires a significant amount of energy;
high initial cost of raw materials;
complex processes of hydrogen storage and transportation;
the necessary infrastructure is absent, requiring significant investment;
unresolved issues of flammability danger.

Thus, based on the analysis, the following conclusions can be made regarding the advantages, disadvantages, and technological problems of applying hydrogen technologies for aviation power units:

– hydrogen is a clean source of energy to support the energy strategies of aircraft power units without harmful emissions;
– hydrogen is more powerful and energy-efficient than fossil fuels;
– hydrogen is a renewable energy source;
– there are no greenhouse gas emissions, reducing the carbon footprint;
– rapid charging of power units;
– long usage time;
– availability of local hydrogen production in remote areas;
– versatility of use: stationary and mobile applications.
– energy supply independence from fossil fuels.

References

1. Lanshin, A.I. (ed.): Aviation Engines and Power Units. TORUS PRESS, 520 p. (2010)
2. Komarov, E.M.: Methods of reducing the emission of harmful substances in combustion chambers of gas turbine engines and gas turbine units. Mech. Eng. Comput. Technol. **5**, 9–29 (2018)
3. The Greenhouse Gas Protocol (GHG Protocol)
4. Mrykin, S.V., Vilchek, M.I., Napadov, K.A.: Method of Structural Diagrams and Reliability Assessment of the Aircraft System (Design Stage), p. 48. Samara State Aerospace University Publishing House, Lab. Workshop. Samara (2012)
5. Volkov, L.I., Shishkevich, A.M.: Reliability of Aircraft, 294 pp. Higher School, Moscow (1975)
6. Varyukhin, A.N., Zakharchenko, V.S., Rakhmankulov, D.Y., Suntsov, P.S., Ovdienko, M.A., Geliev, A.V., Kiselev, I.O., Vlasov, A.V.: Traditional, hybrid and electric power units of local airlines. In: Collection of Scientific Works Aviadvigateli, TsAGI Named After P.I. Baranov, Moscow, vol. 1(14). (2022)
7. Fil, S., Andrii, K., Berbenets, D.: Report: D7.3 Preliminary shaping of the concept and typical flight profile of the aircraft with HTEP. Prelimin. Require. HTEP (2023)
8. Hydrogen energy and fuel cells—a look into the future: Final report of the European Union expert group. (2003)
9. Züttel.: Materials for hydrogen storage. Mater. Today 24–33 (2003)
10. Tarasov, B.P.: Metal hydride systems for reversible hydrogen storage. In: Collection of Abstracts of the Second International Symposium "Safety and Economy of Hydrogen Transport" (IFSSEHT-2003), Alternative Energy and Ecology, pp. 38–39. (2003)
11. Fuel Cell Engine Safety.: U.S. Department of Energy. Energy Efficiency and Renewable Energy. http://www.avt.nrel.gov/pdfs/fcm06r0.pdf
12. Jasionovski, W.J., et al.: Gas distribution equipment in hydrogen service. Int. J. Hydrogen Energy **5**, 323–336 (1980)
13. Wurster, R., Zittel, W.: Hydrogen energy. In: Workshop on Energy Technologies to Reduce CO_2 Emissions in Europe: Prospects, Competition, Synergy. Energy Research Centre of the Netherlands ECN, Petten (1994)
14. Kanilo, P.M.: Toxicity of gas turbine engines and prospects for the use of hydrogen, 140 pp. Kiev: Scientific thought (1982)
15. IRENA.: Renewable Energy Statistics, International Renewable Energy Agency, Abu Dhabi. www.irena.org (2021)

16. IRENA, IEA and REN21: Renewable Energy Policies in a Time of Transition: Heating and Cooling. IRENA, OECD/IEA and REN21 (2020)
17. Kadyk, T., Winnefeld, C., Hanke-Rauschenbach, R., Krewer, U.: Analysis and design of fuel cell systems for aviation. Energies **11**(2), 375 (2018). https://doi.org/10.3390/en11020375
18. Winnefeld, C., Kadyk, T., Krewer, U., Hanke-Rauschenbach, R.: Modeling and designing cryogenic hydrogen tanks for aircraft applications. Energies **11**, 105 (2018)

MIX
Papier aus verantwortungsvollen Quellen
Paper from responsible sources
FSC® C105338

If you have any concerns about our products,
you can contact us on
ProductSafety@springernature.com

In case Publisher is established outside the EU,
the EU authorized representative is:
Springer Nature Customer Service Center GmbH
Europaplatz 3, 69115 Heidelberg, Germany

Printed by Libri Plureos GmbH
in Hamburg, Germany